CATION FLUX ACROSS BIOMEMBRANES

ACADEMIC PRESS RAPID MANUSCRIPT REPRODUCTION

The Proceedings of a Symposium on "Cation Flux across
Biomembranes" Sponsored by the Japan Bioenergetics
Group, Held September 10-13, 1978 at the
Inter-University Seminar House of Kansai in Kobe, Japan

International Symposium on

CATION FLUX ACROSS BIOMEMBRANES

Edited by

Yasuo Mukohata
Department of Biology
Faculty of Science
Osaka University
Toyonaka, Japan

Lester Packer
Membrane Bioenergetics Group
and
Department of Physiology-Anatomy
University of California
Berkeley, California

ACADEMIC PRESS
NEW YORK   SAN FRANCISCO   LONDON   1979
A Subsidiary of Harcourt Brace Jovanovich, Publishers

COPYRIGHT © 1979, BY ACADEMIC PRESS, INC.
ALL RIGHTS RESERVED.
NO PART OF THIS PUBLICATION MAY BE REPRODUCED OR
TRANSMITTED IN ANY FORM OR BY ANY MEANS, ELECTRONIC
OR MECHANICAL, INCLUDING PHOTOCOPY, RECORDING, OR ANY
INFORMATION STORAGE AND RETRIEVAL SYSTEM, WITHOUT
PERMISSION IN WRITING FROM THE PUBLISHER.

ACADEMIC PRESS, INC.
111 Fifth Avenue, New York, New York 10003

United Kingdom Edition published by
ACADEMIC PRESS, INC. (LONDON) LTD.
24/28 Oval Road, London NW1 7DX

Library of Congress Cataloging in Publication Data

International Symposium on Cation Flux across
    Biomembranes, Kobe, Japan, 1978.
    Cation flux across biomembranes.

    Sponsored by Japan Bioenergetics Group.
    1. Membranes (Biology)—Congresses. 2. Cations—
Congresses. 3. Ion flow dynamics—Congresses.
I. Mukohata, Yasuo, Date II. Packer, Lester.
III. Japan Bioenergetics Group. IV. Title.
QH601.I534 1978   574.8'75   79-11934
ISBN 0-12-511050-2

PRINTED IN THE UNITED STATES OF AMERICA

79 80 81 82   9 8 7 6 5 4 3 2 1

CONTENTS

Contributors *xi*
Preface *xix*

PART I: ATP UTILIZING SYSTEMS

Chapter 1: A Perspective on Sodium and Potassium Ion Transport Adenosine Triphosphatase — 3
Robert L. Post

Chapter 2: Detection of Sodium Binding to Na^+-, K^+-ATPase With a Sodium-Sensitive Electrode — 21
Yukichi Hara and Makoto Nakao

Chapter 3: Ouabain-Sensitive and Stoichiometrical Binding of Sodium and Potassium Ions to Na^+-, K^+-ATPase — 29
Hideo Matsui and Haruo Homareda

Chapter 4: Interaction of Inorganic Phosphate with Na^+-, K^+-ATPase: Effect of P_i on the K^+-Dependent Hydrolysis of p-Nitrophenylphosphate — 41
Kazuya Taniguchi, Hideaki Tazawa, Terufusa Tamaki, and Shoichi Iida

Chapter 5: Binding of Na^+ Ions to the Na^+-, K^+-Dependent ATPase — 49
Motonori Yamaguchi

Chapter 6: Rheogenic Na Pump in the Toad Skin — 51
F. Lacaz-Vieira, W. A. Varanda, L. H. Bevevino, and D. T. Fernandes

Chapter 7: Active and Passive Ion Transport by Gastric Vesicles — 53
George Sachs, Edd Rabon, and Gaetano Saccomani

v

Chapter 8:	Reaction Scheme for the Ca-ATPase from Human Red Blood Cells *Alcides F. Rega, Patricio J. Garrahan, Héctor Barrabin, Alberto Horenstein, and Juan P. Rossi*	67
Chapter 9:	Conformation of Various Reaction Intermediates of Sarcoplasmic Reticulum Ca^{2+}-ATPase *Noriaki Ikemoto*	77
Chapter 10:	The Role of Magnesium on the Sarcoplasmic Calcium Pump *Madoka Makinose and Werner Boll*	89
Chapter 11:	Transient Kinetics of Ca^{2+} Transport of Sarcoplasmic Reticulum: A Comparison of Cardiac and Skeletal Muscle *Michihiro Sumida, Taitzer Wang, Frederic Mandel, Arnold Schwartz, and Jeffrey P. Froehlich*	101
Chapter 12:	Occurrence of Two Types of Acid-Stable Phosphoenzyme Intermediate during ATP Hydrolysis by Sarcoplasmic Reticulum *Munekazu Shigekawa and Alfred A. Akowitz*	103
Chapter 13:	Effects of Temperature, Nucleotides, and Metal Cations on the State of Ca^{2+}-, Mg^{2+}-Dependent ATPase of Sarcoplasmic Reticulum Membranes as Studied by Hydrogen–Deuterium Exchange Reaction Kinetics and Saturation Transfer Electron Spin Resonance *Yutaka Kirino, Kazunori Anzai, Taka'aki Ohkuma, and Hiroshi Shimizu*	105
Chapter 14:	Effect of Ca^{2+} Ions on the Reactivity of the Nucleotide Binding Site of the Sarcoplasmic Reticulum Ca^{2+}-, Mg^{2+}-Adenosine Triphosphatase *Masao Kawakita, Kimiko Yasuoka, and Yoshito Kaziro*	119
Chapter 15:	ATP-Induced Ca Binding of Ca-ATPase in the Absence of Added Mg Ion *Jun Nakamura*	125

Contents

Chapter 16:	Entropy-Driven Phosphorylation with Pi of the Transport ATPase of Sarcoplasmic Reticulum *Tohru Kanazawa, Yuichi Takakuwa, and Fumio Katabami*	127
Chapter 17:	Kinetic Properties of Ca^{2+}-, Mg^{2+}-Dependent ATPase of Detergent-Treated Sarcoplasmic Reticulum *Haruhiko Takisawa*	129
Chapter 18:	Proton Relaxation Studies of the Interaction between Manganese (II) and ATPase of Sarcoplasmic Reticulum *Ryoichi Kataoka, Toshiyuki Shibata, and Akira Ikegami*	139
Chapter 19:	Electrogenic Calcium Transport in the Sarcoplasmic Reticulum Membrane *Yves Dupont*	141
Chapter 20:	Initial Rate of Ca Uptake by Fragmented Sarcoplasmic Reticulum from Bullfrog Skeletal Muscle *Yasuo Ogawa, Nagomi Kurebayashi, and Takao Kodama*	161
Chapter 21:	Mechanism of Calcium Release from Fragmented Sarcoplasmic Reticulum: Development of an Optical Method for the Rapid Kinetic Studies *S. Tsuyoshi Ohnishi*	163
Chapter 22:	Ionic Permeability of Sarcoplasmic Reticulum Membrane *Michiki Kasai and Tadaatsu Kometani*	167
Chapter 23:	Mechanism of Cyclic AMP Regulation of Active Calcium Transport by Cardiac Sarcoplasmic Reticulum *Michihiko Tada, Makoto Yamada, Fumio Ohmori, Tsunehiko Kuzuya, and Hiroshi Abe*	179
Chapter 24:	Feedback Regulation of Synaptic Transmission by ATP and Adenosine Derivatives in Mammalian Brain *Yoichiro Kuroda and Kazuo Kobayashi*	191

| Chapter 25: | Structural Studies of Sarcoplasmic Reticulum *in Vitro* and *in Situ*
Sidney Fleischer, Cheng-Teh Wang, Akitsugu Saito, Maria Pilarska, and J. Oliver McIntyre | 193 |

PART II: ATP GENERATING SYSTEMS

Chapter 26:	Flash-Induced Increase of ATPase Activity in *Rhodospirillum rubrum* Chromatophores Margareta Baltscheffsky and Arne Lundin	209
Chapter 27:	Bioenergetics of the Early Events of Bacterial Photophosphorylation Bruno A. Melandri and Assunta Baccarini-Melandri	219
Chapter 28:	Estimation of the Surface Potential in Photosynthetic Membranes Shigeru Itoh, Katsumi Matsuura, Kazumori Masamoto, and Mitsuo Nishimura	229
Chapter 29:	Conversion of Ca^{2+}-ATPase Activity into Mg^{2+}- and Mn^{2+}-ATPase Activities with Coupling Factor Purified from *Rhodospirillum rubrum* Chromatophores Gilbu Soe, Nozomu Nishi, Tomisaburo Kakuno, Jinpei Yamashita, and Takekazu Horio	243
Chapter 30:	Divalent Metal Ions as Modifiers of the Nonlinear Initial Rates of ATPase Activity in Photosynthetic Coupling Factors C. Carmeli, Y. Lifshitz, and M. Gutman	249
Chapter 31:	Changes in Subunit Construction of Chloroplast Coupling Factor 1 with Detachment from the Membrane and Addition of Divalent Cation Yasuo Sugiyama and Yasuo Mukohata	261
Chapter 32:	Effects of Divalent Cations on Membrane ATPase from a Strictly Anaerobic Sulfate-Reducing Bacterium *Desulfovibrio vulgaris* Michiko Takagi, Kunihiko Kobayashi, and Makoto Ishimoto	275
Chapter 33:	Resolution and Reconstitution of Proton Translocating ATPase Nobuhito Sone, Masasuke Yoshida, Hajime Hirata, and Yasuo Kagawa	279

Contents

Chapter 34:	Dissociation and Reconstitution of *Escherichia coli* F_1-ATPase: Analysis of the Defect in Uncoupled Mutants *Masamitsu Futai and Hiroshi Kanazawa*	291
Chapter 35:	Antiport Systems for Calcium/Proton and Sodium/Proton in *Escherichia coli* *Tomofusa Tsuchiya and Keiko Takeda*	299
Chapter 36:	ATP-Linked Cation Flux in *Streptococcus faecalis* *Hiroshi Kobayashi, Jennifer Van Brunt, and Franklin M. Harold*	301
Chapter 37:	Na^+/K^+ Gradient as an Energy Reservoir in Bacteria *Vladimir P. Skulachev*	303
Chapter 38:	On the Proton-Pumping Function of Cytochrome c Oxidase *Mårten Wikström and Klaas Krab*	321
Chapter 39:	Mechanism of H^+ Pump in Rat Liver Mitochondria *G. F. Azzone, T. Pozzan, F. Di Virgilio, and M. Miconi*	331
Chapter 40:	The Stoichiometry of Vectorial H^+ Movements Coupled to Electron Transport and ATP Synthesis in Mitochondria *Albert L. Lehninger, Baltazar Reynafarje, and Adolfo Alexandre*	343
Chapter 41:	Anisotropic Charge Model for H^+-Ejection from Mitochondria *Tomihiko Higuti, Naokatu Arakaki, Makoto Yokota, Akimasa Hattori, Singo Niimi, and Setuko Nakasima*	355
Chapter 42:	Lipophilic Alkylamines as Inhibitors of Coupling and Uncoupling in Beef Heart Mitochondria *Edmund Bäuerlein and Heinz Trasch*	357
Chapter 43:	A Cyanine Dye: Phosphate-Dependent Cationic Uncoupler in Mitochondria *Hiroshi Terada*	365

Chapter 44:	On the Mechanism of Ca^{2+}-Induced Swelling of Rat Liver Mitochondria *Masanao Kobayashi, Yoshiharu Shimomura, Masashi Tanaka, Kazuo Katsumata, Harold Baum, and Takayuki Ozawa*	371
Chapter 45:	The Kinetics of the Reaction of Calcium with Mitochondria at Subzero Temperatures *Britton Chance, Yuzo Nakase, and Fanny Itsak*	377
Chapter 46:	ADP, ATP Transport in Mitochondria as Anion Translocation *Martin Klingenberg*	387
Chapter 47:	Proton and Potassium Translocation by the Proteolipid of the Yeast Mitochondrial ATPase *Richard S. Criddle, Richard Johnston, Lester Packer, Paul K. Shieh, and Tetsuya Konishi*	399
Chapter 48:	Approach to the Membrane Sector of the Chloroplast Coupling Device *Nathan Nelson and Esther Eytan*	409
Chapter 49:	Proton Translocation by Bacteriorhodopsin *Lester Packer, Tetsuya Konishi, Peter Scherrer, Rolf J. Mehlhorn, Alexandre T. Quintanilha, Paul K. Shieh, Irmelin Probst, Chanoch Carmeli, and Janos K. Lanyi*	417

Author Index   *431*
Subject Index   *441*

Contributors

Numbers in parentheses indicate the pages on which authors' contributions begin.

ABE, HIROSHI (179), First Department of Medicine, Cardiology Division, Osaka University School of Medicine, Osaka, Japan

AKOWITZ, ALFRED A. (103), Department of Medicine, University of Connecticut, Farmington, Connecticut

ALEXANDRE, ADOLFO (343), Department of Physiological Chemistry, The Johns Hopkins University School of Medicine, Baltimore, Maryland

ANZAI, KAZUNORI (105), Faculty of Pharmaceutical Sciences, The University of Tokyo, Tokyo, Japan

ARAKAKI, NAOKATU (355), Faculty of Pharmaceutical Sciences, University of Tokushima, Tokushima, Japan

AZZONE, GIOVANNI F. (331), C.N.R. Unit for the Study of Physiology of Mitochondria and Institute of General Pathology, University of Padova, Padova, Italy

BACCARINI-MELANDRI, ASSUNTA (219), Institute of Botany, University of Bologna, Bologna, Italy

BALTSCHEFFSKY, MARGARETA (209), Department of Biochemistry, Arrhenius Laboratory, University of Stockholm, Stockholm, Sweden

BARRABIN, HÉCTOR (67), Departamento de Química Biológica, Facultad de Farmacia y Bioquímica, Universidad de Buenos Aires, Buenos Aires, Argentina

BÄUERLEIN, EDMUND (357), Abteilung Naturstoffchemie, Max-Planck-Institut für medizinische Forschung, Heidelberg, Fed. Rep. Germany

BAUM, HAROLD (371), Department of Biomedical Chemistry, Faculty of Medicine, University of Nagoya, Nagoya, Japan; University of London, London, U.K.

BEVEVINO, L. H. (51), Department of Physiology, Institute of Biomedical Sciences, University of São Paulo, São Paulo, Brazil

BOLL, WERNER (89), Abteilung Physiologie, Max-Planck-Institut für medizinische Forschung, Heidelberg, Fed. Rep. Germany

CARMELI, CHANOCH (249, 417), Department of Biochemistry, The George S. Wise Center of Life Sciences, Tel-Aviv University, Tel-Aviv, Israel; *also* Membrane Bioenergetics Group, University of California, Berkeley, California

Contributors

CHANCE, BRITTON (377), Johnson Research Foundation, University of Pennsylvania, Philadelphia, Pennsylvania

CRIDDLE, RICHARD S. (399), Department of Biochemistry and Biophysics, University of California, Davis, California

DI VIRGILIO, F. (331), C.N.R. Unit for the Study of Physiology of Mitochondria and Institute of General Pathology, University of Padova, Padova, Italy

DUPONT, YVES (141), Laboratoire de Biophysique Moléculaire et Cellulaire, Département de Recherche Fondamentale, C.E.N.G., Grenoble, France

EYTAN, ESTHER (409), Department of Biology, Technion—Israel Institute of Technology, Haifa, Israel

FERNANDES, D. T. (51), Department of Physiology, Institute of Biomedical Sciences, University of São Paulo, São Paulo, Brazil

FLEISCHER, SIDNEY (193), Department of Molecular Biology, Vanderbilt University, Nashville, Tennessee

FROEHLICH, JEFFREY P. (101), Laboratory of Molecular Aging, National Institute of Aging, NIH, Baltimore, Maryland

FUTAI, MASAMITSU (291), Department of Microbiology, Faculty of Pharmaceutical Sciences, Okayama University, Okayama, Japan

GARRAHAN, PATRICIO J. (67), Departamento de Química Biológica, Facultad de Farmacia y Bioquímica, Universidad de Buenos Aires, Buenos Aires, Argentina

GUTMAN, MENACHEM M. (249), Department of Biochemistry, The George S. Wise Center of Life Sciences, Tel Aviv University, Tel Aviv, Israel

HARA, YUKICHI (21), Department of Biochemistry, School of Medicine, Tokyo Medical and Dental University, Tokyo, Japan

HAROLD, FRANKLIN M. (301), Division of Molecular and Cellular Biology, National Jewish Hospital and Research Center, and Departments of Biochemistry and Microbiology, University of Colorado Medical School, Denver, Colorado

HATTORI, AKIMASA (355), Faculty of Pharmaceutical Sciences, University of Tokushima, Tokushima, Japan

HIGUTI, TOMIHIKO (355), Faculty of Pharmaceutical Sciences, University of Tokushima, Tokushima, Japan

HIRATA, HAJIME (279), Department of Biochemistry, Jichi Medical School, Minamikawachi-machi, Tochigi, Japan

HOMAREDA, HARUO (29), Department of Biochemistry, Kyorin University School of Medicine, Mitaka, Tokyo, Japan

HORENSTEIN, ALBERTO (67), Departamento de Química Biológica, Facultad de Farmacia y Bioquímica, Universidad de Buenos Aires, Buenos Aires, Argentina

HORIO, TAKEKAZU (243), Division of Enzymology, Institute for Protein Research, Osaka University, Suita, Osaka, Japan

Contributors

IIDA, SHOICHI (41), Department of Pharmacology, School of Dentistry, Hokkaido University, Sapporo, Japan

IKEGAMI, AKIRA (139), The Institute of Physical and Chemical Research, Wako, Saitama, Japan

IKEMOTO, NORIAKI (77), Department of Muscle Research, Boston Biomedical Research Institute; and Department of Neurology, Harvard Medical School, Boston, Massachusetts

ISHIMOTO, MAKOTO (275), Department of Chemical Microbiology, Faculty of Pharmaceutical Sciences, Hokkaido University, Sapporo, Japan

ITOH, SHIGERU (229), Department of Biology, Faculty of Science, Kyushu University, Fukuoka, Japan

ITSAK, FANNY (377), Johnson Research Foundation, University of Pennsylvania, Philadelphia, Pennsylvania; Department of Biochemistry, Weizmann Institute of Science, Rehovot, Israel

JOHNSTON, RICHARD (399), Department of Biochemistry and Biophysics, University of California, Davis, California

KAGAWA, YASUO (279), Department of Biochemistry, Jichi Medical School, Minamikawachi-machi, Tochigi, Japan

KAKUNO, T. (243), Division of Enzymology, Institute for Protein Research, Suita, Osaka, Japan

KANAZAWA, HIROSHI (291), Department of Microbiology, Faculty of Pharmaceutical Sciences, Okayama University, Okayama, Japan

KANAZAWA, TOHRU (127), Department of Biochemistry, Asahikawa Medical College, Asahikawa, Japan

KASAI, MICHIKI (167), Department of Biophysical Engineering, Faculty of Engineering Science, Osaka University, Toyonaka, Osaka, Japan

KATABAMI, FUMIO (127), Department of Biochemistry, Asahikawa Medical College, Asahikawa, Japan

KATAOKA, RYOICHI (139), The Institute of Physical and Chemical Research, Wako, Saitama, Japan

KATSUMATA, KAZUO (371), Department of Biomedical Chemistry, Faculty of Medicine, University of Nagoya, Nagoya, Japan

KAWAKITA, MASAO (119), The Institute of Medical Science, The University of Tokyo, Tokyo, Japan

KAZIRO, YOSHITO (119), The Institute of Medical Science, The University of Tokyo, Tokyo, Japan

KIRINO, YUTAKA (105), Faculty of Pharmaceutical Sciences, The University of Tokyo, Tokyo, Japan

KLINGENBERG, MARTIN (387), Institut für Physikalische Biochemie der Universität München, München, Fed. Rep. Germany

KOBAYASHI, HIROSHI (301), Research Institute for Chemobiodynamics, Chiba University, Chiba, Japan

KOBAYASHI, KAZUO (191), Department of Neurochemistry, Tokyo Metropolitan Institute for Neurosciences, Fuchu, Tokyo, Japan

KOBAYASHI, KUNIHIKO (275), Department of Chemical Microbiology, Faculty of Pharmaceutical Sciences, Hokkaido University, Sapporo, Japan

KOBAYASHI, MASANAO (371), Department of Biomedical Chemistry, Faculty of Medicine, University of Nagoya, Nagoya, Japan

KODAMA, TAKAO (161), Department of Pharmacology, Juntendo University School of Medicine, Tokyo, Japan

KOMETANI, TADAATSU (167), Department of Biophysical Engineering, Faculty of Engineering Science, Osaka University, Toyonaka, Osaka, Japan

KONISHI, TETSUYA (399, 417), Membrane Bioenergetics Group, University of California, Berkeley, California; Department of Radiochemistry and Biology, Niigata College of Pharmacy, Niigata, Japan

KRAB, KLAAS (321), Department of Medical Chemistry, University of Helsinki, Helsinki, Finland

KUREBAYASHI, NAGOMI (161), Department of Pharmacology, Juntendo University School of Medicine, Tokyo, Japan

KURODA, YOICHIRO (191), Department of Neurochemistry, Tokyo Metropolitan Institute for Neurosciences, Fuchu, Tokyo, Japan

KUZUYA, TSUNEHIKO (179), First Department of Medicine, Cardiology Division, Osaka University School of Medicine, Osaka, Japan

LACAZ-VIEIRA, FRANCISCO (51), Department of Physiology, Institute of Biomedical Sciences, University of São Paulo, São Paulo, Brazil

LANYI, JANOS K. (417), NASA Ames Research Center, Moffett Field, California

LEHNINGER, ALBERT L. (343), Department of Physiological Chemistry, The Johns Hopkins University School of Medicine, Baltimore, Maryland

LIFSHITZ, YAEL (249), Department of Biochemistry, The George S. Wise Center of Life Sciences, Tel Aviv University, Tel Aviv, Israel

LUNDIN, ARNE (209), Department of Biochemistry, Arrhenius Laboratory, University of Stockholm, Stockholm, Sweden

MAKINOSE, MADOKA (89), Abteilung Physiologie, Max-Planck-Institut für Medizinische Forschung, Heidelberg, Fed. Rep. Germany

MANDEL, FREDERIC (101), Department of Pharmacology and Cell Biophysics, University of Cincinnati College of Medicine, Cincinnati, Ohio

MASAMOTO, KAZUMORI (229), Department of Biology, Faculty of Science, Kyushu University, Fukuoka, Japan

MATSUI, HIDEO (29), Department of Biochemistry, Kyorin University School of Medicine, Mitaka, Tokyo, Japan

MATSUURA, KATSUMI (229), Department of Biology, Faculty of Science, Kyushu University, Fukuoka, Japan

McINTYRE, J. OLIVER (193), Department of Molecular Biology, Vanderbilt University, Nashville, Tennessee

Contributors

MEHLHORN, ROLF J. (417), Membrane Bioenergetics Group, University of California, Berkeley, California

MELANDRI, BRUNO A. (219), Institute of Botany, University of Bologna, Bologna, Italy

MICONI, M. (331), C.N.R. Unit for the Study of Physiology of Mitochondria and Institute of General Pathology, University of Padova, Padova, Italy

MUKOHATA, YASUO (261), Department of Biology, Faculty of Science, Osaka University, Toyonaka, Osaka, Japan

NAKAMURA, JUN (125), Biological Institute, Faculty of Science, Tohoku University, Sendai, Japan

NAKAO, MAKOTO (21), Department of Biochemistry, School of Medicine, Tokyo Medical and Dental University, Tokyo, Japan

NAKASE, YUZO (377), Johnson Research Foundation, University of Pennsylvania, Philadelphia, Pennsylvania; Department of Physiology, Wakayama Medical School, Wakayama, Japan

NAKASIMA, SETUKO (355), Faculty of Pharmaceutical Sciences, University of Tokushima, Tokushima, Japan

NELSON, NATHAN (409), Department of Biology, Technion—Israel Institute of Technology, Haifa, Israel

NIIMI, SINGO (355), Faculty of Pharmaceutical Sciences, University of Tokushima, Tokushima, Japan

NISHI, NOZOMU (243), Division of Enzymology, Institute for Protein Research, Suita, Osaka, Japan

NISHIMURA, MITSUO (229), Department of Biology, Faculty of Science, Kyushu University, Fukuoka, Japan

OGAWA, YASUO (161), Department of Pharmacology, Juntendo University School of Medicine, Tokyo, Japan

OHKUMA, TAKA'AKI (105), Faculty of Pharmaceutical Sciences, The University of Tokyo, Tokyo, Japan

OHMORI, FUMIO (179), First Department of Medicine, Cardiology Division, Osaka University School of Medicine, Osaka, Japan

OHNISHI, S. TSUYOSHI (163), Department of Biological Chemistry and Department of Anesthesiology, Hahnemann Medical College, Philadelphia, Pennsylvania

OZAWA, TAKAYUKI (371), Department of Biomedical Chemistry, Faculty of Medicine, University of Nagoya, Nagoya, Japan

PACKER, LESTER (399, 417), Membrane Bioenergetics Group, University of California, Berkeley, California

PILARSKA, MARIA (193), Department of Molecular Biology, Vanderbilt University, Nashville, Tennessee

POST, R. L. (3), Department of Physiology, Vanderbilt University Medical School, Nashville, Tennessee

POZZAN, T. (331), C.N.R. Unit for the Study of Physiology of Mitochondria and Institute of General Pathology, University of Padova, Padova, Italy

PROBST, IRMELIN (417), Membrane Bioenergetics Group, University of California, Berkeley, California

QUINTANILHA, ALEXANDRE T. (417), Membrane Bioenergetics Group, University of California, Berkeley, California

RABON, EDD (53), Laboratory of Membrane Biology, University of Alabama Medical Center, Birmingham, Alabama

REGA, ALCIDES F. (67), Departamento de Química Biológica, Facultad de Farmacia y Bioquímica, Universidad de Buenos Aires, Buenos Aires, Argentina

REYNAFARJE, BALTAZAR (343), Department of Physiological Chemistry, The Johns Hopkins University School of Medicine, Baltimore, Maryland

ROSSI, JUAN P. (67), Departamento de Química Biológica, Facultad de Farmacia y Bioquímica, Universidad de Buenos Aires, Buenos Aires, Argentina

SACCOMANI, GAETANO (53), Laboratory of Membrane Biology, University of Alabama Medical Center, Birmingham, Alabama

SACHS, GEORGE (53), Laboratory of Membrane Biology, University of Alabama Medical Center, Birmingham, Alabama

SAITO, AKITSUGU (193), Department of Molecular Biology, Vanderbilt University, Nashville, Tennessee

SCHERRER, PETER (417), Membrane Bioenergetics Group, University of California, Berkeley, California

SCHWARTZ, ARNOLD (101), Department of Pharmacology and Cell Biophysics, University of Cincinnati College of Medicine, Cincinnati, Ohio

SHIBATA, TOSHIYUKI (139), The Institute of Physical and Chemical Research, Wako, Saitama, Japan

SHIEH, PAUL K. (399, 417), Membrane Bioenergetics Group, University of California, Berkeley, California

SHIGEKAWA, MUNEKAZU[1] (103), Department of Medicine, University of Connecticut, Farmington, Connecticut; Department of Biochemistry, Asahikawa Medical College, Asahikawa, Japan

SHIMIZU, HIROSHI (105), Faculty of Pharmaceutical Sciences, The University of Tokyo, Tokyo, Japan

SHIMOMURA, YOSHIHARU (371), Department of Biomedical Chemistry, Faculty of Medicine, University of Nagoya, Nagoya, Japan

SKULACHEV, VLADIMIR P. (303), Department of Bioenergetics, A. N. Bolozersky Laboratory of Molecular Biology and Bioorganic Chemistry, Moscow State University, Moscow, U.S.S.R.

---

[1]*Present Address: Department of Biochemistry, Asahikawa Medical College, Asahikawa, Japan.*

SOE, GILBU (243), Division of Enzymology, Institute for Protein Research, Suita, Osaka, Japan

SONE, NOBUHITO (279), Department of Biochemistry, Jichi Medical School, Minamikawachi-machi, Tochigi, Japan

SUGIYAMA, YASUO (261), Department of Biology, Faculty of Science, Osaka University, Toyonaka, Osaka, Japan

SUMIDA, MICHIHIRO (101), Department of Pharmacology and Cell Biophysics, University of Cincinnati College of Medicine, Cincinnati, Ohio; Department of Biochemistry, Ehime University School of Medicine, Ehime, Japan

TADA, MICHIHIKO (179), First Department of Medicine Cardiology Division, Osaka University School of Medicine, Osaka, Japan

TAKAGI, MICHIKO (275), Department of Chemical Microbiology, Faculty of Pharmaceutical Sciences, Hokkaido University, Sapporo, Japan

TAKAKUWA, YUICHI (127), Department of Biochemistry, Asahikawa Medical College, Asahikawa, Japan

TAKEDA, KEIKO (299), Department of Microbiology, Faculty of Pharmaceutical Sciences, Okayama University, Okayama, Japan

TAKISAWA, HARUHIKO (129), Department of Biology, Faculty of Science, Osaka University, Toyonaka, Osaka, Japan

TAMAKI, TERUFUSA (41), Department of Pharmacology, School of Dentistry, Hokkaido University, Sapporo, Japan

TANAKA, MASASHI (371), Department of Biomedical Chemistry, Faculty of Medicine, University of Nagoya, Nagoya, Japan

TANIGUCHI, KAZUYA (41), Department of Pharmacology, School of Dentistry, Hokkaido University, Sapporo, Japan

TAZAWA, HIDEAKI (41), Department of Pharmacology, School of Dentistry, Hokkaido University, Sapporo, Japan

TERADA, HIROSHI (365), Faculty of Pharmaceutical Sciences, University of Tokushima, Tokushima, Japan

TRASCH, HEINZ (357), Abteilung Naturstoffchemie, Max-Planck-Institut für Medizinische Forschung, Heidelberg, Fed. Rep. Germany

TSUCHIYA, TOMOFUSA (299), Department of Microbiology, Faculty of Pharmaceutical Sciences, Okayama University, Okayama, Japan

VAN BRUNT, JENNIFER (301), Division of Molecular and Cellular Biology, National Jewish Hospital and Research Center; Departments of Biochemistry and Microbiology, University of Colorado Medical Center, Denver, Colorado

VARANDA, W. A. (51), Department of Physiology, Institute of Biomedical Sciences, University of São Paulo, São Paulo, Brazil

WANG, CHENG-TEH (193), Department of Molecular Biology, Vanderbilt University, Nashville, Tennessee

WANG, TAITZER (101), Department of Pharmacology and Cell Biophysics, University of Cincinnati College of Medicine, Cincinnati, Ohio

WIKSTRÖM, MÅRTEN (321), Department of Medical Chemistry, University of Helsinki, Helsinki, Finland

YAMADA, MAKOTO (179), First Department of Medicine Cardiology Division, Osaka University School of Medicine, Osaka, Japan

YAMAGUCHI, MOTONORI (49), Department of Biology, Faculty of Science, Osaka University, Toyonaka, Osaka, Japan

YAMASHITA, JINPEI (243), Division of Enzymology, Institute for Protein Research, Suita, Osaka, Japan

YASUOKA, KIMIKO (119), The Institute of Medical Science, The University of Tokyo, Tokyo, Japan

YOKOTA, MAKOTO (355), Faculty of Pharmaceutical Sciences, University of Tokushima, Shomachi, Tokushima, Japan

YOSHIDA, MASASUKE (279), Department of Biochemistry, Jichi Medical School, Minamikawachi-machi, Tochigi, Japan

PREFACE

An international symposium on "Cation Flux across Biomembranes" sponsored by the Japan Bioenergetics Group was convened September 10–13, 1978 at the Inter-University Seminar House of Kansai in Kobe, Japan. The unique feature of the symposium was that eighty of the leading investigators concerned with ATP-utilizing and ATP-generating systems associated with cation fluxes across membranes met to discuss biochemical mechanisms in depth and their relation to cation transport functions. Three types of membrane systems were reported on.

ATP-UTILIZING SYSTEMS

1. The plasma membrane, associated with the ATP dependent Na^+-K^+ transport system, which draws upon most of the cell's energy for cation fluxes. Also considered was the novel plasma membrane H^+-K^+ ATPase associated with the HCl secretion mechanism across the plasma membrane of gastric mucosa.
2. The sarcoplasmic recticulum membrane associated with Ca^{++} transport, which plays a key role in excitation–contraction coupling in muscle.

ATP-GENERATING SYSTEMS

3. The inner membranes of mitochondria, chloroplasts, and bacteria associated with H^+ fluxes generated by oxidation–reduction reactions, and their coupling to secondary ion flows and oxidative and photosynthetic phosphorylation. H^+ transport associated with the photoreaction cycle of bacteriorhodopsin, the light energy converted in halobacteria was also considered.

It is noteworthy that shortly after this symposium was held, the Nobel Prize in Chemistry was awarded to Dr. Peter Mitchell, who has contributed much to providing the Chemiosmotic Theory as a framework for integrating the relationship between cation fluxes and energy generating and utilizing systems. Viewed in this manner, this volume is both a timely and appropriate forum to summarize an unprecedented amount of experimental work that followed upon some of the earlier hypotheses for the relationship between ATP coupled systems and ion transport. A remarkable similarity in basic biochemical mechanisms is emerging, despite the fact that nature chose to so organize these membrane systems to result in functionally different phenomena as revealed by the direction and specificity of cation fluxes.

The success of this symposium is largely due to contributions made to the organizing committee by several organizations. Without them it would not have been possible to gather in Kobe the distinguished scientists who took part in the symposium. We wish to thank:

The Ministry of Education, Science, and Culture of Japan

Commemorative Association for the Japan World Exposition (1970)

Japan Society for the Promotion of Science

Eisai Co., Ltd.

Suntory Ltd.

Kowa Co., Ltd.

Yoshikawa Oil & Fat Co., Ltd.

Minophagen Pharmaceutical Co., Ltd.

Wako Pure Chemicals Co., Ltd.

Toyo Kagakusangyo Co., Ltd.

The organizing committee also thanks the participants in this symposium for taking the time, without exception, to contribute to this volume, a feat almost without precedent. Thanks are also due to several persons who served with us on the Editorial Advisory Committee: Lester Packer, Sidney Fleischer, Martin Klingenberg, Robert Post, Albert Lehninger, and Yuji Tonomura.

The organizing committee would also like to acknowledge Professor Emeritus Shiro Akabori, former president of Osaka University and honorary chairman of the Japan Bioenergetics Group, and Professor Ryo Sato, director of the special project research group on Biomembranes in Japan for having lent their support to this symposium.

The Japan Bioenergetics Group was founded in 1974 by the late Professor Saburo Muroaka and Professors Yuji Tonomura and Yasuo Mukohata. It presently has approximately 300 members in Japan and is affiliated with the joint IUB-IUPAB Interunion Bioenergetics Group. The Japan Bioenergetics Group yearly sponsors three-day symposia on topics on the frontier of bioenergetics.

Research in the field of bioenergetics is now experiencing considerable momentum and is taking on new directions. It is hoped that this volume will help to focus on some of the major unsolved problems in which the challenges in the future lie.

The organizing committee:

> Makoto Nakao, *Chairman*
> Yasuo Kagawa
> Michiki Kasai
> Yasuo Mukohata

PART I

ATP-UTILIZING SYSTEMS

A PERSPECTIVE ON SODIUM AND POTASSIUM ION TRANSPORT ADENOSINE TRIPHOSPHATASE[1]

Robert L. Post

Department of Physiology,
Vanderbilt University Medical School,
Nashville, Tennessee, USA

I. INTRODUCTION

In this paper I review my current thinking about the sodium and potassium ion-transport adenosine triphosphatase.

I will cite few references to specific articles. For such information I refer the reader to the many recent reviews which are available (Albers, 1976; De Weer, 1975; Garrahan and Garay, 1976; Glynn and Karlish, 1975; Jorgensen, 1975; Mullins, 1972; Schwarz et al., 1975; Skou, 1975; Whittam and Chipperfield, 1975). A particularly attractive and most recent one is that by Cavieres (1978). In 1979 Academic Press will publish the proceedings of The 2nd International Conference on the Properties and Functions of Na,K-ATPase. This conference was organized by the Institute of Biophysics of the University of Aarhus, Denmark. It will be a rich resource.

When I am thinking of transport studies with intact cells, I will refer to "the pump" and when I am thinking of enzymatic studies in preparations of broken membranes, I will refer to "the enzyme". Actually they are different aspects of the same thing.

---

[1] Supported by grant No. 5R01 HL-01974 from the National Heart, Lung, and Blood Institute of the NIH.

II. SIGNIFICANCE

A. Physiological Function of the Pump

The function of the pump is to replace intracellular sodium ions with extracellular potassium ions in animal cells. By doing this it generates a concentration gradient of each ion across the plasma membrane. These gradients are extremely useful for many cell functions including volume regulation, electrical signaling, fluid secretion or excretion, and active transport of sugars, amino acids, calcium ions, or hydrogen ions. The enzyme also generates directly an outward electric current which makes the inside of the cell more negative electrically than it would be otherwise. This electrogenic activity is important for the function of certain sensory receptors, for an example.

B. Biological Distribution

The enzyme is present in most animal cells (but not in the dog erythrocyte, for an exception). It is found in large amounts in organs which transport large amounts of sodium and potassium. Most specifically with reference to mammals, the major organs are the kidney and the brain (Bonting, 1970). Other rich sources are the electric organ of the electric eel, the rectal gland of the dogfish, and the salt gland of the duck. A good source which can be grown in culture is the brine shrimp.
In general the enzyme has similar properties wherever it is found. However, there is good evidence for significant species differences. For example, with respect to inhibition by particular cardioactive steroids, the rat is (almost notoriously) relatively insensitive to ouabain. In this article I ignore such differences. I take my ideas on transport reactions mostly from studies on erythrocytes and squid axons. I take my ideas on reactions of the enzyme from studies on broken membranes from the rich sources mentioned above.

C. Cardioactive Steroids

Cardioactive steroids, such as the digitalis glycosides or, most commonly, ouabain, are specific inhibitors of this pumping enzyme. In complex preparations, such as cells, tissues, or whole animals, these steroids are most useful in marking out phenomena due specifically to the actions of the pump. Conversely, the pump is almost certainly the receptor for the pharmacological and therapeutic actions of these drugs. These drugs increase the force of the beat of the failing heart. There is often an associated diuresis relieving the congestion of congestive heart failure.

III. THE PUMP

A. Stoichiometry and Sidedness

When operating at, or reasonably close to, its capacity, this pump or active ion-transport system transports 3 sodium ions outward and 2 potassium ions inward across the plasma membrane per one molecule of intracellular ATP hydrolyzed to intracellular ADP and intracellular inorganic phosphate. (The sidedness of the hydrolytic water molecule is not known.) The reaction is catalyzed by intracellular magnesium ion. Extracellular magnesium or ATP appears to have no effect. Ouabain inhibits only from the extracellular solution.

The inequality of positive charge transported is not compensated by transport of some other ion by the pump. The pump really is electrogenic and moves one positive charge outward per cycle. The cell compensates for the charge imbalance by moving ions across the membrane by other pathways.

The stoichiometry is a little flexible. The pump can operate at other than the optimum stoichiometry but at a reduced rate. The best example of this is "uncoupled sodium transport". It appears as follows. When sodium and potassium ions are both absent from the extracellular solution, there remains a net transport of sodium ions outward at a rate of about 1 to 10% of capacity. There is also an associated splitting of ATP at a rate of about 2 to 3 sodium ions per molecule of ATP. (A corresponding sodium-dependent ATPase activity in the absence of potassium ions can also be observed in partially purified preparations of the enzyme in broken membranes.)

B. Energy Requirements

In human erythrocytes under physiological conditions it is possible to estimate whether the energy supplied by the hydrolysis of ATP is sufficient to meet the requirements of the pump. The reasoning is as follows. Both sodium and potassium ions move against concentration gradients. For sodium ion the ratio of the final concentration on the outside to the initial concentration on the inside is about 10 and for potassium ion the transport ratio is about 25 in the other direction. At 37°C about 1.4 kcal/mol are required for transport against a concentration ratio of 10-fold. Accordingly 4.2 kcal/mol are required for transport of 3 sodium ions and 3.7 kcal/mol are required for 2 potassium ions, since the ratio is higher. In addition 1 positive charge is transported outward per cycle against a membrane potential of 10 millivolts, the inside being negative relative to the outside. This electrical work requires 0.2 kcal/mol. Thus the energy requirement to do the work of transport is about 8.5 kcal/cycle. The standard free energy of hydrolysis of ATP is about 7 kcal/mol under standard conditions. That is, with concentrations of substrates and products at 1 M. Since the physiological concentration of inorganic phosphate is about 1,000-fold lower than this, 3 times 1.4 or 4.2 kcal/mol more are available under physiological conditions. In addition the concentration of ATP is greater than that of ADP. Thus 11.2, or more, kcal/mol are available to drive each cycle of the pump. The energy supply is adequate.

C. Kinetics with Respect to Sodium and Potassium

Since this pump is a transport enzyme, I consider intracellular sodium ion and extracellular potassium ion as substrates for translocation. Correspondingly I consider extracellular sodium ion and intracellular potassium ion as products. Ions on the same side of the membrane are "cis" to each other and ions on opposite sides are "trans". The kinetic patterns are more or less symmetrical with respect to the two sides of the membrane but the ion affinities are quite different.

1. *Net Transport.* On each side of the membrane the substrate ion is inhibited competetively by the cis product on the same side of the membrane. Specifically, intracellular potassium ion is a competetive inhibitor of

intracellular sodium ion. Correspondingly extracellular sodium ion is a competitive inhibitor of extracellular potassium ion. Thus the product inhibition is reciprocal.

The kinetics shows a complexity; it is sigmoid. On each side of the membrane, as the concentration of the inhibitory cis product ion is reduced, the kinetics with respect to the cis substrate ion approaches a simple hyperbola. And as the concentration of the product is increased, the kinetics becomes more sigmoid. Such behavior seems compatible with the polyvalent stoichiometry.

The trans effects are relatively simple, provided that substrate or product concentrations are sufficiently high to saturate the pump, which is the case in the usual isotonic solutions. In this case a deficiency of a trans substrate or an excess of a trans product only reduces the maximal transport rate with respect to the cis substrate, that is, the inhibition pattern is noncompetitive. This noncompetitive pattern has been interpreted as evidence for "simultaneous" as opposed to "consecutive" transport of sodium and potassium (Garrahan and Garay, 1976). I will discuss this later.

2. <u>Reversible In=Out Exchange Translocation.</u> The translocation step for each substrate ion can become reversible provided that the other substrate ion is absent. That is, there can be an in=out exchange of intracellular sodium ion for extracellular sodium ion in the absence of extracellular potassium ion. Correspondingly there can be an exchange of extracellular potassium ion for intracellular potassium ion in the absence of intracellular sodium ion. The further requirements are the following.

a. <u>Sodium ion.</u> "Sodium-sodium exchange", as it is called, requires a high concentration of extracellular sodium ion and the presence of intracellular ATP and ADP. Presumably an ATP=ADP exchange reaction is involved since analogs of ATP which cannot donate their terminal phosphate group are not able to substitute for ATP. Under optimum conditions the rate of exchange is about 30% of sodium efflux in the presence of extracellular potassium.

b. <u>Potassium ion.</u> "Potassium-potassium exchange", as it is called, requires the presence of intracellular inorganic phosphate <u>and</u> ATP. In this case the ATP does not need to donate its phosphate group since nonhydrolyzable analogs are effective substitutes. Presumably binding of

ATP to the pump is sufficient. Under optimum conditions the rate is about 10% of the capacity of the pump for net inward transport of potassium ion.

D. Affinities for Sodium and Potassium Ions

Nevertheless, even with all these complexities it is possible to recognize that the pump has a higher affinity for intracellular sodium ion than for intracellular potassium ion and also a higher affinity for extracellular potassium ion than for extracellular sodium ion. To be more specific, taking values from the exchange reactions I think of the half-maximal concentrations as in Table I.

Thus the selectivity for potassium and against sodium is much greater with respect to the extracellular solution than is the selectivity for sodium and against potassium with respect to the intracellular solution.

These relative affinities of the transport system in intact membranes are helpful in assigning sidedness to reactions of the ions with the enzyme in preparations of broken membranes

E. Substrate Specificity

Outward net transport accepts only sodium ion. Of all the other inorganic monovalent cations only lithium ion seems able to replace sodium ion. Demonstration of the outward transport of lithium ion, which is very slow, requires an absence of competing ions on the inside of the cell. For net inward transport the enzyme accepts the following ions as congeners of potassium:- ammonium, rubidium, cesium, thallous and even lithium ion. Lithium ion is more active as a congener of potassium ion than of sodium ion.

TABLE I. Approximate Half-maximal Concentrations, mM.

Ion	Sidedness	
	In	Out
Na^+	1	100
K^+	10	0.1

Just as the active sites for net outward transport are sharply tuned to sodium ion, so are the active sites for net inward transport sharply tuned against it. This remarkably contrasting selectivity is found not only for net transport but also in all the partial reactions of both the pump and the enzyme. In fact, it is a basis for classifying the partial reactions into two contrasting categories, as I will show later.

IV. THE ENZYME

A. The Sodium and Potassium Adenosine Triphosphatase of Broken Membranes

The plasma membrane of the cell can be broken and isolated without harming the pump embedded within it. After this treatment the pump can continue to split ATP and presumably can continue to transport sodium ions outward and potassium ions inward. That is, more precisely, it can transport sodium ions from the face of the membrane which was cytoplasmic or intracellular before the membrane was broken to the face of the membrane which was external or extracellular; and it can also transport potassium ions from the face which was external before the membrane was broken to the face which was cytoplasmic. This transport activity can no longer be detected experimentally by depletion of ionic substrate from the solution on one side of the membrane or accumulation of ionic product on the other. This is because the breaks in the membrane allow sodium and potassium ions to diffuse freely from one face of the membrane to the other and thus maintain a constant concentration of both ions at both faces. However, the ATPase activity of the pump remains detectable and can be distinguished from other ATPase activities, which may be present in a membrane preparation, by a requirement for sodium, potassium, and magnesium ions simultaneously, together with a sensitivity to inhibition by cardioactive steroids. Since it is possible to detect the activity of the transport system in a broken membrane preparation, it has been possible to purify it and to analyze its components.

B. Reconstitution

Evidence that the pump is still undamaged after extensive purification has been obtained by reconstituting the purified enzyme into tight phospholipid vesicles. Such reconstituted pumps transport sodium and potassium. The stoichiometry of the ions and the sidedness with respect to ATP and ouabain are those found in the intact cell. Thus the sodium and potassium ATPase and the pump are different activities of the same system.

In these reconstituted preparations the pump molecules have a scrambled or random orientation. However, in the experiments, the ATP is supplied only from the outside of the vesicles. Thus only those molecules with a reversed or "inside out" orientation respond and pump sodium inward and potassium outward. When such molecules are in tight vesicles, they are also protected from ouabain, which cannot get inside the vesicles and reacts only with the external aspect of the pump. In leaky vesicles all pump molecules are inhibited by ouabain. Thus addition of ouabain inhibits the ATPase activity of all but the pumping molecules.

C. The Phosphoenzyme

In the presence of sodium and magnesium ions the enzyme accepts the terminal phosphate group of ATP to form a phosphoenzyme. Under these conditions there is also an associated ATP=ADP exchange activity, which is consistent with reversibility of the transphosphorylation. This exchange activity is low at sodium concentrations up to 0.15 M because most of the phosphoenzyme changes to a form lacking the ability to react with added ADP. The rate is about two-thirds of the rate of the sodium-dependent ATPase activity in the absence of potassium. Transient kinetic studies have shown that the phosphoenzyme, with two reactive states, qualifies as an intermediate in the ATPase activity. In the presence of potassium and magnesium ions the enzyme accepts a phosphate group from inorganic phosphate and there is an active exchange of ^{18}O between water and inorganic phosphate under these conditions.

In both cases the active site of phosphorylation is the same beta-carboxyl group of a specific aspartyl residue. The same phosphate group at the active site is at a high energy level when it exchanges with the terminal phosphate group of ATP, and at a low energy level when it exchanges with inorganic phosphate. Accordingly I have proposed that the binding of monovalent cations controls the energy level

of the phosphate group. Binding of sodium ions raises the
level whereas binding of potassium ions lowers it.
Presumably these changes in the energy level of the
phosphate group are mediated by changes in the conformation
of an active center in which the active site of
phosphorylation is located. The form catalyzing ATP=ADP
exchange is called "E1" and the form catalyzing P_i=HOH
exchange is called "E2".

D. Composition

The enzyme is composed of two polypeptide chains and
phospholipid. The large chain, "catalytic subunit", or
"alpha chain" carries the active site phosphate group and
has been labeled covalently with derivatives of ouabain.
Its molecular weight is about 100,000. The small chain,
"glycopeptide", or "beta chain" is a glycopeptide. It has a
molecular weight of about 40,000 with respect to protein.
Acidic phospholipids such as phosphatidyl serine seem to be
necessary for activity. Crosslinking experiments indicate
that the native enzyme probably consits of 2 alpha subunits
and 2 beta subunits.

E. Conformations and their Changes

The two kinds of reactivity of the active site phosphate
group suggest two conformations of the enzyme. There is
more direct evidence available. The enzyme with sodium
bound to it is structurally different from the same enzyme
with potassium bound to it. Differences are observed in the
rate of loss of activity of the enzyme during digestion with
trypsin, in the sizes of the peptides released from the
alpha subunit during graded digestion with trypsin, and in
the intrinsic fluorescence of the native enzyme.

Furthermore, each conformation can be produced in the
absence of the specific monovalent cation which stabilizes
that conformation when it is acting alone. Specifically, a
combination of sodium ion, magnesium ion and a low
concentration of ATP produces the conformation found with
potassium ion alone. Alternatively addition of a high
concentraion of ATP in the presence of potassium ion
produces the conformation found with sodium ion alone. Not
only are there two conformations, but these conformations
are also interconvertible through the action of ATP.

F. Reaction Sequence of the Enzyme

The following scheme shows a reaction sequence. I believe it is essentially correct but I am still thinking about it. The scheme is derived in part from the kinetics of the whole enzyme, transient kinetics of the phosphoenzyme (in both broken membranes and in tight homogeneously oriented vesicles derived from erythrocyte membranes), binding studies, particularly of formycin nucleotides (Karlish et al., 1978a), and studies of the effects of ligands on the conformation as deduced from the results of tryptic digestion (Jorgensen, 1975b), nucleotide binding (Karlish et al. (1978b), or from intrinsic fluorescence (Karlish and Yates, 1978).

The symbols (E1) and (E2) refer to conformers considered more specifically later (Table III). (E-P) refers to the phosphoenzyme. P_i is inorganic phosphate. The sidedness of the ligands with respect to the cell membrane is shown by their location relative to the main loop of the reaction sequence, which is shown by solid arrows. Ligands inside the loop react from the intracellular solution and ligands outside the loop react from the extracellular solution. The stoichiometry with respect to sodium and potassium and the actions of magnesium ion are not shown for simplicity. Magnesium ion is required for transphosphorylation from ATP or from P_i and remains bound to the phosphoenzyme more tightly than the phosphate group (Fukushima and Post, 1978). More data are needed also to interpret other effects. The change in the reaction pathway associated with "uncoupled sodium transport" in the absence of extracellular potassium ion is shown by the dotted arrows. I do not show details of the transition from (E1-P) to (E2-P) since I am not clear about the extent to which the binding of sodium ion interacts with that of ADP in this process.

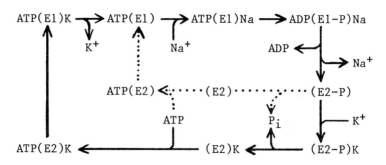

There is a significant feature of the kinetics not shown explicitly in this scheme. The half-maximal concentration for ATP decreases in the absence of potassium ion. In the presence of K^+ it is about 0.1 mM and in the absence it is about 0.2 μM. This change reflects a strong antagonism between the binding of K^+ and the binding of ATP. The stability of the (E2)K complex appears to be so great as to suggest an "occluded" conformation of the carrier. For release of K^+ to the intracellular solution, binding of ATP and presumably transformation of (E2)K into (E1)K is required. In the absence of K^+ the corresponding transformation occurs more easily.

V. CLASSIFICATION OF PARTIAL REACTIONS

Partial reactions remain when one or more ligands are removed from a reaction system or possibly when an enzyme is poisoned or modified by a suitable agent or treatment. I believe that such reactions are useful for understanding the complete reaction sequence. I do not regard a simple deficiency of one or more ligands as likely to produce changes so pathological as to render a partial reaction

TABLE II. Monovalent Cations on (Na,K)pump ATPase

Function or property	Monovalent cation	
	Na^+	K^+
Net transport	outward	inward
In=out exchange dependency	ADP + ATP	P_i + ATP
Molecular exchange	ATP=ADP	P_i=HOH
Phosphate donor	ATP	P_i
ATP binding	tight	weak
p-Nitrophenyl phosphatase activity	absent	present
Inhibition by		
DCCD[a]	present	absent
$F^- + Mg^{2+}$	absent	present
$Be^{2+} + Mg^{2+}$	absent	present
$VO_3^- + Mg^{2+}$	absent	present

[a] DCCD is dicyclohexylcarbodiimide.

inappropriate for understanding the complete system in the native enzyme. I consider it useful to classify these reactions.

A most striking feature of the partial reactions of this transport enzyme is the way they separate into two contrasting groups, one dependent on Na^+ uniquely, and the other dependent on K^+ or its congeners (Table II). When I consider these contrasts in relation to chemical evidence for changes in conformation induced by the binding of Na^+ or K^+ (see above), I do favor an alternation between these conformations as part of the reaction sequence.

VI. WORKING HYPOTHESES

"I have given up the search for truth;
now I am looking for a working hypothesis."
                                            -- Anonymous

A. Molecular Mechanism of Translocation

Like any other enzyme a pump should have an active site for each translocated substrate. I am calling such a site a carrier, whatever the molecular mechanism.

The pump is embedded in a lipid bilayer membrane and has a cytoplasmic aspect on one face and an external aspect on the other. I think of the catalytic subunit of the pump as containing deep inside itself 3 carriers or binding sites for monovalent cations. I think of these sites as having two access pathways (inward and outward) for all of them (or at least pathways which open and close synchronously if different for each site). These access pathways allow the carriers to accept ions from, or to release them into, the fluid on either side of the membrane. Each access pathway can be open or shut. They give the carriers access either to the cytoplasm (intracellular fluid), or to the external medium (extracellular fluid), or to neither, but never to both simultaneously. (In the last case the pump would become a pore (compare Hille and Schwarz, 1978). If a carrier has no access to either solution, i.e. if both its pathways are shut, then it is in an occluded conformation. The mechanism of transport thus becomes 1) acceptance by a carrier of an ion from the solution in contact with one face of the membrane, 2) occlusion of the ion, 3) release of the ion into the solution in contact with the other face of the membrane, and 4) return of the pump to its original state. This model of active transport is thus a sort of "molecular peristalsis".

B. Conformers

I think of the smallest functional unit of the (Na,K)pump as consisting of one catalytic subunit and one glycopeptide. I will call this pair a monomer. Correspondingly I will call the native enzyme a dimer. I think of each monomer of the pumping enzyme as having two principal conformations (Table III).

1. <u>The (E1) conformer.</u> I will consider (E1) first. In the dephosphoform the carriers have access to the cytoplasmic aspect of the pump and accept 3 sodium ions at a half-maximal concentration of about 1 mM. The pump also accepts ATP at a half-maximal concentration of about 0.2 uM. It also accepts one magnesium ion. There is little interaction energy of binding between these ligands. As long as any one of the binding sites is empty or incorrectly filled, little happens. But when the pump is fully loaded and becomes ATP·Mg·(E1)·Na$_3$, then the sodium ions become occluded on their carriers and (reversible) transphosphorylation takes place yielding ADP·Mg·(E1-P)·Na$_3$. This transphosphorylation modifies the occluded pump in two important ways. 1) The occluded carriers now obtain open access pathways to the external aspect of the pump. 2) The affinity of the enzyme for Na$^+$ is tremendously reduced; the half-maximal concentration rises to about 0.5 M in phosphoenzyme experiments

TABLE III. Proposed Conformers of (Na,K)pump ATPase

Characteristic	Conformer designation	
	(E1)	(E2)
Favored monovalent cation	Na$^+$	K$^+$
Active sites for translocation	3	2
Favored carrier sidedness	cytoplasmic	external
ATP affinity	tight	weak
Molecular exchange reaction	ATP=ADP	P$_i$=HOH
Translocation ability	good	poor

(Post et al., 1975). Thus there is a change in carrier sidedness and a tremendous negative interaction of binding between the phosphate group on the enzyme and the bound sodium ions. I have phosphorylated the enzyme from ATP in 3 M NaCl and found it to be relatively quite stable. All of it was in the form of (E1-P), i.e. sensitive to added ADP and almost insensitive to saturating concentrations of added K^+ during a chase experiment (not shown). So I think that (E1-P) can still be stable after it has released its ADP provided the sodium ions still remain bound to it. I do not know to what extent ADP can stabilize (E1-P) after the sodium ions have come off. Yamaguchi and Tonomura (1978) have evidence for binding of ADP to some form of the phosphoenzyme in 0.1 M NaCl. Under physiological conditions the phosphoenzyme rapidly loses most of its bound Na^+ or ADP or both and becomes (E2-P).

In this model I have permitted (E1-P) to translocate sodium ions without requiring conversion to (E2-P). This permission implies a conformational change which is subsidiary to the pricipal change. I derive this permission from the conformational changes of hemoglobin in which the R and T conformations can show subsidiary changes in structure according to the ligands bound to them (Perutz, 1978). In any case I think of an equilibrium between (E1-P) and (E2-P) in the absence of monovalent cations and nucleotides as being strongly in favor of (E2-P). I no longer think that free Mg^{2+} has much effect on this equilibrium.

    2. The (E2) conformer. Now I consider (E2). The most prominent feature is the lack of translocation. For in=out exchange of K^+, binding of ATP is required, but binding of ATP converts (E2) into (E1) according to tryptic digestion patterns. Therefore (E2) cannot complete inward translocation by itself. (E2-P) accepts potassium ions from the extracellular fluid to become (E2-P)·K_2, which engages in P_i=HOH exchange. At a low concentration of P_i this leads to release of P_i and conversion of (E2-P)·K_2 to (E2)·K_2, which I think of as an occluded conformation. Presumably if this complex is phosphorylated again from P_i, the bound K^+ can be released again into the extracellular fluid. In order to release K^+ into the intracellular fluid, binding of ATP is required as indicated above. Possibly Mg·ATP is more effective than free ATP since it is supposed to be the true substrate.

Another feature is the change in the number of carrier sites from 3 in (E1) to 2 in (E2). I suppose that two of the carrier sites in (E1) are altered, principally with respect to their affinity for monovalent cations. The third site is altered more. It can still bind ions, and in this way its occupant can modify the kinetics; but the occupant must get off the site before translocation can take place since this carrier now translocates only in the empty state (Cavieries, 1977).

In case there is no potassium ion in the extracellular fluid, spontaneous dephosphorylation allows all the carrier sites to occlude in the empty state. Binding of ATP to the potassium-free (E2) takes place more easily than to the corresponding potasium-complexed (E2)·K_2. This converts the enzyme to (E1) and allows the carriers to release K^+ into and accept Na^+ from the intracellular solution.

3. <u>A Similarity between (E1) and (E2)</u>. In both conformers, formation of a phosphoenzyme, with magnesium ion bound to it, may open the external access pathway between the carriers and the external solution. This pathway may remain closed in the dephosphoenzyme.

VII. DISCUSSION

A. Hybrid Occupancy of the Carriers

Hybrid occupancy is a developing topic. It may help in the interpretation of many peculiar phenomena. Suppose the carrier sites are occupied partly by sodium ions and partly by potassium ions. What then? Whatever conformer is heterogeneously occupied in this way should be under some strain. But it must translocate only slowly, if at all, or the stoichiometry of sodium to potassium transport would not persist. It might do unexpected things. Hybrid occupancy might be a factor in some of the following phenomena.

1. Fukushima and Tonomura (1975) have observed that addition of K^+ and ADP to the phosphoenzyme results in a partial reversal of phosphorylation with formation of ATP -- which remains bound to the enzyme! Perhaps addition of K^+ to a complex of ADP with disodium (E1-P) allows a partial reversal of the forward reaction with formation of an occluded conformation of ATP·(E1)·Na_2K. This complex might be occluded not only with respect to the monovalent cations but also with respect to the ATP.

2. Intracellular potassium ion inhibits in=out exchange of sodium at low concentrations of intracellular sodium ion but at high concentrations of sodium it stimulates the exchange. This paradoxical effect may be an indicator of hybrid occupancy. Perhaps intracellular potassium ion attacks a disodium (E1) complex and accelerates displacement of one of the sodium ions into the intracellular solution.

B. Subunit Interactions

Functional relationships between the subunits have not been demonstrated directly but may be important for correlating data obtained from the enzyme or pump operating in a steady state with data obtained in partial reactions. They have been offered in various models (Cavieres, 1977). Interaction between monomers may reconcile the demands of transport kinetics experiments for simultaneous translocation of sodium and potassium with the reaction sequence presented above. (Actually it would be enough if the steady-state ratio of one conformer to the other were independent of the rate of the reaction.) One reconciliation hypothesis combines an alternation of conformations in the reaction sequence of each monomer together with simultaneous transport of sodium and potassium. This is done by arranging the monomers to operate 180 degrees out of phase with each other. In this way when one monomer, (E1), is binding sodium ions and is being phosphorylated from ATP, the other monomer, (E2), is binding potassium ions and is releasing inorganic phosphate into the medium. Then the (E1) monomer becomes an (E2) monomer and _vice versa_ and the cycle continues. Available evidence suggests that any such interaction is not very strong. It will be interesting to see if more direct evidence can be developed.

REFERENCES

Albers, R.W. (1976). In "The Enzymes of Biological Membranes" (A.N. Martonosi, ed.), pp. 283-301. Plenum, New York.

Bonting, S.L. (1970). In "Membranes and Ion Transport" (E.E. Bittar, ed.), pp. 257-363. Wiley, New York.

Cavieres, J.D. (1977) In "Membrane Transport in Red Cells" (J.C. Ellory and V.L. Lew, eds.). Academic Press, New York.
De Weer, P. (1975). In "MTP International Reviews of Science, Physiology", (C.C. Hunt, ed.), vol. 3, pp. 231-278. University Park Press, Baltimore.
Fukushima, Y., and Post, R.L. (1978). J. Biol. Chem. 253:6853-6862.
Fukushima,Y., and Tonomura, Y. (1975). J. Biochem. 77:533-541.
Garrahan, P.J., and Garay, R.P. (1976). Current Topics in Membranes and Transport 8:29-97.
Glynn, I.M., and Karlish, S.J.D. (1975). Annu. Rev. Physiol. 37:13-55.
Hille, B., and Schwartz, W. (1978). J. Gen. Physiol. 72:409-442.
Jorgensen, P.L. (1975a). Q. Rev. Biophys. 7:239-274.
Karlish, S.J.D., and Yates, D.W. (1978). Biochim. Biophys. Acta 527:115-130.
Karlish, S.J.D., Yates, D.W., and Glynn, I.M. (1978a) Biochim. Biophys. Acta 525:230-251.
Karlish, S.J.D., Yates, D.W., and Glynn, I.M. (1978b) Biochim. Biophys. Acta 525:252-264.
Mullins, L.J. (1972) In "Role of Membranes in Secretory Processes" (L. Bolis, ed.), pp. 182-202. North Holland, Amsterdam.
Perutz, M.F. (1978). Scientific American 239:92-125.
Post, R.L. (1977). In "Biochemistry of Membrane Transport" FEBS Symposium No. 42 (G. Semenza and E. Carafoli, eds.) pp. 352-362. Springer Verlag, New York.
Post, R.L. (1978). In "Molecular Specialization and Symmetry in Membrane Function" (A.K. Solomon and M. Karnovsky, eds.) pp. 212-221. Harvard University Press, Cambridge.
Post, R.L., Toda, G., Kume, S., and Taniguchi, K. (1975). J. Supramolecular Structure 3:479-497.
Schwartz, A., Lindenmayer, G.E., and Allen, J.C. (1975). Pharmacol. Rev. 27:3-14.
Skou, J.C. (1975) Q. Rev. Biophys. 7:401-434.
Whittam, R., and Chipperfield, A.R. (1975) Biochim. Biophys. Acta 415:149-171.

DETECTION OF SODIUM BINDING TO Na$^+$,K$^+$-ATPase
WITH A SODIUM SENSITIVE ELECTRODE

Yukichi Hara[1]
Makoto Nakao[2]

Department of Biochemistry
Tokyo Medical and Dental University
Tokyo, Japan

I. INTRODUCTION

It is well known that Na$^+$,K$^+$-ATPase plays an important role in the active transport of Na$^+$ and K$^+$ across the cell membrane. To clarify the mechanism of the sodium pump, it is necessary to know the number of Na$^+$ and K$^+$ bound to and released from the enzyme and to know how they are bound and released. The affinities of the cation binding sites of Na$^+$, K$^+$-ATPase for Na$^+$ and K$^+$ are expected to change successively or simultaneously in active transport along the cycle of ATPase reaction. It is therefore probable that the affinities for these ions vary with a ligand ATP, which probably causes the modulation of the enzyme forms.

It was reported by Matsui et al.(1,2) and Kaniike et al.(3) that the binding of ^{22}Na was not affected by ATP but the binding of ^{42}K was reduced by ATP. If this is the case, what brings about the Na$^+$ binding to the enzyme? In order to elucidate this point, the effect of ATP on the Na$^+$ binding was examined. A cation sensitive electrode system may be useful for these binding experiments because the data obtained by the isotope method can not exclude the effect of isotope

---

The abbreviations used are: Na$^+$,K$^+$-ATPase, Na$^+$ and K$^+$-stimulated adenosin triphosphatase: EDTA, ethylenediaminetetraacetic acid.

[1,2]Supported by a grants from the Ministry of Education, Science and Culture of Japan.

exchange reactions between free radioactive ions and the cold ones which bound to the enzyme previously. A potassium selective electrode was used by Hastings (4).

In the present paper, we describe a technique by which a small change of Na^+ concentration could be measured using a newly devised highly sensitive Na^+-selective electrode system, and some results. The ATP dependent Na^+ binding observed here needed the presence of K^+ in the absence of Mg^{++}.

II. METHODS

A. Enzyme Preparations

Pig brain NaI enzyme was prepared with a slight modification of the method of Nakao et al.(5). The enzyme was washed three times with a solution containing 0.3mM NaCl, 25mM imidazole-HCl (pH 7.3), 0.1mM EDTA, and various concentrations of KCl, and finally suspended in the same solution to make approx. 5mg/ml of protein concentration. The specific activity of this enzyme was around 250 μmoles pi/mg/h, and concentrations of the enzyme were calculated from the levels of phosphorylation from $AT^{32}P$. Protein concentrations were determined by the method of Lowry et al.(6).

B. Na^+ Sensitive Electrode System

A small Na^+ sensitive glass electrode (6∅) was purchased from Toko Chemical Lab. Co. The ion selectivity constant for potassium against sodium was 1.5×10^{-1} at 0.3mM NaCl, 25mM imidazole (pH 7.3), and 0.1mM EDTA. A polyacrylamide gel bridge, containing the same salt concentration as that of the sample solution, was placed between the sample solution and a Ag/AgCl/KCl reference electrode in order to constitute a double junction reference electrode. The sample solution was placed in a small vessel purchased from Radiometer Co. and continuously stirred with a disc type magnetic stirrer bar.

The sample chamber and electrodes were kept in a copper cylindrical case. A heat-exchangeable gum hose in which temperature controlled water (25C°) was circulated was wound about the cylindrical case. This equipment was set up in an aluminum case in order to shield it electrostatically. The drift of temperature in the copper cylinder was within 0.01°C

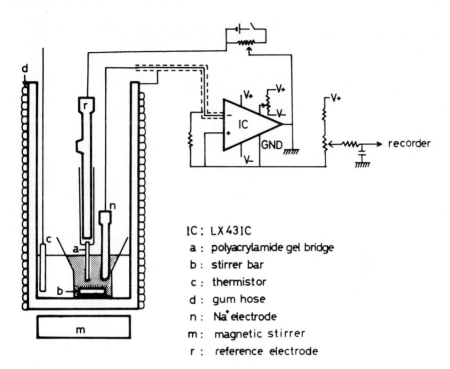

FIGURE 1. Diagram of the experimental arrangement for the detection of sodium concentration changes. A polyacrylamide gel bridge (a) was placed between a reference electrode (r) and the sample solution (dotted area). (n) & (c) represent a small Na^+-sensitive glass electrode and a thermister rod, respectively. The sample solution was stirred with a magnetic disk (b). These were surrounded by a copper cylinder. The electric potential was amplified by the inside-out follower amplifier. All experiments were carried out at 25C°.

during each experiment. The amplifier was a varactor-bridge type operational amplifier (LX 431C CR BOX Tokyo). The voltage was recorded with a pen recorder on the full scale of 1mV. The noise voltages referred to input with and without electrodes were 2 and 5 μv respectively.

C. Calibration and Sodium Binding

2 or 2.5ml of sample solutions were used for each experiment. After the electrodes were dipped into the sample solution, it took 2 hours to stabilize the output voltages. When the calibration was carried out, a small portion (20-

25 μl) of the solutions with the same ionic composition as the sample solution except for the sodium concentration was added to the sample solution to make a small change in the sodium concentration.

When the ATP dependent Na^+ binding was measured, 20μl of a solution having the same ionic composition as the enzyme solution and containing 12-20mM ATP-Tris was added twice to the enzyme solution. The voltage change with the second addition of ATP was subtracted from the voltage change with the first addition, in order to exclude the nonspecific effects of the addition of ATP solution (the change of pH, ionic strength, etc.) on the Na^+ sensitive electrode. From this difference, the decrease in the sodium concentration was calculated on the basis of the calibrations carried out in the same sample solution. This decrease in the sodium concentration was considered to be due to the ATP-dependent Na^+ binding. The number of sodium ions bound to enzyme was calculated from the concentration of the active sites of enzyme and the degree of the decrease in the sodium concentration.

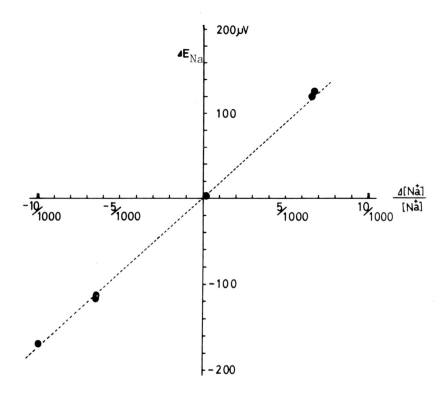

FIGURE 2. The plot of the voltage change against the change in the sodium concentration.

III. RESULTS

A. Calibrations in the Absence of the Enzyme

The sensitivity of this measurement system was checked as follows. 2.0ml of 0.3mM NaCl, 0.2mM KCl, 0.1mM EDTA, and 25mM imidazole (pH 7.3) were placed in the sample chamber. After the drift of the Na^+-sensitive electrode potential became small, 20 μl of the solutions with 0, 0.1, 0.3 and 0.5mM NaCl were added successively to the solution. It took approx. 10 min to stabilize output voltages after each addition to the solution. The plot of the voltage changes against the changes in sodium concentration was practically

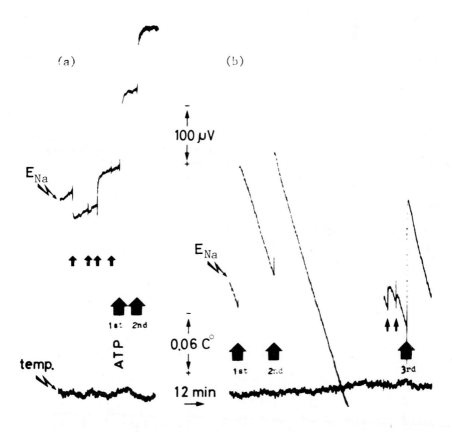

FIGURE 3. ATP-dependent sodium bindings in the presence of 0.15mM KCl (a) and in the presence of 0.5mM KCl (b). The large and small arrows represent the additions of ATP and the solutions for calibration, respectively.

linear because the changes were very small (Fig.2), although there should be a non-linear relationship between them.

B. Sodium Binding in the Absence of Potassium Ion

When the enzyme solution contained 0.1-1mM NaCl, the ATP dependent Na^+ binding was hardly detected, although a very small amount of binding lower than 0.1mole Na^+/mole enzyme was observed in a few cases.

C. Sodium Binding in the Presence of Sodium Ion and Potassium Ion

The Na^+ binding was measured at 0.3mM NaCl and 0-1mM KCl, 25mM imidazole, and 0.1mM EDTA. Fig. 3 a and b show voltage changes with additions of 180μM ATP in the presence of 0.3mM NaCl, 0.15mM KCl, and 1.26μM Na^+,K^+-ATPase, or 0.3mM NaCl, 0.5mM KCl, and 1.78μM Na^+,K^+-ATPase. Although the base lines drifted, the differences between the voltage changes with the first addition of ATP and those with the second were clearly observed in both cases (Fig. 3 a and b). No difference between the voltage changes with the second addition and with the third was noticed in the latter experiment (Fig. 3 b). While the phosphate determination was carried out after the experiment, hydrolysis of ATP during the experiment was not detected. Table (I) shows a control experiment which was carried out without the enzyme protein. There was no difference recognized among the voltage changes with each addition.

Exp I		0 μM Na^+,K^+-ATPase	
			Voltage change
	ATP	1st addition	-250μv
		2nd	-250
Exp II		3.03 μM Na^+,K^+-ATPase	
	ATP	1st addition	-240μv
		2nd	-160
		3rd	-160

TABLE1. Effect of ATP on the voltage change of the sodium sensitive electrode in the presence or absence of the enzyme. The sample solutions contain 0.3mM NaCl, 0.08mM KCl, 25mM imidazole-HCl (pH 7.3), and 0.1mM EDTA with or without 3.03μM Na^+,K^+-ATPase. 20μl of 13mM ATP solutions were added two or three times to 2.5ml of enzyme solution.

Detection of Sodium Binding

To clarify the effect of K^+ on Na^+ binding, ATP-dependent Na^+ binding was measured at various concentrations of KCl as shown in Fig. 4. The ATP dependent Na^+ binding increased with the increasing K^+ concentrations. The maximum binding, 1.2 moles Na^+/mole enzyme, was obtained at 0.3-0.5mM KCl. These ATP dependemt Na^+ bindings were reduced to less than 5% by previous heat treatment or ouabain treatment of the enzyme preparation. Na^+ binding was also observed when the ATP was replaced by ADP (91μM), and the amount was as much as 70% of that with ATP (107μM).

IV. DISCUSSION

Small changes of sodium ion concentration could be measured successfully with a ion sensitive electrode system which consisted of a small Na^+-sensitive glass electrode, a double junction reference electrode, and a low-noise operational amplifier with a varactor diode bridge. ATP-dependent

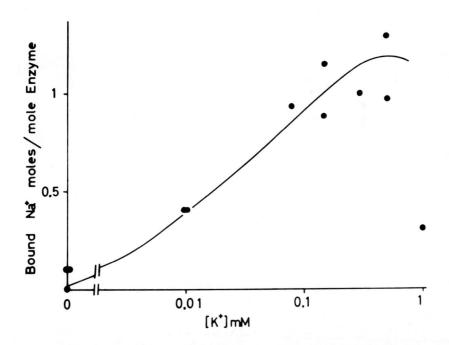

FIGURE 4. Effect of KCl on the ATP-dependent sodium binding. The sodium binding was measured in the presence of 1-3μM Na^+,K^+-ATPase, 0.3mM NaCl, 25mM imidazole-HCl (pH 7.3), 0.1mM EDTA, and various concentrations of KCl.

Na^+ binding to pig brain NaI enzyme was detected by means of this system. The electrode system responds to the activity of sodium ions rather than to their concentrations. In order to overcome this inconvenience, calibrations in terms of concentration were carried out in the enzyme solution before and/or after the ATP additions.

The decrease in sodium ion concentrations observed was due to the Na^+,K^+-ATPase, because this decrease was sensitive to ATP and reduced to almost zero by the treatment of the enzyme with ouabain or heat-treatment prior to use. The active transport of cations into membranous vesicles might have been involved in the change in the Na^+ concentration. However, the added ATP was not hydrolyzed in the absence of Mg^{++} and no increment in inorganic phosphate was observed. Therefore, the observed change in the Na^+ concentration was not considered to be due to the active transport of Na^+ into the vesicles.

The ATP dependent binding of sodium ions increased with the increase in KCl concentration. Therefore a potassium ion was necessary for the ATP-dependent sodium binding. On the other hand, it has been reported that potassium competes with ATP for binding to Na,K-ATPase (1,7). The replacement of potassium ions by sodium ions has been reported by Post et al.(8). Therefore, it is possible that the ATP-dependent sodium binding may occur with the removal of K^+ from the enzyme.

REFERENCES

1) Matsui,H., Hayashi,Y., Homareda,H., and Kimimura,M. (1977) B.B.R.C. 75:373.
2) Matsui,H., and Homareda,H., (1978). Proceedings of the 4th Japan Bioenergetics Group Meeting 4:120.
3) Kaniike,K., and Lindenmayer,G.E., (1976). J.Biol. Chem. 251:4794.
4) Hastings,D.F., (1977). Anal.Biochem. 83:416.
5) Nakao,T., Nakao,M.,Kawai,K., Fujihira,Y., Hara,Y., and Fujita,M. (1973). J.Biochem. 73:781.
6) Lowry,O.H. Rosenbrough,N.J., Farr,A.L., and Randall,R.J., (1951). J.Biol.Chem. 193:265.
7) Nørby,J.G., and Jensen,J.,(1974). Anals of the New York Academy of Sciences 242:158.
8) Post,R.L., Hegyvary,C., and Kume,S., (1972). J.Biol.Chem. 247:6530.

OUABAIN-SENSITIVE AND STOICHIOMETRICAL BINDING OF
SODIUM AND POTASSIUM IONS TO Na^+,K^+-ATPase

Hideo Matsui[1]
Haruo Homareda

Department of Biochemistry
Kyorin University School of Medicine
Mitaka, Tokyo, Japan

INTRODUCTION

Membrane bound Na^+,K^+-ATPase is considered the molecular pump which performs sodium and potassium transport across cell membrane by the energy liberated from ATP hydrolysis (1). Interaction of Na^+ and K^+ to Na^+,K^+-ATPase is a central problem of the cation transport, and has been mainly studied kinetically by testing the effect of the cations to ATP hydrolysis (2-4), and to partial reactions such as phosphorylation and dephosphorylation of the enzyme (5-7), ATP-ADP exchange (8, 9) and nucleotide binding (10-14). Although the direct investigation of the cation binding to the enzyme is important to the understanding of molecular mechanism of the ion pump, these experiments are few (15) because of the following difficulties. 1) Number of the specific binding site is far less than that of the nonspecific binding sites. 2) Difference of affinity for the cation between the specific site and nonspecific site is not sufficiently high (two figures or so). 3) Particularly, affinity of the specific Na^+ binding is low (dissociation constant is greater than one tenth mM). To overcome these disadvantage, use of the highly active enzyme preparation is necessary by increase of concentration of the specific sites (16-19). The specific

---

[1]Supported by grant from the Ministry of Education of Japan.

K^+ binding was demonstrated as ouabain-sensitive K^+ binding from our laboratory by centrifugation method with the purified enzyme and $^{42}K^+$ (18). Determination of the specific Na^+ binding was more difficult than of K^+ binding because of the much lower affinity of Na^+ than of K^+. This paper reports that the specific Na^+ binding is also demonstrated as the ouabain-sensitive binding. Characteristics of the ouabain-sensitive Na^+ and K^+ binding are in well agreement with the kinetical observations on the effect of Na^+ and K^+.

METHODS

Enzyme Preparations. The Na^+,K^+-ATPase was prepared from canine kidney outer medulla by the method developped in our laboratory (20). Specific activities of the enzyme preparations were about 1,000-1,500 µmol $Pi \cdot h^{-1} \cdot mg^{-1}$.

$^{22}Na^+$ *and* $^{42}K^+$ *Binding.* The ouabain-sensitive $^{22}Na^+$ and $^{42}K^+$ bindings were determined by the centrifugation method reported previously (18) with slight modification. One half milligram of the enzyme was preincubated with or without 0.2 mM ^3H-ouabain in a total volume of 0.5 ml containing 1 mM EDTA/imidazole alone at 37° for 10 min. The enzyme suspension was cooled in an ice bath. Then, $^{22}NaCl$ or ^{42}KCl together with Imidazole/HCl buffer pH 8.0 (for Na^+ binding) or pH 7.0 (for K^+ binding) at 0° and other ligands, if necessary, were added to the enzyme mixture in a final volume of 1 ml. The reaction mixture was centrifuged at 40,000 rpm 20 min. The precipitate separated from supernatant was dissolved in 0.3 ml of 0.2N NaOH and transferred into counting vial. Radioactivities of ^{22}Na or ^{42}K and of ^3H were counted simultaneouly by a Beckman liquid scintillation spectrometer. The ouabain-sensitive Na^+ or K^+ binding was defined as the difference of radioactivity between the absence and the presence of ouabain.

^3H-*Ouabain Binding.* The radioactivity of ^3H-ouabain recovered in the precipitate was counted simultaneously with that of ^{22}Na or ^{42}K as mentioned above. This ^3H-ouabain is total level of ouabain which consists of the specifically bound ouabain, nonspecifically bound ouabain and unbound ouabain contained in water space of the precipitate (21). The latter two were determined by the control run contained 100 mM Na^+, 100 mM K^+ and 5 mM Mg^{2+} together with ^3H-ouabain and omitted preincubation. The specific ouabain binding, therefore, was calculated from the total level of ^3H-ouabain by subtracting the value of control run. The experiments were done in duplicate for K⁺ binding and in triplicate for Na⁺ binding. Standard deviations of the each determination were usually within 3 % of the observed values.

RESULTS

Ouabain-Sensitive K⁺ Binding. When centrifugation method is applied to the measurement of the cation binding to Na⁺,K⁺-ATPase, the precipitate fraction contains not only the specifically bound cation to the enzyme, but also non-specifically bound cation and unbound cation which is contained in the water space of the precipitate. Table I displays these relationship. At 0.2 mM KCl, 16.4 nmol (per mg protein) of total ^{42}K⁺ found in the precipitate consisted of 4 nmol of ouabain-sensitive specific ^{42}K binding, about 8 nmol of nonspecific binding which was displaced by 5 mM Mg^{2+} or 100 mM Na , and 4 nmol of unbound ^{42}K⁺ which was determined after addition of high level of cation to displace the specific and binding. The ouabain-sensitive ^{42}K⁺ binding was replaced by congeners of K⁺. The replacement order of the congeners for K⁺ binding was the same to their relative sequence of apparent affinity for ATP hydrolysis and dephosphorylation of the phosphoenzyme (Fig. 1.)(4, 7).

TABLE I. Ouabain-Sensitive ^{42}K⁺ Binding to Na⁺,K⁺-ATPase with various ligands

Exp.	Additions	^{42}K⁺ (nmol·mg⁻¹)		
		−Ou	+Ou	Δ
I[a]	None	16.4	12.8	3.6
	0.2 mM ATP	13.9	11.9	2.0
	5 mM MgCl₂	8.1	4.0	4.1
	" + 0.2 mM ATP	6.0	4.2	1.8
	" + 0.2 mM Pi	5.6	4.3	1.3
	100 mM NaCl	4.2	—	—
	" + 5 mM MgCl₂	3.9	—	—
II[b]	None	4.2	0.5	3.7
	10 mM NaCl	0.4	0.4	0

[a] Concentration of ^{42}KCl was 0.2 mM. Specific ouabain binding was 2.5 nmol·mg⁻¹.
[b] Concentration of ^{42}KCl was 0.02 mM. Specific ouabain binding was 2.2 nmol·mg⁻¹.

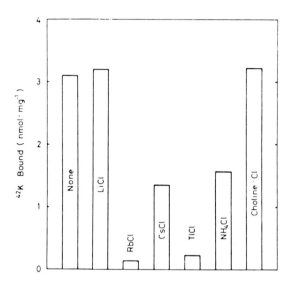

Fig. 1. Displacement of ouabain-sensitive K⁺ binding by congeners of K⁺. Concentration of ^{42}KCl was 20 µM and those of congeners of K⁺ were 0.3 mM.

Stoichiometry of the ouabain-sensitive K⁺ binding was 2 mol of bound K⁺ per mol of bound ouabain in the presence (18) or absence of Mg^{2+} (Fig. 2). Although an apparent dissociation constant (K_d) was approximately 50 µM in the presence of 5 mM Mg^{2+} as reported previously (18), it decreased to 4.1 µM in the absence of Mg^{2+} (Fig. 2).

As shown in Fig. 2, Scatchard plot of the ouabain-sensitive K⁺ binding showed an unusual curved line. This pattern indicates that the two binding sites to K⁺ are not equivalent but cooperative. Assuming that after binding of one K⁺ to the first site the affinity of the second site to K⁺ increases from zero, i.e. sequential binding of two potassium ions, the following relationships are proposed,

$$E + M \rightleftharpoons EM \qquad [E][M]/[EM] = K_1 \qquad (1)$$

$$EM + M \rightleftharpoons EMM \qquad [EM][M]/[EMM] = K_2 \qquad (2)$$

$$[E_t] = [E] + [EM] + [EMM] \qquad (3)$$

where K_1 and K_2 denote K_d's for the first and the second potassium binding respectively, [M] is free potassium concentration, and $[E_t]$, [E], [EM] and [EMM] are concentrations of total, free, one potassium bound and two potassium bound

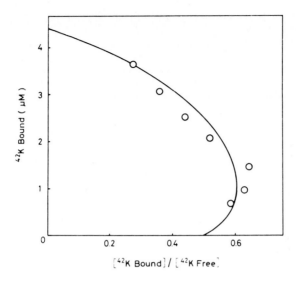

Fig. 2. Scatchard type plot of ouabain-sensitive K^+ binding without Mg^{2+}. The ouabain-sensitive $^{42}K^+$ binding measured at various concentrations of K^+ (O) was plotted for Scatchard type analysis in which the X-axis and Y-axis for the regular Scatchard plot (22) were inverted. The specific ouabain binding was 2.2 nmol·mg^{-1}. The solid line represents the best fit of equation (5) to the observed values with K_1 = 4.4 µM and K_2 = 3.8 µM. The apparent K_d of ouabain-sensitive K^+ binding was 4.1 µM which was $\sqrt{K_1K_2}$.

enzyme respectively. Then, we obtain the total bound potassium $[M_B]$ as :

$$[M_B] = [EM] + 2[EMM] = \frac{[E_t][M](K_2 + 2[M])}{K_1K_2 + K_2[M] + [M]^2} \quad (4)$$

To obtain Scatchard type plot, substitution of $[M_B]/[M]$ and $[M_B]$ by X and Y respectively gives a second order equation.

$$K_1K_2X^2 + K_2XY + Y^2 - [E_t]K_2X - 2[E_t]Y = 0 \quad (5)$$

If $K_2 < 4K_1$ in equation (5), the theoretical curve shows a part of ellipse. When K_1 = 4.4 µM, K_2 = 3.8 µM were applied, the calculated curve was well fit for the observed values (Fig. 2). In equation (4) substitution of $[M_B]$ by $[E_t]$, which means half maximal binding, gives $\sqrt{K_1K_2}$ for the apparent K_d.

TABLE II. Ouabain-Sensitive $^{22}Na^+$ Binding to Na^+,K^+-ATPase with various ligands

Additions	$^{22}Na^+$ (nmol·mg^{-1})		
	-Ou	+Ou[b]	Δ
None[a]	31.2	24.1	7.1
0.2 mM KCl	22.3	22.9	-0.6
0.2 mM ATP[c]	25.6	20.8	4.8
0.2 mM Pi	28.1	24.1	4.0
50 mM Choline Cl	16.0	11.9	4.1
5 mM MgCl$_2$	9.7	9.7	0
" + 100 mM NaCl 100 mM KCl	6.8	6.7	0.1

[a]Concentration of $^{22}NaCl$ was 0.5 mM.
[b]Specific ouabain binding was 3.7 nmol·mg^{-1}.
[c]About 10 mM Tris/HCl was contained.

Ouabain-Sensitive Na^+ Binding. In the experiment of Na^+ binding, the binding was also ouabain-sensitive as same as the K^+ binding was (Table II). Since the affinity of Na^+ to the enzyme is rather low and therefore a higher concentration (0.5 mM) of Na^+ than K^+ has to be used for the detection of Na^+ binding, the ratio of ouabain-sensitive bound $^{22}Na^+$ to total radioactivity in precipitate was less than one fourth because of the increase of nonspecific $^{22}Na^+$ bound and unbound $^{22}Na^+$. Nevertheless, the distinct differences of the values between the presence and the absence of ouabain were observed. The ouabain-sensitive Na^+ binding was specifically inhibited by low concentration of K^+ (0.2 mM) at which the nonspecific Na^+ binding did not show any decrease. Contrary to the K^+ binding, ATP had no effect to the ouabain-sensitive Na^+ binding. The specific Na^+ binding reached nearly a saturation level at the concentrations more than 0.5 mM of NaCl in the reaction mixture, while the nonspecific binding was increased linearly with the NaCl concentrations increased (Fig. 3). Although precise determination of the saturation level is difficult, repeated experiments suggests that two mol of Na^+ are bound to one unit of enzyme bound one mol of ouabain. Scatchard plot of the experimental data did not show a straight line, but fit a curved line calculated from the equation (5) with $K_1 = 1$ mM and $K_2 = 0.07$ mM (Fig. 4). The results indicate that the Na^+ binding is also cooperative as K^+ does.

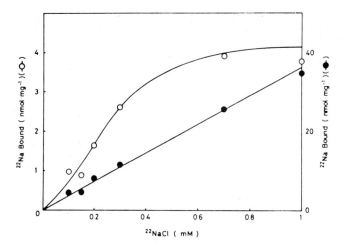

Fig. 3. Saturation curve of ouabain-sensitive $^{22}Na^+$ binding. Ouabain-sensitive $^{22}Na^+$ bindings (O) were measured at concentrations of Na^+ indicated, and its apparent K_d was 0.26 mM. The specific ouabain binding was 2.2 nmol·mg^{-1}. The ^{22}Na recovered in the precipitate fraction of ouabain-pretreated enzyme, i.e. the ouabain-insensitive $^{22}Na^+$ binding plus unbound $^{22}Na^+$ contained in the water space of the precipitate, (●) was also plotted.

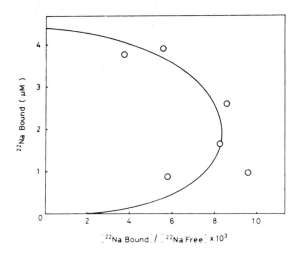

Fig. 4. Scatchard type plot of ouabain-sensitive $^{22}Na^+$ binding. The data of ouabain-sensitive $^{22}Na^+$ binding in Fig. 3 were plotted for Scatchard type analysis (O). Solid line represents the best fit of equation (5) to the observed values with K_1 = 1 mM and K_2 = 0.07 mM.

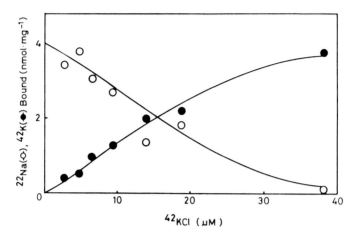

Fig. 5. Simultaneous measurement of $^{22}Na^+$ binding and ^{42}K binding by titration of K^+ against Na^+. The enzyme was incubated with a constant concentration (0.5 mM) of Na^+ and an increasing concentration of K^+. As the two radiation energy of $^{22}Na^+$ and $^{42}K^+$ could not be detected differentially, the radioactivity of $^{22}Na^+$ was determined after decay of ^{42}K (10 days later), and the radioactivity of ^{42}K was calculated by subtraction of ^{22}Na count from the total count. These procedure for simultaneous determination of $^{22}Na^+$ and $^{42}K^+$ binding considerably increased the experimental error.

Relation Between K^+ Binding Site and Na^+ Binding Site. The quantitative survey of K_d for Na^+ binding and K_i for Na^+ ihhibition to K binding, and of K_d for K^+ binding and K_i for K^+ inhibition to Na^+ binding is necessary to consider the relation between Na^+ binding site and K^+ binding site. At present, sufficient data for each constant are not available. However, a double labeling experiment with $^{22}Na^+$ and $^{42}K^+$ would provide an information that the Na^+ site and K^+ site exist simultaneously or not. Fig. 5 is a preliminary result of this type of experiments. Numbers fo binding sites both for Na^+ and for K^+ were not additive. The result, no simultaneous existence of the both site, is incompatible with the half-of-the sites model (22, 23).

DISCUSSION

The apparent K_d for K^+ binding in the absence of Mg^{2+} was 4.1 μM (Fig. 2) which was one tenth of the K_d in the presence of Mg^{2+} (18). This value of K_d is far less than the $K_{0.5}$'s for K^+ activation to ATP hydrolysis (3, 4) and dephosphorylation of phosphoenzyme (7, 25), and for K^+ inhibition to ouabain binding (26). It is also ten times as low as the K_d measured by differential potentiometric titration with K^+ selective electrode (19). These discrepancies might be not due to the difference of ion binding sites, but due to the difference of ligand conditions such as Na^+, Mg^{2+}, ATP, Pi, and buffer and to the difference of experimental temperature. The decrease of temperature increases the affinity of K^+ to the enzyme (26).

The K_d for Na^+ binding, 0.26 mM is the same value as the $K_{0.5}$ for Na^+ to ATPase activation which was determined by extrapolation to the absence of Mg^{2+} and K^+ by Robinson (27). Contrary to K^+, the Na^+ activation was not affected by the change of reaction temperature (26). Kaniike et al. (17) reported a high affinity Na^+ binding with K_d of 0.3 mM. They applied the centrifugation mehtod to the Na^+ binding experiment, however, the K_d for Na^+ and number of binding sites were obtained by simple application of Schatchard plot. As shown in Fig. 3 and 4, the specific and ouabain-sensitive Na^+ binding exhibited a cooperativity and Scatchard plots did not give a straight line. Therefore, the K_d and the number of high affinity binding sites which they presented is not necessarily equal to those of the ouabain-sensitive binding we presented here. Ostroy et al. studied for Na^+ binding by NMR method (16). They showed an increase of Na^+ binding by the addition of ATP, however, the value of binding is too high to explain the stoichiometrical relationship.

The Na^+ and K^+ observed as the ouabain-sensitive binding described in this report would be the same cations that stimulate the Na^+,K^+-ATPase activity and are transported by the enzyme, in view of the following specificities of the binding. 1) The high affinity of the ouabain-sensitive binding is explainable to the various $K_{0.5}$'s for Na^+ and K^+ obtained from kinetical experiments. 2) The order of inhibition to K^+ binding by congeners of K^+ is the same to that of increase in the rate of hydrolysis of the phosphoenzyme made from ATP (7, 25). 3) The specific inhibition to Na^+ binding by low concentration of K^+ (Table II). 4) The effect of ATP, inhibition of K^+ binding but no effect to Na^+ binding, shows a reverse correlation with the effect of cations to ATP binding (10-14). 5) Although the stoichiometry between the ouabain-sensitive Na^+ binding and the ouabain binding is not yet conclusive at present, the results obtained suggest that the ratio of bound Na^+ to bound K^+ and to bound ouabain is 2:2:1 instead of 3:2:1. The study of the cation binding to the Na^+,K^+-ATPase in equilibrium and the study of the cation transport in erythrocytes in dynamic state are quite different phases or different levels of phenomena for the molecular mechanism of ion pump. Therefore, this binding ratio of 2:2:1 may not be completely incompatible with the stoichiometry of transport of 3:2:1 (28, 29).

The measurement of simultaneous binding of Na^+ and K^+ in a double labeling experiment did not show an additive binding of Na^+ and K^+. This result has an apparent similarity to the result of simultaneous phosphorylation experiment from ATP and Pi in which the level of phosphoenzyme was not additive but constant (30). These results are obviously against the half-of-the-sites model. If the enzyme operates in a half-of-the-sites system, the coupling between oligomers should not be tight to resolve the incompatibility.

ACKNOWLEDGMENTS

We are grateful to Dr. Y. Hayashi, Dr. M. Taguchi and Miss Y. Matsumura for their skilful cooperation to the purification of the enzyme.

REFERENCES

1. Glynn, I.M., and Karlish, S.J.D. (1975). Ann. Rev. Physiol. 37:13.
2. Skou, J.C. (1957). Biochim. Biophys. Acta 23:394.
3. Skou, J.C. (1960). Biochim. Biophys. Acta 42:6.
4. Rossi, B., Gache, C. and Lazdunski, M. (1978). Eur. J. Biochem. 85:561.
5. Albers, R.W., Fahn, S. and Koval, G.J. (1963). Proc. Natl. Acad. Sci. U.S.A. 50:474.
6. Post, R.L., Sen, A.K. and Rosenthal, A.S. (1965). J. Biol. Chem. 240:1473.
7. Post, R.L., Hegyvary, C. and Kume,S. (1972). J. Biol. Chem. 247:6530.
8. Fahn, S., Koval, G.J. and Albers, R.W.(1966). J. Biol. Chem. 241:1882.
9. Banerjee, S.P. and Wong, S.M.E. (1972). J. Biol.Chem. 247:5409.
10. Nørby, J.G. and Jensen, J. (1971). Biochim. Biophys. Acta 233:104.
11. Hegyvary, C. and Post, R.L. (1971). J. Biol. Chem. 246:5234.
12. Robinson, J.D. (1976). Biochim. Biophys. Acta 429:1006
13. Karlish, S.J.D., Yates, D.W. and Glynn, I.M. (1978). Biochim. Biophys. Acta 525:230.

14. Karlish, S.J.D., Yates, D.W. and GLynn, I.M. (1978). Biochim. Biophys. Acta 525:252.
15. Ahmed, K. and Judah, J.D. (1966). Biochim. Biophys. Acta 112:58.
16. Ostroy, F., James, T.L., Hoggle, J.H., Sarrif, S. and Hokin, L.E. (1974). Arch. Biochem. Biophys. 162:421.
17. Kaniike, K., Lindenmayer, G.E., Wallick, E.T., Lane, L.K. and Schwartz, A. (1976). J. Biol. Chem. 251:4794.
18. Matsui, H., Hayashi, Y., Homareda, H. and Kimimura, M. (1977). Biochem. Biophys. Res. Commun. 75:373.
19. Hastings, D.F. (1977). Anal. Biochem. 83:416.
20. Hayashi, Y., Kimimura, M., Homareda, H. and Matsui, H. (1977). Biochim. Biophys. Acta 482:185.
21. Matsui, H. and Schwartz, A. (1968). Biochim. Biophys. Acta 151:655.
22. Scatchard, G. (1949). Ann. N.Y. Acad. Sci. 51:660.
23. Stein, W.D., Lieb, W.R., Karlish, S.J.D. and Eilam, Y. (1973). Proc. Natl. Acad. Sci. U.S.A. 70:275.
24. Repke, K.R.H., Schön, R., Henke, W., Schönfeld, W., Streckenbach, B. and Dittrich, F. (1974). Ann. N.Y. Acad. Sci. 242:203.
25. Post, R.L., Taniguchi, K. and Toda,G. (1974). Ann. N.Y. Acad. Sci. 242:80.
26. Inagaki, C., Lindenmayer, G.E. and Schwartz, A. (1974) J. Biol. Chem. 249:5135.
27. Robinson, J.D. (1977). Biochim. Biophys. Acta 482:427.
28. Sen, A.K. and Post, R.L. (1964). J. Biol. Chem. 239:345.
29. Garrahan, P.J. and Glynn, I.M. (1967). J. Physiol. 192:217.
30. Post, R.L. (1977). In "Biochemistry of Membrane Transport" (Semenza, G. and Carafoli, E. eds.), FEBS-Symposium No. 42. p.352. Springer-Verlag, Berlin, Heidelberg, New York.

INTERACTION OF INORGANIC PHOSPHATE WITH Na^+,
K^+-ATPase: EFFECT OF P_i ON THE K^+ DEPENDENT
HYDROLYSIS OF P-NITROPHENYLPHOSPHATE[1]

Kazuya Taniguchi
Hideaki Tazawa
Terufusa Tamaki
Shoichi Iida

Department of Pharmacology
School of Dentistry
Hokkaido University Sapporo Japan

I. INTRODUCTION

Na^+, K^+-ATPase accepts a phosphate group from ATP or P_i to form phosphoenzyme (Post et al.,1975). Treatment of K^+ sensitive phosphoenzyme formed from P_i with high concentration of Na^+ developed sensitivity to ADP with concomitant loss of sensitivity to K^+ and synthesis of ATP (Taniguchi and Post 1975). The results indicate that ATP is broken down via ADP sensitive phosphoenzyme (E_1P) and K^+ sensitive phosphoenzyme (E_2P). There appeared to be at least four reactive states of phosphoenzyme which equilibriate with inorganic phosphate. The site of attachment of the phosphate group is the same and the active site was determined to be a β-aspartyl phosphate of 10,000 dalton of peptide of Na^+, K^+-ATPase (Post et al., 1975 ; Degani et al., 1974 ; NISHIGAKI et al., 1974). It is very important to estimate the energy level of phosphoenzyme to clarify the mechanism of the energy transduction in Na^+, K^+-ATPase. The energy level of E_1P is assumed to be 1 Kcal/mol lower than the free energy of hydrolysis of ATP (Post et al., 1973). In this paper thermodynamic constant of K^+ complexed phosphoenzyme ($KMgE_2P$) formation of pig kidny Na^+, K^+-ATPase was estimated from the apparent dissociation constant for Mg^{2+}

[1]Supported by Grant-in-Aid for Scientific Research (149001) by Japan Ministry of Education during 1976 to 1978 .

and P_i. $KMgE_2P$ is assumed to appear as the final acid stable phosphoenzyme in the reaction sequence of Na^+, K^+-ATPase (Taniguchi & Post, 1975). These apparent dissociation constants were obtained kinetically from concentration dependency of p-nitrophenylphosphate (PNPP), Mg^{2+} and P_i on K^+ dependent PNPP hydrolysis (Nagai et al., 1966 ; Post et al., 1972).

II. RESULTS

A. K^+ Dependent PNPPase Activity in the Presence of P_i

A working hypothesis of PNPP hydrolysis and $KMgE_2P$ formation in the presence of Mg^{2+}, P_i and K^+ is presented in scheme 1. The step of PNPP hydrolysis with rate constant k_3 is assumed to be irreversible (Hobbs and De Weer, 1976) and other steps are assumed to be in rapid equilibrium.

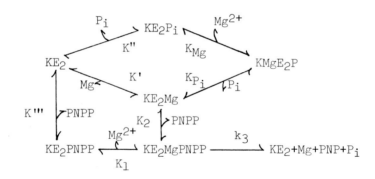

Scheme I

Where K', K'', K''', K_1, K_2, K_{P_i} and K_{Mg} are dissociation constants for coresponding reactions shown in the scheme. PNPPase activity, V is given :

$$V = V_m^{ap}\left(1 + \frac{K_m^{ap}}{[PNPP]}\right)^{-1} \quad \text{where} \quad V_m^{ap} = k_3 Et\left(1 + \frac{K_1}{[Mg]}\right)^{-1}$$

$$\text{and} \quad K_m^{ap} = K_2\left[1 + \frac{[P_i]}{K_{P_i}}\left(1 + \frac{K_{Mg}}{[Mg]}\left(1 + \frac{K''}{[P_i]}\right)\right)\right]\left(1 + \frac{K_1}{[Mg]}\right)^{-1}$$

Apparent K_m for PNPP and V_m in the presence of 40 mM K^+ and 100 mM Tris-HCL at pH 7.4 in the absence or presence of 0.5, 1, 2 and 4 mM of P_i were calculated from the activity in the presence of 1, 2, 4, 8 and 16 mM of PNPP using the direct linear plot of Eisenthal and Cornish-Bowden (1974) with an aid of a computor. K_1 and k_3E_t at 0°C were calculated to be 2.0 mM and 15 μmoles PNP/mg/hr from V_m^{ap} values in the presence of 0.525, 1.25, 2.4, 4.9 and 9.9 mM of Mg^{2+}. Data pairs of K_m^{ap}/V_m^{ap} and P_i concentrations at constant Mg concentration gave single straight line with coefficient of determination about 0.99. Values of K_2, K', K_{P_i} and K_{Mg} were calculated to be 9.3, 1.2, 1.8 and 0.32 mM from the reciplocal values of slope and intercept and Mg^{2+} concentrations by the method described above. Thus all the dissociation constants in the scheme 1 were calculated.

where $K' = K_{Mg} \cdot K''/K_{P_i}$ and $K''' = K_2 K'/K_1$

B. Temperature Dependency of Binding of Mg and P_i to Potassium Bound Enzymes

Values of apparent dissociation constant (K_d) obtained at representative temperatures are shown in Table I. The enzyme increased the affinity for Mg several fold by binding of P_i. Affinity for Mg of KE_2 and KE_2P_i was decreased at higher temperature. The enzyme increased the affinity for P_i several fold by binding of Mg. Affinity for P_i of KE_2 and KE_2Mg was not changed remarkably with temperature.

The ratio of affinity of KE_2 to P_i and to Mg^{2+} was about 6 at 0° and decreased to 3 at 40°.

Table I. Apparent Dissociation Constants[a] for Mg and P_i

Complex	KE_2Mg	$KMgE_2P$	KE_2P_i	$KMgE_2P$
Dissociation form	$Mg+KE_2$	$Mg+KE_2P_i$	P_i+KE_2	P_i+KE_2Mg
Temperature °C				
0	1.2	0.32	7.0	1.8
6.1	1.1	0.36	5.7	1.9
20	1.1	0.38	5.2	1.8
25	1.4	0.38	6.4	1.8
34	1.9	0.61	6.0	1.9
40	2.2	0.60	7.0	1.9

[a] Units in mM

C. Thermodynamic Constants of Enzyme Ligand Complexes and Potassium Complexed Phosphoenzyme (KMgE$_2$P) Formation.

From the values of apparent K_d at different temperatures (Table I), apparent standard free energy change $\Delta G°$ at 0° and 20°, enthalpy change $\Delta H°$ and entropy change $\Delta S°$ were calculated for the formation of enzyme ligand complexes and KmgE$_2$P formation. The folowing equations were used:

$$\Delta G° = -RT\ln(10^3 \cdot K_d^{-1}), \quad \Delta S° = (\Delta H° - \Delta G°) \cdot T^{-1}$$

$\Delta H°$ was estimated from the slope of van't Hoff plot. Van't Hoff plot of binding of Mg^{2+} to KE$_2$P$_i$ and KE$_2$ are shown in Fig. 1. The plot gave two intersecting straight lines for both cases at around 20°C.

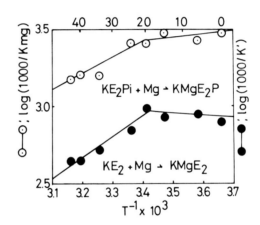

Fig. 1. Van't Hoff plot of apparent association constants. Each points were estimated from the apparent dissociation constants for Mg and Pi at temperatures between 0° and 43° as described in the text. Unit in the upper abscissa is °C.

Thermodynamic constant estimated from the data shown in the Table I, Fig. 1 and from other experiments are shown in Table II.

Binding of Mg to KE$_2$ and KE$_2$P$_i$ was coupled with -3.6 and -4.3 Kcal/mol of $\Delta G°$ with increase in 14 and 12 e.u./mol of $\Delta S°$ respectively with little change in $\Delta H°$ below 20°. Above 20°, -4 and -4.6 Kcal/mol of $\Delta G°$ were estimated for these reactions but the driving force was -5.7 and -4.8 Kcal/mol of $\Delta H°$ respectively.

Binding of P$_i$ to KE$_2$ and KE$_2$Mg was coupled with -2.7 and -3.4 Kcal/mol of $\Delta G°$ with increase in 18 and 13 e.u./mol of

Table II. Thermodynamic Constants of Enzyme Ligand Complex Formation

Reactant	Temperature, °C					
	0 ~ 20			20 ~ 43		
	$\Delta G_0°$ Kcal/mol	$\Delta H°$ Kcal/mol	$\Delta S°$ e.u./mol	$\Delta G_{20}°$ Kcal/mol	$\Delta H°$ Kcal/mol	$\Delta S°$ e.u./mol
Mg + KE_2	-3.6	0.2	14	-4.0	-5.7	-5.8
Mg + KE_2P_i	-4.3	-0.9	12	-4.6	-4.8	-0.7
P_i + KE_2	-2.7	2.1	18	-3.0	-1.8	4
P_i + KE_2Mg	-3.4	0.2	13	-3.7	-0.2	11

$\Delta G_0°$, $\Delta G_{20}°$: Standard free energy changes at 0° and 20°.

$\Delta S°$ respectively below 20°. Above 20°, the similiar values of $\Delta G°$ were estimated for these reactions but the reactions were favored by both increase in $\Delta S°$ and slight decrease in $\Delta H°$.

Binding of Mg to KE_2 and subsequent binding of P_i and vice versa should form $KMgE_2P$ (Post et al., 1975). These data suggested that $KMgE_2P$ formation was coupled with -7 to -7.7 Kcal/mol of $\Delta G°$ with increase in 27 to 30 e.u./mol of $\Delta S°$ and decrease in 5.9 to 6.6 Kcal/mol of $\Delta H°$ above and below 20° respectively.

III. DISCUSSION

Present results suggested that binding of Mg^{2+} and P_i and vice versa to KE_2 caused -7 to -8 Kcal/mol of $\Delta G°$ which seems to be consistent with a hypothesis of the presence of low energy acyl phosphate. Thermodynamic constant estimated from apparent dissociation constants for Mg and P_i from acid stable phosphoenzyme (unpublished data) also seems to support the conclusion. The standard free energy change of aspartyl phosphate formation from free aspartic acid and P_i is reported to be 11.5 Kcal/mol in aqueous solution (Atkinson & Morton, 1960). The great difference between $\Delta G°$ of water soluble aspartyl phosphate formation and phosphoenzyme formation from P_i might be due to solvation effects of the ligands which have bean ignored (George et al., 1970). The solvation energy for physiological ligands of Na^+, K^+-ATPase are similiar to or much greater in magnitude than strong covalent bond energies, e.g. Na^+ 97, K^+ 77 and Mg^{2+} 259 Kcal/l (Halliwell and Nyburg,

1963). When those solvated ligands are moved to an active center of Na^+, K^+-ATPase, the binding of a ligand may induce the change in the aqueous environment of active site to nonaqueous one which might cause the desolvation of ligands. Na^+, K^+-ATPase would accept phosphate group from inorganic phosphate to form an aspartyl phosphate.

Driving force of Mg binding to KE_2P_i and KE_2 changed abruptly from increase in ΔS° to decrease in ΔH° at around 20°. Abrupt change in the apparent activation energy of Na^+, K^+-ATPase around 20° required phospholipids (Taniguchi and Iida, 1972). Importance of phospholipids in the phosphorylation reactions has been shown (Taniguchi and Tonomura, 1971; Taniguchi and Post, 1975). These data seem to suggest that Mg binding to Na^+, K^+-ATPase induces some conformational change in which phospholipids molecules participate. The data also suggest the importance of both enthalpy and entropy in energy transduction of Na^+, K^+-ATPase. Abrupt changes in ΔH° and ΔS° of P_i binding to Mg complexed Ca^{2+}, Mg^{2+}-ATPase of sarcoplasmic reticulum at around 18° has been reported recently (Kanazawa and Katabami., 1978), though the driving force for the reaction was the increase in entropy between 0° and 38°.

In the present experiments, a $KMgE_2P$ was aasumed to be present during the turn over of the enzyme. The thermodynamic constants for its formation were estimated from the temperature dependency of the apparent dissociation constants obtained from competitive inhibition of P_i to K^+-PNPPase activity. Another method to estimate the apparent dissociation constants is to measure the amount of acid stable $KMgE_2P$ under various Mg and P_i concentrations. The latter experiments are now in progress using the same lot of enzyme preparation described above. If thermodynamic constants obtained by two different methods agree well, it means that the acid stable $KMgE_2P$ or potassium magnesium enzyme phosphate complex which is indistinguishable to acid stable $KMgE_2P$ is present during turn over of the enzyme.

IV. SUMMARY

Thermodynamic constant for potassium complexed phosphoenzyme formation from P_i and Mg was estimated from apparent dissociation constants for Mg and P_i. These dissociation constants were obtained kinetically from competitive inhibition of P_i to potassium dependent p-nitrophenyl phosphatase activity. The phosphoenzyme formation was coupled with -7 to -8 Kcal/mol of standard free energy change with increase in entropy and decrease in enthalpy respectively below and above 20°.

ACKNOWLEDGMENTS

We wish to thank Drs. Y. Tonomura and R.L. Post for their helpful discussions and critical comments with respect to this work, and to K. Sato for the computer analysis.

REFERENCES

Atkinson, M.R., and Morton, R.K. (1960). in Comparative Biochemistry II (Florkin, M., and Mason, H.S, eds), p 1. Academic Press, New York and London.
Degani, C., Dahms, A.S., and Boyer, P. (1974). Ann. N.Y. Acad. Sci. 242:77.
Eisenthal, R., and Cornish-Bowden, A. (1974). Biochem. J. 139:715.
George, P., Witonsky, R.J., Trachtman, M., Wu, C., Dorwart, W., Richman, L., Richman, W., Shurayh, F., and Lentz, B. (1970). Biochem. Biophys. Acta 223:1.
Halliwell, H.F., and Nyburg, S.C. (1963). Trans. Farady Soc., 59:1126.
Hobbs, A.N., and De Weer, P. (1976). Arch. Biochem. Biophys. 173:386.
Kanazawa, T., and Katabami, H. (1978). This symposium.
Nagai, K., Izumi, F., and Yoshida, H. (1966). J. Biochem. 59:295.
Nishigaki, I., Chen, A., and Hokin, L.E. (1974). J. Biol. Chem. 249:4911.
Post, R.L. Hegyvary, C., and Kume, S. (1972). J. Biol. Chem. 247:6530.
Post, R.L., and Kume, S. (1973). J. Biol. Chem. 248:6993.
Post, R.L., Kume, S., and Rogers, F.N. (1973). in Mechanisms In Bioenergetics (Azzone, G.F., Ernster, L., Papa, S., Quagliariello, E., and Silliprandi, N., eds). p 203. Academic Press, New York and London.
Post, R.L., Toda, G., and Rogers, F.N. (1975). J. Biol. Chem. 250:691.
Taniguchi, K., and Tonomura, Y. (1971). J. Biochem. 69:543.
Taniguchi, K., and Iida, S. (1972). Biochem. Biophys. Acta. 274:536.
Taniguchi, K., and Post, R.L. (1975). J. Biol. Chem., 250:3010.

BINDING OF Na$^+$ IONS TO THE Na$^+$, K$^+$-DEPENDENT ATPase

Motonori Yamaguchi

Department of Biology
Osaka University
Toyonaka, Osaka, Japan

Binding of Na$^+$ ions to the Na$^+$, K$^+$-ATPase has been examined by Kaniike et al. and Homareda et al., using the ultracentrifugation method (1,2). However, it is very difficult to measure by this method the binding of Na$^+$ ions to the enzyme during the ATPase reaction. I devised a new method, by which the binding can be measured rapidly. The reaction mixtures, containing ^{22}NaCl, ^{3}H-Glucose and the enzyme, were incubated at 0°C, and then rapidly filtrated on two piled membrane filters (∅ 0.4 μm). The enzyme was completely trapped on the upper filter, and the amount of ^{22}NaCl in the filtrate trapped on the lower filter was determined by counting ^{22}Na. The volume of the filtrate was determined by counting ^{3}H-Glucose. Non-specific binding was removed by the addition of 75 mM choline-Cl.

I Na$^+$ binding without the ATPase reaction

(i) The amount of ^{22}Na ions bound to the enzyme is about 8.0 μmoles/g protein in 0.8 mM ^{22}NaCl and no Mg^{2+} ions. (ii) Na$^+$ ions bound to the enzyme can be displaced by the ions as follows; Rb$^+$, K$^+$, NH$_4^+$, Li$^+$, Cs$^+$ (in the order of effectiveness). (iii) Na$^+$ ions bound to the enzyme can be displaced by Mg^{2+} ions. (iv) The amount of Na$^+$ ions displaced by K$^+$ or Mg^{2+} ions is 3.5 times as that of the maximum amount of a phosphorylated intermediate. (v) In the presence of EDTA, Na$^+$ binding is unaffected by ATP.

II Na$^+$ binding during the ATPase reaction

(i) Na^+ ions bound to the enzyme in the presence of 0.8 mM NaCl and 0.1 mM $MgCl_2$ are rapidly released by adding ATP. They rebind to the enzyme 2 min later. (ii) EP is rapidly formed and decomposed within 30 sec. Na^+-dependent P_i-liberation also stops within 30 sec.

These results suggest that the Na^+, K^+-ATPase has at least three binding sites for Na^+ ions, and the affinity of Na^+ ion to the enzyme changes by its reaction with ATP.

The author wishes to thank Professor Y. Tonomura for his valuable suggestions and encouragement.

1. Kaniike, K., Lindenmayer, G. E., Wallick, E. T., Lane, L. K., and Schwartz, A. (1976). J. Biol. Chem. 215, 4794-4795.
2. Homareda, H., Hayashi, Y., and Matsui, H. (1977). 3rd Annual Meeting of Japan Bioenerg. Group, Tokyo, 41-43.

RHEOGENIC Na PUMP IN THE TOAD SKIN

F.Lacaz-Vieira[1]
W.A.Varanda
L.H.Bevevino
D.T.Fernandes

Department of Physiology
Institute of Biomedical Sciences
University of São Paulo
Sao Paulo, Brazil

The intimate nature of the mechanism involved in the Na transport across amphibian skins is still deeply controvertial particularly regarding the Na-K coupling at pump level in the basolateral membrane of the epithelial cells. Transient alterations of 42 K unidirectional fluxes across the isolated short-circuited skins have been shown to be a powerful tool in the study of Na and K movements across the tissue and their relationships (Varanda and Lacaz-Vieira,1979). In this study, experimental results have been compared to computer simulated predictions. Experiments were performed in the short-circuited state. K efflux transients observed upon the action of ouabain are compared to the theoretical predictions obtained by simulation based on a 2 intraepithellial series compartment model which adequately describes K efflux. Inhibition of the Na pump located in the basolateral membrane of the epithelial cells was simulated by using differential equations which describe the model, having as initial boundary conditions those obtained after 42-K efflux had reached the steady state. Numerical integration was carried out by the forth order Runge-Kutta method. Two alternative approaches were used. The first assumed the existence of a non-rheogenic Na pump in the basolateral membrane with the electrical potential difference across this barrier given by a diffusion potential. Simulation of pump inhibition predicts a monotonic decline of the rate of 42-K discharge into the outer bathing solution, J_{21}^K (see insert, curve a). This behavior is not

---

[1] Supported by grant from FAPESP - 76/1376 and 76/0006

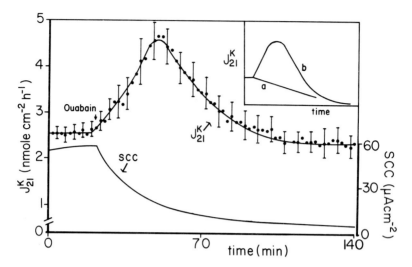

Fig.1 Effect of ouabain on the short-circuit current (SCC) and on the rate of 42-K discharge into the outer bathing solution, J_{21}^K. Insert: computer predictions regarding J_{21}^K. See text for details.

in agreement with the experimental results. The second model assumed a rheogenic Na pump as the primary cause of the electrical potential difference across the inner barrier and, due to the short circuited condition, also across the outer barrier. Simulation of Na pump inhibition predicts transient increase in J_{21}^K, followed by a decline with time (see insert, curve b). This transient behavior predicted by the rheogenic Na pump model was observed experimentally (see fig.1) except for the final steady state value which did not fall below the control steady-state value. Reasons for this discrepancy are unknown. Increase in the K permeability of the outer barrier due to small cell swelling could possibly be a reason. The results do not lead to a clear conclusion regarding the intimate nature of the rheogenic Na pump, whether it is a pure Na pump or a Na-K pump with a Na/K ratio significantly greater than 1. However, as for the transient increase in J_{21}^K induced by ouabain to be significant, the contribution of the rheogenic component of the electrical potential difference has necessarely to be of large magnitude.

REFERENCE

Varanda,W. and Lacaz-Vieira,F. (1979).Submitted to publication.

ACTIVE AND PASSIVE ION TRANSPORT BY GASTRIC VESICLES

George Sachs
Edd Rabon
Gaetano Saccomani

Laboratory of Membrane Biology
University of Alabama Medical Center
Birmingham, Alabama

I. INTRODUCTION

Various cations are transported actively across biomembranes, by direct coupling to a metabolic process. H^+ is translocated by bacteria, mitochondria and chloroplasts in an electrogenic (uniport) process, Na^+ and K^+ by a plasma membrane $Na^+ + K^+$ ATPase, Ca^{++} by sarcoplasmic reticulum or plasma membrane ATPase. In this paper, we shall review the ATPase derived from gastric parietal cells (1). This enzyme was described as a K^+ activated, ouabain insensitive enzyme (2) and later shown to be capable of H^+ transport (3,4). We shall refer to this enzyme as an $H^+ + K^+$ ATPase. What makes this H^+ transporting enzyme interesting is its role in producing the largest H^+ gradient in vertebrates, namely $10^{6.6}$. As the following data will show, the enzyme acts as an electroneutral exchanger of H^+ for K^+. While this raises certain physiological difficulties, it is an excellent model for ion translocation mechanisms, especially H^+.

In this paper, we will review general characteristics of the enzyme and its transport function. For methodological details, the referenced papers provide the methods.

---

[1]Supported by NIH grant AM 15878; NSF PCM 77-08951 and NSF PCM 78-09208.

II. GENERAL CHARACTERISTICS

1. **PURIFICATION.** A combination of differential and density gradient centrifugation and free flow electrophoresis results in a 40-fold enrichment of the K^+-activated ATPase from hog gastric mucosa. Associated with this, on single dimensional SDS gels, a single group of 105,000 M_r polypeptides is left (5). There is evidence however that this group of peptides is not homogeneous. For example, isoelectric focussing patterns show considerable heterogeneity (Figure 1) and tryptic hydrolysis shows a varying resistance of the peptides. Carbohydrate staining shows further that this region contains a glycoprotein.

2. **KINETICS.** The K_M for the H^+-K^+ ATPase with respect to ATP is $\sim 10^{-4}$ M. The enzyme in intact vesicles shows an ionophore stimulation (6) and ionophores do not affect the K_A for K^+ ($\sim 10^{-3}$ M) but increase the V_{max} (Figure 2). Thus, the K^+ site is not modified, but the accessibility is. The H^+-K^+ ATPase has a transition temperature of 28.1 C, and an E_A of 10.7 Kcal above and 31.1 Kcal below this temperature. With labelled ATP, a phosphoprotein is formed, with a 105,000 M_r (7) discharged by K^+. The pH dependence of the phosphorylation/dephosphorylation is shown in Figure 3. Since optimal ATPase is found at pH 7.4, the dephosphorylation is rate limiting. The phosphoenzyme is chased by CDTA, K^+ and ATP, and ADP. The rate of dephosphorylation appears adequate to account for ATPase activity. Turnover number (V/EP) is 10 sec^{-1} at 20°C.

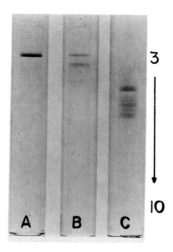

FIGURE 1. SDS gel of free flow purified fraction before (A) and after (B) trypsin treatment for 15 min at 30°C at a 100/1 ratio, and gel isoelectric focussing with pH gradient 3-10 (C).

3. PHOSPHOLIPIDS. As monitored by phospholipase A_2 digestion, these are required for ATPase activity. Intact vesicles show progressive inhibition with respect to ATPase, but not pNPPase, and phospholipids PE $\geq$ PC > PS restore activity. In lyophilized preparations, both enzyme activities are affected (ATPase > pNPPase) but only ATPase activity is reconstituted with phospholipid addition. The loss of phospholipid in the two types of preparation allows a provisional assignment of asymmetry to the bilayer (Figure 4) (8).

4. PEPTIDE COMPOSITION. Tryptic digestion destroys enzyme activity in a biphasic reaction (9) (K_I = 2.88 x 10^{-3}; K_{II} = 0.34 x 10^{-3} sec^{-1}). Associated with this, there is an increase of phosphoenzyme, but a constant quantity of P-E is chased by K^+. Specifically ATP and ADP protect activity in the slow phase and also 60% of the 105,000 M_r region. 40% of the peptide is degraded to 78,000 M_r and 30,000 M_r peptides. In the absence of ligands, 87K and 47K peptides are phosphorylated. Cross-linking with Cu^{++} phenanthroline produces higher M.W. trimers. 35% of the peptide is resistant to trypsin. Irradiation is also consistent with at least a dimeric structure for the ATPase and transport activity (10) and this is modelled in Figure 5. The overall enzyme activity is thus:

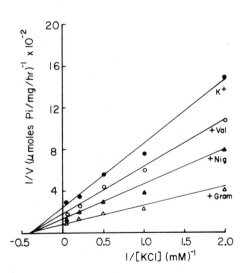

FIGURE 2. Lineweaver-Burk plot of 1/v vs 1/k as a function of different ionophores.

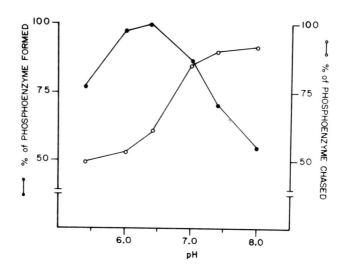

FIGURE 3. The pH dependence of formation of E-Pi and discharge of E-Pi by 20 mM K^+ at 4° C.

$$ATP + E \xrightleftharpoons{Mg^{++}} ATP - E$$
$$ATP - E \xrightleftharpoons{Mg^{++}} ADP - E - Pi$$
$$ADP - E - Pi \xrightleftharpoons{} E - Pi_I + ADP$$
$$E - Pi_I \xrightleftharpoons{H^+} E - Pi_{II}$$
$$E - Pi_{II} \xrightarrow{K^+} E + Pi$$

III. H^+ TRANSPORT

1. <u>BASIC PROPERTIES</u>. The addition of small quantities of ATP to vesicles in KCl solution results in transient uptake of H^+. Increasing time of preincubation in KCl, or the addition of valinomycin accelerates and increases H^+ uptake (12). With disappearance of ATP, the H^+ gradient dissipates, with a $t\frac{1}{2}$ of 300 sec. From

$$P = \ln 2 \cdot r/3 \, t_{\frac{1}{2}} \qquad P_{H^+}, \, _{OH^-} = 10^{-8} \text{ cm/sec}$$

The leak of H^+ is reduced in SO_4^{--} solutions, and not much increased by SCN^-, indicating that the major leak component is as HCl or $K^+:H^+$ antiport. The K^+ requirement for transport appears to be intravesicular, or at least dependent on access

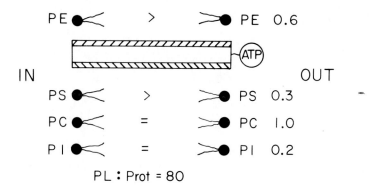

FIGURE 4. The arrangement of glycerophospholipids associated with the K^+-ATPase as determined by phospholipase A_2 action. The relative ratios are also shown.

of K^+ to a pathway present on the inner surface of the ATPase external group of vesicles. Preincubation for varying lengths of time gives a family of curves, eventually resulting in an overshoot of the H^+ gradient (Figure 6). Addition of valinomycin to the steady state increases H^+ gradient to a new level in the presence of ATP. Two interpretations are suggested, an H^+ uniport coupled to a K^+ conductance or a $H^+:K^+$ exchange pump. Modelling the former alternative shows that no overshoot can be anticipated. With an $H^+:K^+$ exchange pump where K^+ is required internally, osmotic sensitivity of the signal would depend on K^+ entry being affected. Thus, with preincubation, osmotic sensitivity is progressively lost, substantiating the $H^+:K^+$ exchange hypothesis (11).

2. ELECTRONEUTRAL EXCHANGE. The exchange could be neutral or electrogenic. To distinguish this, uptake of lipid permeable anions or cations (^{14}C-SCN^-, or diethyloxodicarbocyanine) can be used. The addition of ATP does not induce measurable movements of either, but the addition of valinomycin to the former, tetrachlorsalicylanilide (TCS) to the latter, by inducing K^+ or H^+ diffusion potentials, reveals responses of either probe (Figure 7). Nigericin inhibits gradients of either ion, additionally demonstrating lack of a pump potential. There are no discernible osmotic effects of ATP addition when sucrose con-

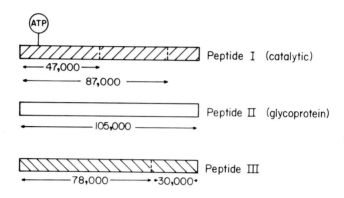

FIGURE 5. The structure of the 105,000 M_r region as revealed by trypsin hydrolysis of catalytic peptide (I); glycopeptide (II) that is trypsin resistant; and tryptic hydrolysis of ligand-insensitive peptide (III).

taining vesicles are mixed with isotonic KCl. This would also not be anticipated if net flow of HCl occurred. The K_A for intravesicular K^+ for transport is considerably larger than the K_A for ATPase activation (without or with ionophore) showing a difference between transport site and activation site (12).

3. <u>SIZE of GRADIENT</u>. At pH 6.1, the change in external pH (ΔH_o) can be monitored directly using pH electrodes or bromcresol green (BCG) and changes of internal pH (ΔH_i) using weak base probes (9-aminoacridine, ^{14}C-aminopyrine) potential probes + TCS (DOCC, di-SC$_3$-(5)) or acridine orange. The ΔH_o in our best preparations is 100 nmol/µl vesicle volume. The calculated ΔH_i ranges from 1-50 nmol/µl, depending on the probe. Although it is easy to dismiss these data as due to probe deficiencies, it is also intuitive that the ΔpH is considerably less than 5.3 at pH 6.1 and 6.6 at pH 7.4 (11), not due to buffering artefacts, unknown leaks or non-uniformity of vesicles. A different explanation has to be sought such as H^+ binding eg. 18 mol/mol catalytic subunit.

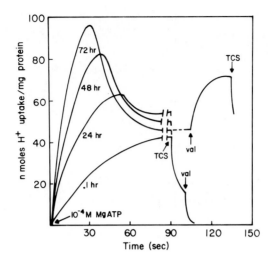

FIGURE 6. The effect of increasing times of KCl incubation on H^+ uptake by gastric vesicles showing the development of an overshoot, and the requirement of both TCS and valinomycin for dissipation.

IV. K^+ TRANSPORT

1. GENERAL ASPECTS. Uptake of $^{86}Rb^+$ into an osmotically sensitive space is slow, the P at 4° C being 8.3×10^{-12} cm/sec at 22° C, 1.1×10^{-9} and at 37° C, 4×10^{-9} cm/sec (13). From the P value, and the area of 1 mg of vesicles, 600 cm^2, the flux of Rb^+ is 6 nmol/mg/min in the absence of val-SCN$^-$. Clearly therefore, in unequilibrated vesicles, K^+ entry rate limits an $H^+:K^+$ exchange. An impermeant anion will even further restrict the pump. In contrast to the slow net flux of cation, a rapid cation:cation exchange is present, giving an isotopic permeability of 10^{-8} cm/sec. The slow K^+ entry that is also anion dependent argues against (a) a specific carrier system in unenergized vesicles and (b) unless the rate is changed, against the K^+ leak path being adequate to account for secretory rates observed in vivo, since a mammalian stomach can secrete of the order of 40 µE/effective cm^2/min, many times that of the K^+ movement, either measured directly in the stomach, or in this vesicle preparation.

2. ACTIVE TRANSPORT. Following equilibration of the vesicles, addition of ATP results in active K^+ efflux (13). The

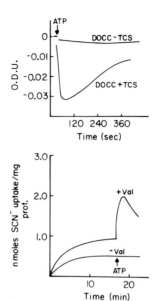

FIGURE 7. The effect of the addition of ATP on the uptake of a lipid permeable cation (DOCC) ± TCS, or on a lipid permeable anion, SCN⁻, ± valinomycin.

quantity effluxed increases with increasing purification of the vesicles from microsomal to gradient to free flow electrophoretic fraction. The data with valinomycin and ANS⁻ (14) confirms the presence of a K^+ gradient simultaneous with an H^+ gradient (Figure 8). Since ionophores do not prevent formation of the gradient, it is not secondary to development of a vesicle potential. Equally, since electrogenic ionophores do not block H^+ uptake, this is not due to an electrogenic K^+ efflux. All alkali cations show this phenomenon, the sequence being $K^+ = Rb^+ > Cs^+ >> Na^+$, as for the proton transport, in vesicles or in tissue. A Hill plot as a function of $\log [K^+]$ gives a slope of about 2, both for transport and ATPase activity (Figure 9).

3. K^+ CONDUCTANCE. This is very low. Application of a pH pulse (5.5-8.5) to vesicles equilibrated with acridine orange results in a deflection of O.D. This dissipates with a $t_{½}$ of 1.1×10^3 sec, whether TCS (a protonophore) is present or not. The addition of KCl or K_2SO_4 accelerates the dissipation rate 10-fold, with or without TCS indicating the presence of a $K^+:H^+$ antiport. Valinomycin in the absence of TCS further accelerates dissipation to a $t_{½}$ of 50 sec, showing the presence of an endogeneous H^+ conductance. In the presence of TCS, the dissipation $t_{½}$ is 25 sec (Figure 10). Thus, the vesicles contain no K^+ conductance, an H^+ conductance and a $K^+:H^+$ antiport.

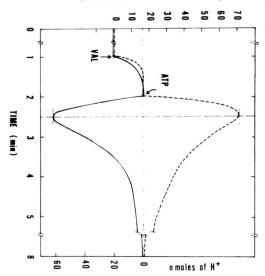

FIGURE 8. The relationship between development of a pH gradient (upper curve) and the valinomycin depend on ANS fluorescence enhancement with the addition of ATP.

V. Cl^- TRANSPORT

1. <u>GENERAL ASPECTS.</u> Cl^- influx has the same time constants as Rb^+ uptake. As for the latter, there is a slight increase of Cl^- rate with the addition of valinomycin. Addition of ATP after Cl^- equilibration, in contrast to the cation data, does not change Cl^- distribution. Preliminary data however, with ATP addition to the microsomal fraction of rabbit, did show a Cl^- uptake (15). An active Cl^- component could have been lost with the purification of the hog vesicles.

2. <u>SELECTIVITY.</u> Anion effects are seen on H^+ uptake. The sequence for H^+ transport (initial velocity) is $I > NO_3^- > Br^- > Cl^- > SO_4^{--}$, isethionate. This anion selectivity sequence corresponds to that of intact tissue as well as liposomes. Pre-equilibration of vesicles with K^+-anion still shows anion effects, since maintenance of the H^+ gradient requires anion entry to compensate for the H^+ anion leak. The leak is reduced in SO_4^{--}, and ageing, trypsin or chymotrypsin treatment induce a Cl^- conductance (16).

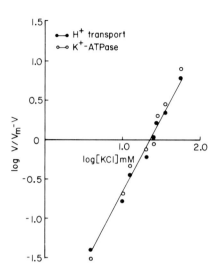

FIGURE 9. A Hill plot of H^+ transport and ATPase activity of $[K^+]$ giving a slope of 2.

VI. NET ION FLUXES

Apart from isotopic techniques, ion transport can be measured in a stop-flow spectrophotometer by light scattering changes following application of osmotic gradients. Pathways for ions can be further defined by the use of lipid permeable co-ions (eg. SCN^-, $TPMP^+$). A gradient of urea results in shrinking of the vesicles with a $t_{1/2}$ < 1 sec and reswelling with a $t_{1/2}$ of 5 sec, giving a $P = 4.6 \times 10^{-7}$ cm/sec. A KCl gradient gives the same shrinking rate, but a very slow reswelling, that is temperature dependent (Figure 11). Surprisingly, valinomycin only minimally accelerates the reswelling phase, although a large increase in H^+ uptake rate is observed. This effect prior to these stop-flow measurements would have been explained by increased K^+ entry into the water space of the vesicles for subsequent $K^+:H^+$ exchange. Since valinomycin apparently does not induce net flow of K^+.val $.Cl^-$, the effect of valinomycin on H^+ uptake could be due to K^+.val cycling within the membrane itself. ATP does not alter these osmotic properties, hence the KCl permeability is maintained at low levels, insufficient to account for the H^+ rate. An additional factor must be present such as Ca^{++} (17), or net flow of K^+ across the active membrane does not occur.

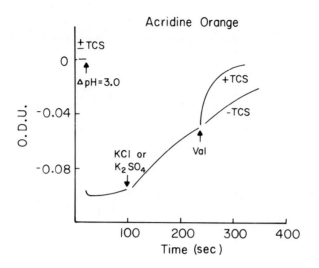

FIGURE 10. The use of acridine orange ± TCS to monitor an applied pH gradient showing the initial slow dissipation, accelerated by K^+ due to $K^+:H^+$ antiport, and accelerated by valinomycin only in the presence of TCS.

VII. MODEL OF TRANSPORT

From all the above, a KCl entry step that is rate limiting and passive combined with an $H^+:K^+$ exchange that is active and electroneutral is supported (Figure 12). The relevant equations are:

$$dH_i / dt = k_o (E - \ln H_i/H_o + \ln K_i/K_o)H_o - K_{DH} (H_i - H_o)$$

$$dK_i / dt = -k_o''' (E - \ln H_i/H_o + \ln K_i/K_o)K_I - K_{DK} (K_i - K_o)$$

where E is ATP driving force, k_o is pump rate constant, K_{DH}, DK are diffusion constants. At steady state, with 1:1 coupling,

$$K_{DH} (H_i - H_o) = K_{DK} (K_i - K_o)$$

and hence

$$H_i = K_{DK} (K_i - K_o) + K_{DH} H_o / K_{DH}$$

These equations predict the characteristics shown in Figure 6,

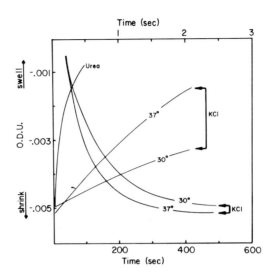

FIGURE 11. A stop-flow study of gastric vesicle permeability to H_2O (30, 37° C, shrinkage) (top axis) and to urea and KCl (swelling, 30, 37° C) (bottom axis).

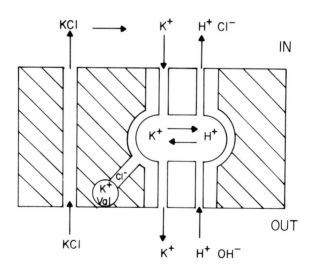

FIGURE 12. A model of $K^+:H^+$ exchange in gastric vesicles consistent with the net flux data and indicating the possibility of intramembranal cycling.

namely an overshoot in H_i as a function of $[K_i]$.

VIII. RELATIONSHIP TO INTACT TISSUES

In spite of the internal consistency of the data for the vesicles, there are certain differences from those predicted which must be emphasized, namely (a) continued transport in the presence of SO_4^{--}, glucuronate, isethionate; (b) lack of SCN^- effect in vesicles, in spite of large concentrations, although frog transport is sensitive; (c) a large deficit in the anticipated gradient; (d) no adequate explanation for the absolute O_2 or complex I dependence of acid secretion. These problems must be solved before the vesicles as isolated above can be accepted as a complete model for acid secretion.

REFERENCES

1. Saccomani, G., Crago, S., Mihas, A.A., Dailey, D., and Sachs, G., Acta Physiol. Scand. Special Suppl. 293 (1978).
2. Ganser, A.L., and Forte, J.G., Biochim. Biophys. Acta. 307: 169 (1973).
3. Lee, J., Simpson, G., and Scholes, P., Biochem. Biophys. Res. Comm. 60:825 (1974).
4. Sachs, G., Rabon, E., Saccomani, G., and Sarau, H.M., Ann. N.Y. Acad. Sci. 264:456 (1975).
5. Saccomani, G., Stewart, H.B., Shaw, D., Lewin, M., and Sachs, G., Biochim. Biophys. Acta 465:311 (1977).
6. Ganser, A.L., and Forte, J.G., Biochim. Biophys. Res. Comm. 54:690 (1973).
7. Saccomani, G., Shah, G., Spenney, J.G., and Sachs, G., J. Biol. Chem. 250:4802 (1975).
8. Chang, H.H., Spisni, A., Sachs, G., Spitzer, H.L., and Saccomani, G., in preparation (1978).
9. Saccomani, G., Dailey, D., and Sachs, G., J. Biol. Chem. submitted (1978).
10. Chang, H.H., Saccomani, G., Rabon, E., Schackmann, R., and Sachs, G., Biochim. Biophys. Acta 464:313 (1977).
11. Rabon, E., Chang, H.H., and Sachs, G., Biochemistry 17: 3345 (1978).
12. Sachs, G., Chang, H.H., Rabon, E., Schackmann, R., Lewin, M., and Saccomani, G., J. Biol. Chem. 251:7690 (1976).
13. Schackmann, R., Schwartz, A., Saccomani, G., and Sachs, G., J. Membr. Biol. 32:361 (1977).

14. Lewin, M., Saccomani, G., Schackmann, R., and Sachs, G., J. Membr. Biol. 32:301 (1977).
15. Soumarmon, A., and Racker, E., Proc. Johnson Found. Symp. in press (1978).
16. Rabon, E., Kajdos, I., and Sachs, G., Biochim. Biophys. Acta submitted (1978).
17. Michelangeli, F., and Proverbio, F., Acta Physiol. Scand. Special Suppl. 399 (1978).

REACTION SCHEME FOR THE Ca-ATPase FROM
HUMAN RED BLOOD CELLS

Alcides F. Rega
Patricio J. Garrahan
Héctor Barrabin
Alberto Horenstein
Juan P. Rossi

Departamento de Química Biológica
Facultad de Farmacia y Bioquímica
Universidad de Buenos Aires
Buenos Aires, Argentina

I. INTRODUCTION

Under normal conditions the concentration of Ca ion in the cytoplasm of human red cells is less than 10^{-7} M while that in plasma is $1.5 \cdot 10^{-3}$ M. It is clear from these values that even disregarding the low membrane potential of red cells, the distribution of Ca between intra and extracellular media in circulating red cells is far from equilibrium. This condition is maintained by a system that actively pumps Ca out of the cells. ATP hydrolysis is the energy source for this process which can therefore be considered to proceed according to the following equation:

$$n\ Ca_i^{2+} + ATP_i \longrightarrow n\ Ca_0^{2+} + ADP_i + Pi_i$$

where i and 0 mean inside and outside respectively and n is a stoichiometry factor whose value (1 or 2) remains to be ascertained (see Schatzmann and Roelofsen, 1977).

Supported by grants from the Consejo Nacional de Investigaciones Científicas y Técnicas (CONICET), Argentina and from the Programa de Desarrollo Científico y Tecnológico de la Organización de los Estados Americanos.

A.F.R. and P.J.G are established investigators of the CONICET.

II. PROPERTIES OF THE MEMBRANE PREPARATION USED FOR THESE STUDIES

Fragmented membranes from human red blood cells were used (Garrahan et al., 1969). Recent studies have demonstrated that depending on the procedure used for isolation of the membranes, Ca-ATPase will be either combined or free of cytoplasmic activator (Bond and Clough, 1973; Farrance and Vincenzi, 1977). Procedures based on hypotonic hemolysis in media containing calcium-chelating agents render Ca-ATPase devoid of activator, while procedures based on hemolysis in either isotonic solutions without chelators or hypotonic solutions with low amounts of Ca render membranes with the activator bound to the Ca-ATPase.

Results in Fig. 1a allow to compare the response to Ca of Ca-ATPase of red cells lysed in a 150 mM NaCl solution with that of membranes prepared by lysis in a 1 mM EDTA; 30 mM TrisHCl (pH 7.4) solution. Membranes from cells lysed in 150 mM NaCl (Isotonic) possess higher Ca-ATPase activity than membranes from cells lysed in EDTA containig solutions (EDTA). As judged by the concentration needed for half-maximum ATPase activity, Ca-ATPase from "Isotonic" membranes has a higher apparent affinity for Ca than that from "EDTA" membranes. Reciprocal plot of the data in Fig. 1a (Fig. 1b) stresses the differences between the Ca-ATPase in the two membrane preparations. The experimental points of the activation curve of the enzyme from "Isotonic" membranes fall on a straight line from whose intercept at the abcisa an apparent dissociation constant of $0.6\ \mu M$ can be calculated. The curve of the enzyme from "EDTA" membranes shows a break suggesting the presence of two populations of enzyme: one with high ($14\ \mu M$) and the other with low ($2.1\ \mu M$) apparent dissociation constant for Ca. This complicated behaviour of the Ca-ATPase from membranes prepared in EDTA containing solutions was first described by Schatzmann in 1973.

Combination of the enzyme with the activator depends on Ca. In fact, after deprivation of Ca by exposure to CDTA "Isotonic" membranes show a Ca-ATPase whose behaviour is not different than that of "EDTA" membranes (Fig 1a and b). On the other hand incubation of "EDTA" membranes with activator results in membranes whose Ca-ATPase shows an apparent dissociation constant for Ca almost identical to that of the enzyme of "Isotonic" membranes.

These results confirm previous findings by others (Scharff and Foder, 1977; Farrance and Vincenzi, 1977) and indicate that i) maximum activation and apparent affinity for Ca of the Ca-ATPase depend on whether the enzyme is associated or not with the cytoplasmic activator; ii) binding of the activator to the

Reaction Scheme from Human Red Blood Cells

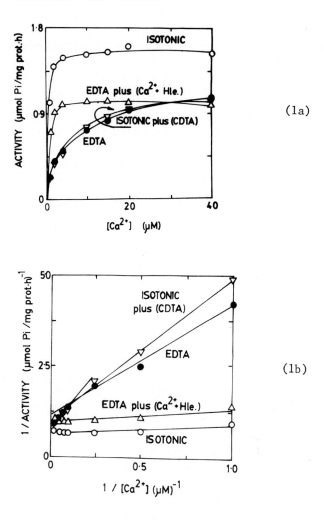

FIGURE 1a and b. Activation by Ca of Ca-ATPase from human red cell membranes prepared by different procedures. To obtain isotonic membranes a high-speed pellet of cells suspended in 150 mM NaCl was frozen at -70°C and then submitted to one wash with 20 volumes of 150 mM TrisHCl (pH 7.7 at 37°C) and three washes with 30 mM TrisHCl (pH 7.7 at 20°C). EDTA membranes were prepared as described previously (Garrahan et al., 1969). EDTA plus (Ca^{2+} + Hle) are EDTA membranes that were incubated for 20 min at 37°C in a cell-free hemolysate diluted 5 times with a 30 uM $CaCl_2$ solution. Isotonic plus (CDTA) membranes are Isotonic membranes after incubation for 20 min at 37°C in a 1 mM CDTA; 15 mM TrisHCl (pH 7.7 at 20°C) solution. Ca-ATPase activity was measured as described before (Richards et al., 1978). (Unpublished results).

ATPase needs Ca at low concentration; and iii) the response to Ca of the Ca-ATPase from membranes that have been exposed to EDTA during isolation is that of Ca-ATPase free of activator. Whether in the intact cell the activator is combined or not with the Ca-ATPase remains an open question (Scharff and Foder, 1978).

Exposure of membranes to EDTA during isolation does not result in irreversible damage of the Ca-ATPase. It should be mentioned however that when "activated" Ca-ATPase is to be assayed, the assay medium should not contain chelating componds. For this reason the experiments presented here were performed with fragmented membranes prepared by hypotonic lysis in EDTA containing solutions and hence devoid of activator since to study the effects of divalent cations on partial reactions, the use of strong chelating agents is necessary.

III. PROPERTIES OF THE PHOSPHOENZYME

As in other transport ATPases, hydrolysis of ATP by the Ca-ATPase in human red cells is a multi-stage process involving the formation of a phosphoenzyme (EP) followed by its hydrolysis (Knauf et al., 1974; Katz and Blostein, 1975; Rega and Garrahan, 1975). The idea that EP is indeed the phosphorylated ATPase is supported by the following:

EP has a high molecular wheight and its phosphate moiety is covalently bound.

EP undergoes rapid turnover.

The sensitivity of the phosphate bond to hydroxylamine and pH is that characteristic of acylphosphates.

The K_d for Ca during phosphorylation (7.0 μM) is very close to that for ATPase activity (9.2 μM).

The K_m for ATP during phosphorylation (1.6 μM) is very close to the K_m of the ATPase measured under similar conditions (2.5 μM).

Although maximum activation of the Ca-ATPase requires Mg, the reaction of formation of EP takes place even in media prepared without $MgCl_2$ (Rega and Garrahan, 1975), that is media containing 1 μM or less contaminating Mg. This is consistent with the finding that though at a lower rate, Ca-dependent ATP hydrolysis and hence binding of ATP at the active center of the Ca-ATPase takes place in the absence of added Mg (Richards et al. 1978). However, at 0°C Mg lowers the half-time of formation of the phosphoenzyme and doubles the steady-state level of phosphoenzyme (Garrahan and Rega, 1978). It will be shown later that these effects of Mg during the phosphorylation reaction are not the only cause of the large increase in ATPase activity elicited by Mg.

The peptide region around the phosphate in the Ca-ATPase from human red cells has the same electrophoretic behaviour than that of the Ca-ATPase from sarcoplasmic reticulum of rat skeletal muscle and that of the Na,K-ATPase of human red cells. This is shown in the experiment of Fig. 2. Each enzyme was phosphorylated with ($^{32}P\gamma$)ATP in media containing the cations indicated in the figure. After denaturation with TCA they were digested with pronase to a limit tripeptide carying radioactive phosphorus. The limit peptides were submmited to high-voltage electrophoresis at pH 2 and the radioactivity on the paper was measured. Results in Fig. 2 show that in red cells phosphorylated in media with either Na or Ca there is a peak of radiactivity which is absent in red cells phosphorylated in media with Mg alone, and whose mobility is the same regardless of

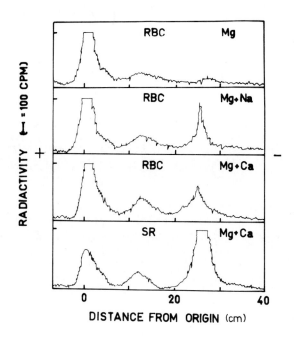

FIGURE 2. The distribution of radiactivity after high-voltage electrophoresis at pH 2 of pronase digests of red cell membranes (RBC) phosphorylated in media containing either Mg, Mg plus Na or Mg plus Ca, and of sarcoplasmic reticulum from rat skeletal muscle phosphorylated in medium containing Mg plus Ca. Isolation and electrophoresis of the limit peptides were performed as described by Bastide et al.(1973). (Unpublished results).

whether the media contained Na or Ca. Other experiments showed that the Na- or Ca-dependent peaks are sensitive to hydroxylamine. The mobility of these Ca- or Na-dependent phosphopeptides is not different to the mobility of the Ca-dependent phosphopeptide from sarcoplasmic reticulum. These results extend to Na,K-ATPase and Ca-ATPase from red cell membranes previous findings of Bastide et al.(1973) who compared the Na,K-ATPase from guinea pig kidney with the Ca-ATPase from rabbit muscle. It seems therefore that, as judged by the electrophoretic mobility of the limit phosphopeptide, the aminoacid sequence around the active site of the Ca-ATPase from red cells is similar to that of the active site of the Na,K-ATPase from red cells and of the Ca-ATPase from sarcoplasmic reticulum.

IV. THE HYDROLYSIS OF THE PHOSPHOENZYME

The rate of hydrolysis of EP is independent of Ca (Rega and Garrahan, 1975).

In the experiment shown in Fig. 3 phosphoenzyme made in the absence of Mg was chased with EGTA in a control medium, in a medium containing eithe Mg or ATP and in a medium with ATP plus Mg. The rate of hydrolysis of EP in a medium containing either ATP or Mg alone is not very different to that of the control. But when ATP and Mg are present together the reaction EP + H_2O ⇌ E + Pi proceeds at a much higher rate, suggesting that rapid hydrolysis of EP requires high concentrations of ATP and Mg.

In the experiment in Fig. 4 the enzyme was phosphorylated in the absence of Mg and 5 s before stopping the reaction, enough Mg was added to raise its concentration in the reaction medium to 0.5 mM. Phosphorylation was stopped by the addition of CDTA or CDTA plus ATP. It can be seen that even under conditions in which, due to the presence of CDTA, all the Mg is chelated, ATP still accelerates dephosphorylation, indicating that Mg and ATP are not needed simultaneously for rapid dephosphorylation. This result was taken as indicative that, as proposed before (Rega and Garrahan, 1975), Mg drives EP from a state of low reactivity towards water to a state which, provided ATP is also present, has high reactivity towards water (E'P). As judged by the concentration of $MgCl_2$ needed for half-maximum acceleration of the phosphorylation reaction in the presence of excess ATP, the apparent dissociation constant for Mg during the reaction of formation of E'P from EP is 0.08 mM. Although E'P could represent a class of phosphoprotein with Mg either boung with high energy or sequestered, the fact is that E'P does not need Mg in the medium to undergo rapid hydrolysis.

ATP accelerates dephosphorylation in the absence of Ca and

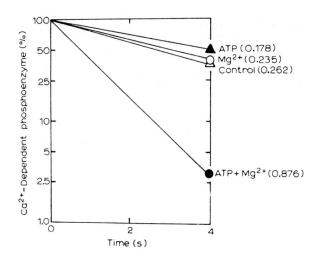

FIGURE 3. Dephosphorylation of phosphoenzyme made in the absence of Mg. All media contained 30 mM EGTA. The figures in brackets are rate constants in s^{-1}. ATP and Mg were 1 and 1.5 mM respectively (Garrahan and Rega, 1978).

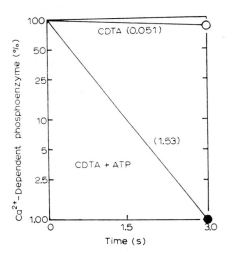

FIGURE 4. Dephosphorylation of phosphoenzyme in medium without Mg. CDTA and ATP were 20 and 1 mM respectively. The figures in brackets are rate constants in s^{-1} (Garrahan and Rega, 1978).

at much higher concentrations than those necessary for phosphorylation (Rega and Garrahan, 1975). The effects of ATP therefore can not be attributed to combination of the nucleotide at the catalytic center of the Ca-ATPase or to an additional Ca-dependent phosphorylation. It seems therefore that during dephosphorylation of the Ca-ATPase, ATP binds to E'P at a site which is different from the catalytic site.

Recent studies (Richards et al., 1978) have demonstrated that the Ca-ATPase from red cells has two sites for ATP. One has high affinity for ATP (Km 2.5 µM) and represents the catalytic site of the Ca-ATPase. The other has a lower affinity for ATP (Km 145 µM) and, provided Mg is present, its occupation largely increases the turnover of the enzyme. It is likely that this is the site at which ATP combines to accelerate the hydrolysis of E'P.

V. REVERSAL OF THE REACTION OF ATP HYDROLYSIS

Provided Ca is present, at high ADP the phosphorylation reaction is reversible both in the presence and in the absence of Mg (Rega and Garrahan, 1978). This means that EP as well as E'P are able to donate its phosphoryl group to ADP to form ATP. It seems therefore that formation of EP and E'P are reactions that take place with little change in bond energy and hence that EP and E'P have energy-rich phosphate bonds.

After showing that the reactions related to phosphorylation are reversible, the reversal of the overall reaction of ATP hydrolysis was attempted in red cells (Rossi et al., 1978). For this purpose cells that have been depleted of most of their ATP were loaded with (^{32}P) orthophosphate and treated with iodoacetamide to block glycolysis. The cells were incubated for short periods of time in media with and without 2 mM $CaCl_2$ and the amount of radiactivity in the ATP measured with the results shown in Table I. In media without Ca there is labelling of ATP probably because of residual glycolytic activity. In media with Ca the specific activity of ATP is almost doubled. The extra incorporation of radioactive phosphorus into ATP observed in the presence of Ca is absent when the cells are incubated in media containing either the ionophore A23187, which abolishes the Ca gradient across the membrane or $LaCl_3$ which inhibits the Ca-pump (Quist and Roufogalis, 1975). The effects of the ionophore and of $LaCl_2$ guive evidence that the extra incorporation of radioactive phosphorus from Pi into ATP observed in the presence of Ca is driven by the calcium gradient and is catalyzed by the Ca-pump. This result strongly suggests that, provided there is a calcium concentration gradient across the membrane, the overall reac-

TABLE I. The incorporation of $(^{32}P)P_i$ into ATP in starved and phosphate loaded cells.

Condition	Relative specific activity $\dfrac{cpm/mol\ ATP}{cpm/mol\ P_i}$
EGTA	0.175
Calcium	0.322
EGTA + 5µM A23187	0.130
Calcium + 5µM A23187	0.147
EGTA + 100µM LaCl$_3$	0.180
EGTA + 100µM LaCl$_3$	0.162

The concentrations of EGTA and calcium were 0.1 and 2 mM respectively. (Rossi et al.,1978).

tion of ATP hydrolysis by the Ca-ATPase can be reversed.

VI. A REACTION SCHEME

Results presented in so far suggest that hydrolysis of ATP by the Ca-ATPase from red cell membranes is reversible and proceeds through a series of reactions that depend on Ca, Mg and ATP as follows.

$$E + ATP \xrightleftharpoons{Ca,\ Mg} E{\sim}P + ADP \qquad (1)$$

$$E{\sim}P \xrightarrow{Mg} E{\sim}^!P \qquad (2)$$

$$E{\sim}^!P + H_2O \xrightleftharpoons{ATP} P_i + E' \qquad (3)$$

$$E' \xrightleftharpoons{} E \qquad (4)$$

Although Mg is not required, it increases the rate of reaction (1). When Mg is absent E∼P undergoes hydrolysis at a low rate and Ca-ATPase activity is about 5% of maximum. At high Mg and low ATP reaction (1) is rapid, the level of phosphoenzyme in the steady state is maximum and E∼'P is formed (reaction 2). Because of the low ATP concentration reaction (3) proceeds at a low rate. Although higher than in medium without Mg, at low ATP and high Mg, Ca-ATPase represents about 15% of maximum. The need of high ATP concentration for rapid hydrolysis of E∼'P (reaction 3) may well be the cause of this low ATPase activity. Under conditions in which all the efectors are non-limitant the rate of formation of E∼P and E∼'P are high. The rate of hydrolysis of E∼'P is maximum and so is the

ATPase activity. It is known that all the effects of ATP and Mg in the Ca-pump are excerted from the inner surface of the cell membrane. In fresh human red cells the intracellular concentration of ATP ranges from 1 to 1.5 mM (Bartlett, 1959) and that of Mg is about 0.7 mM (Gerber et al., 1973). These concentrations are enough for ATP and Mg to exert fully their effects on the Ca-ATPase during catalysis of the reactions shown here.

REFERENCES

Bartlett, G.R. (1959). J. Biol. Chem. 234:459.
Bastide, F., Meissner, G., Fleischer, S. and Post, R.L. (1973). J. Biol. Chem. 248:8385.
Bond, G.H. and Clough, D.L. (1973). Biochim. Biophys. Acta 323:592.
Farrance, M.L. and Vincenzi, F.F. (1977). Biochim. Biophys. Acta 471:49.
Garrahan, P.J., Pouchan, M.I. and Rega, A.F. (1969). J. Physiol. 204:305.
Garrahan, P.J. and Rega, A.F. (1978). Biochim. Biophys. Acta 513:59.
Gerber, G., Berger, H., Janig, G.R. and Rapoport, S.M. (1973). Eur. J. Biochem. 38:553.
Katz, S. and Blostein, R. (1975). Biochim. Biophys. Acta 314:324.
Knauf, P.A., Proverbio, F. and Hoffman, J.F. (1974). J. Gen. Physiol. 63:324.
Quist, E.E. and Roufogalis, B.D. (1975) FEBS Lett. 50:135.
Rega, A.F. and Garrahan, P.J. (1975). J. Membrane Biol. 22:313.
Rega, A.F. and Garrahan, P.J. (1978). Biochim. Biophys. Acta 507:182.
Richards, D.E., Rega, A.F. and Garrahan, P.J. (1978). Biochim. Biophys. Acta 511:194.
Rossi, J.P.F.C., Garrahan, P.J. and Rega, A.F. (1978). J. Membrane Biol. (In the press).
Scharff, O. and Foder, B. (1977). Biochim. Biophys. Acta 483:416.
Scharff, O. and Foder, B. (1978). Biochim. Biophys. Acta 509:67.
Schatzmann, H.J. and Roelofsen, B. (1977). In "Biochemistry of Membrane Transport" (G. Semenza and E. Carafoli, ed.), p. 389 Springer-Verlag, Berlin.
Schatzmann, H.J. (1973). J. Physiol. 235:551.

CONFORMATION OF VARIOUS REACTION INTERMEDIATES OF SARCOPLASMIC RETICULUM CA^{2+}-ATPASE

Noriaki Ikemoto[1]

Department of Muscle Research, Boston Biomedical Research Institute; and Department of Neurology, Harvard Medical School, Boston, Massachusetts

I. INTRODUCTION

Upon binding of Ca^{2+} to the high affinity α-sites ($K_a \simeq 3 \times 10^6$ M^{-1}) (1,2) of the purified ATPase of sarcoplasmic reticulum maximal activation of both formation and decomposition of the phosphorylated intermediates (EP) is observed (1,3-8). Further binding of Ca^{2+} to the low affinity γ-sites ($K_a \simeq 1 \times 10^3$ M^{-1}) (1,2) produces inhibition of ATP-hydrolysis, whereas there is little or no change in the steady-state concentration of EP (9,10).

These phenomena fit well the mechanism shown in Fig. 1. Upon the addition of ATP to the enzyme in which the α-sites have bound Ca but the γ-sites are empty (Fig. 1, scheme 2) the enzyme is capable of carrying out the whole reaction cycle. The initial step of the reaction is obviously the formation of an enzyme-ATP complex; subsequently ATP is cleaved resulting in the sequential formation of three types of acid-stable EP. The first two forms of EP have emerged from the finding that the α- or transport-sites show dynamic decrease in their affinity for Ca^{2+} as the reaction proceeds from the first form of EP to the second form (11,12), permitting the enzyme to release Ca^{2+}, presumably at the interior side of the membrane. The third form of EP is based upon recent reports that appreciable amounts of ADP-insensitive EP become measurable in the absence

---

[1]Supported by grants from NIH (AM16922), the National Science Foundation, the American Heart Association, and the Muscular Dystrophy Association of America.

of added K^+ (13), and that in the absence of Ca^{2+}, phosphate drives back the enzyme reaction to form an acid-stable EP (14, 15). Finally the enzyme returns to the original form after the liberation of phosphate from the enzyme. If ATP is added to the enzyme in the presence of high [Ca^{2+}], at which concentration both α- and γ-sites are saturated with Ca^{2+} (scheme 3), it appears that the reaction does not proceed beyond the second form of EP. At the low extreme of [Ca^{2+}], at which both α- and γ-sites are empty (scheme 1), there is no formation of EP or Pi-liberation, but an enzyme-ATP complex appears to be formed (16).

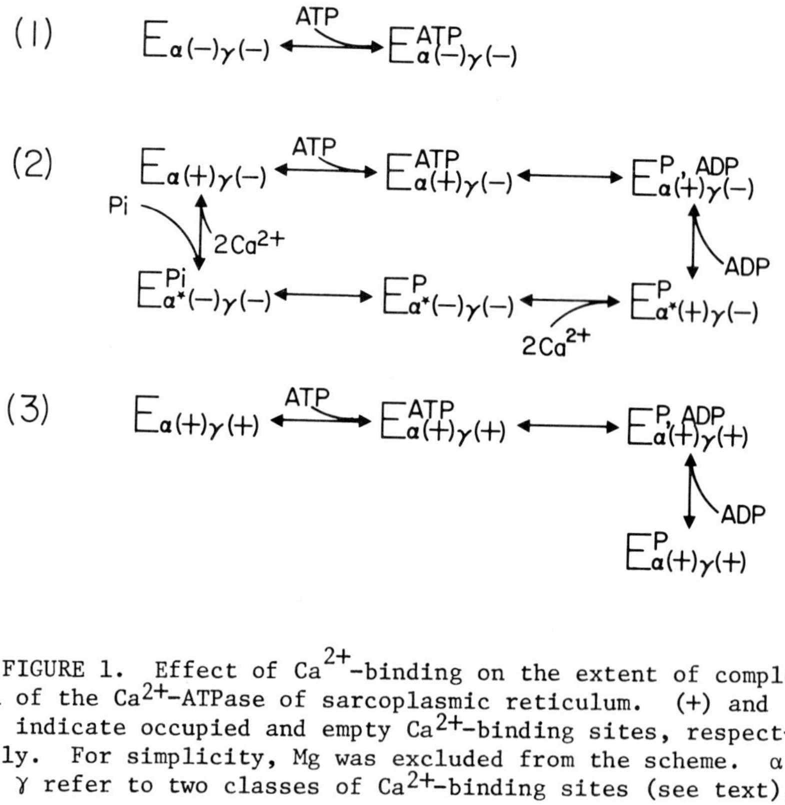

(1) $E_{\alpha(-)\gamma(-)} \xrightleftharpoons{ATP} E^{ATP}_{\alpha(-)\gamma(-)}$

(2) $E_{\alpha(+)\gamma(-)} \xrightleftharpoons{ATP} E^{ATP}_{\alpha(+)\gamma(-)} \rightleftharpoons E^{P,ADP}_{\alpha(+)\gamma(-)}$

$Pi \uparrow \downarrow 2Ca^{2+}$ $\downarrow ADP$

$E^{Pi}_{\alpha^*(-)\gamma(-)} \leftarrow E^{P}_{\alpha^*(-)\gamma(-)} \rightleftharpoons E^{P}_{\alpha^*(+)\gamma(-)}$
   $\qquad\qquad\qquad\qquad 2Ca^{2+}$

(3) $E_{\alpha(+)\gamma(+)} \xrightleftharpoons{ATP} E^{ATP}_{\alpha(+)\gamma(+)} \rightleftharpoons E^{P,ADP}_{\alpha(+)\gamma(+)}$

$\qquad\qquad\qquad\qquad\qquad\qquad\downarrow ADP$

$\qquad\qquad\qquad\qquad\qquad E^{P}_{\alpha(+)\gamma(+)}$

FIGURE 1. Effect of Ca^{2+}-binding on the extent of completion of the Ca^{2+}-ATPase of sarcoplasmic reticulum. (+) and (-) indicate occupied and empty Ca^{2+}-binding sites, respectively. For simplicity, Mg was excluded from the scheme. α and γ refer to two classes of Ca^{2+}-binding sites (see text).

Attempts have been made by many workers to monitor the enzyme conformations corresponding to various reaction intermediates (17-33). The usefulness of SH-directed reagents in studies of enzyme conformation has already been recognized in several reports (27,30-33). This paper deals with studies on the conformational changes of the purified ATPase of sarcoplasmic reticulum induced by a) Ca^{2+}-binding, b) nucleotide-binding, and c) formation of EP with the use of stopped flow fluorometry for following the reaction of -SH groups with S-mercuric N-dansyl cysteine (Dns.Cys.Hg).

II. EXPERIMENTAL PROCEDURES

A. Preparations

Fragmented sarcoplasmic reticulum was prepared from rabbit skeletal fast-twitch (white) muscles (2). The Ca^{2+}-ATPase was purified from the fragmented sarcoplasmic reticulum solubilized with Triton X-100 (2) with some modifications (34). Dns.Cys.Hg was prepared according to Leavis and Lehrer (35) with the modification that HgCl and HCl rather than $HgNO_3$ and HNO_3 were used in the reaction with N-dansyl cysteine (34).

B. Stopped Flow Studies

The reaction of Dns.Cys.Hg with the purified ATPase was studied in a Durrum stopped flow apparatus equipped with a Jarrell Ash 0.25 m excitation monochromator under the conditions described in the figure legends. Increase of the dansyl fluorescence (λ_{em} = 500 nm, λ_{ex} = 335 nm) resulting from the binding of Dns.Cys.Hg with thiols was monitored with a photomultiplier through a cut-off filter which excluded light below 430 nm.

III. RESULTS AND DISCUSSION

A. Reaction of Dns.Cys.Hg with Thiols

In the native enzyme twelve thiols per 10^5 dalton of the purified ATPase peptide are titrated with Nbs_2 while 16 thiols are reactive in the presence of sodium dodecyl sulfate. Increasing amounts of the reagent were added to the enzyme and the number of blocked thiols was determined. As shown in Fig. 2, the number of blocked thiols is equal to moles of the added reagent up to 12 mol of reagent per 10^5 g enzyme, above which no further reaction takes place. These results show that the 12 free thiols described above react stoichiometrically with Dns.Cys.Hg. Furthermore, the fluorescence intensity of the enzyme-bound reagent is proportional to the number of blocked thiols. Thus, one can determine the time course of thiol-blocking by monitoring the increase of fluorescence intensity by means of stopped flow spectroscopy.

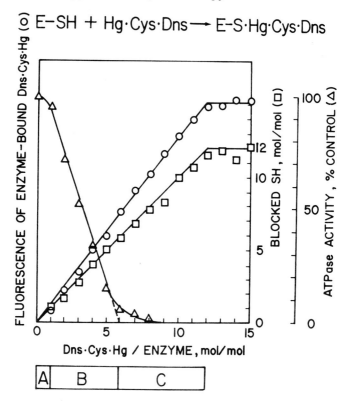

FIGURE 2. The number of thiols blocked with Dns.Cys.Hg, the increase of fluorescence intensity (λ_{em} = 500 nm, λ_{ex} = 335 nm), and the inhibition of ATPase activity as a function of the amount of Dns.Cys.Hg added to the purified ATPase. Various amounts of Dns.Cys.Hg were mixed with 2 μM purified Ca^{2+}-ATPase of sarcoplasmic reticulum in a solution containing 0.1 M KCl, 5 mM $MgCl_2$, 1 mM EGTA, 0.9 mM $CaCl_2$ (free [Ca^{2+}] = 1.5 x 10^{-6} M), 20 mM Tris-maleate, pH 7.0, at 22°. For the determination of the number of blocked thiols, a 1 ml sample treated with Dns.Cys.Hg was mixed with a 1 ml solution containing 2 mM EDTA, 0.32% sodium dodecyl sulfate, 0.4 mM Nbs_2, 200 mM Tris-HCl, pH 8.0. After incubation for 60 min at 22°, A_{412} was determined (36). The fluorescence intensity was determined at the time when the emission intensity at 500 nm reached maximum, which varied in the range of 30 s - 3 min depending upon the amount of Dns.Cys.Hg. For the ATPase assay, the reaction was started by diluting a 0.1 ml fraction of the enzyme-Dns.Cys.Hg mixture with a 1.0 ml solution con-

taining 0.1 M KCl, 5 mM $MgCl_2$, 5 mM ATP, 1 mM EGTA, 0.9 mM $CaCl_2$, 0.1 M Tris-maleate, pH 7.0, at 25°; the reaction was stopped by 6.7% trichloroacetic acid and the liberated Pi was determined colorimetrically (37).

The mode of inhibition of ATPase activity as a function of the number of blocked thiols shown in Fig. 2 permits one to classify the 12 free thiols into three classes: A, B and C. Blocking of the most reactive thiol has little or no effect on the enzyme activity (class A). Blocking of 5 more thiols results in almost complete inhibition of ATPase activity (class B). The remaining least reactive 6 thiols form class C. The type of experiment shown in Fig. 2 has also been carried out at 1×10^{-9} M and 1×10^{-2} M Ca^{2+}. The results were basically the same as those shown in Fig. 2, indicating that the conclusions described above - viz. (a) stoichiometric reaction of Dns.Cys.Hg with free thiols, (b) proportionality between the fluorescence intensity and the number of blocked thiols, and (c) number of the members of each thiol class - are not affected by the Ca^{2+} concentration.

The rate constants of the reaction of Dns.Cys.Hg with individual thiol classes have been determined according to the method of analysis shown in Ref. 34. We have found that the members of class C, SH_7 through SH_{12}, have about the same rate constant (k_C). The kinetic constants of members of class B differ and reliable calculation was possible only for the most reactive member of class B (SH_2). Therefore, its constant (k_2) was used as a representative value of the class B (k_B).

B. Conformational changes induced by Ca^{2+}-binding

We have determined the kinetic constants of individual thiol classes in the presence of various [Ca^{2+}]. The strategy of this experiment is that in a first step various conformational states of the enzyme are created by controlling [Ca^{2+}], and in a second step Dns.Cys.Hg was allowed to react with the thiol classes described above. The results are as shown in Fig. 3. In this figure the axes A, B and C (inset) represent the kinetic constants of classes A, B and C, respectively; and 1 unit of these axes corresponds the values of k_A, k_B and k_C at pCa = 9 (viz., $k_A = 2 \times 10^6$ M^{-1} s^{-1}; $k_B = 3 \times 10^4$ M^{-1} s^{-1}; $k_C = 1.4 \times 10^3$ M^{-1} s^{-1}) respectively.

The k_A value is virtually independent of [Ca^{2+}], whereas k_B and k_C show appreciably larger changes depending upon [Ca^{2+}] (Fig. 3). Upon increasing [Ca^{2+}] in the range of $8 \geq$ pCa ≥ 5 activation of enzyme activity takes place, the constant of class B increases, and that of class C decreases. Upon further increase of [Ca^{2+}], the former decreases and the latter increases in parallel with Ca^{2+}-binding to the low affinity γ-sites and inhibition of ATPase activity.

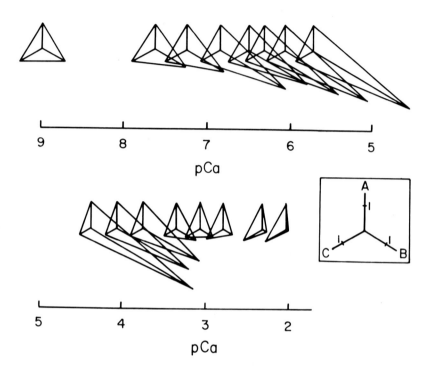

FIGURE 3. Diagrammatic representation of the rates of Dns.Cys.Hg incorporation into each of the three thiol classes as a function of pCa. The axes A, B and C represent classes A, B and C, respectively (for further details, see text). For the determination of k_1 (K_A) and k_2 (k_B), 5.0 μM Dns.Cys.Hg and for that of k_{7-12} (k_C), 17.1 μM Dns.Cys.Hg reacted with 2.5 μM purified ATPase in a solution containing 0.1 M KCl, 5 mM $MgCl_2$, 1 mM EGTA, various concentrations of $CaCl_2$, 20 mM Tris-maleate, pH 7.0 with the use of a stopped flow apparatus at 22°. The concentrations of free Ca^{2+} were calculated as described previously (1). The values of k_A, k_B and k_C were calculated as described elsewhere (34).

The fact that, in both activating and inhibitory ranges of [Ca^{2+}], changes of k_B and k_C take place in a reciprocal fashion suggests that Ca^{2+}-binding has different effects on

Conformation of Various Reaction Intermediates

the enzyme conformation in different domains of the enzyme molecule. The simplest interpretation of this phenomenon would be that Ca^{2+}-binding to the α-sites produces a conformation of the enzyme in which the region containing class B becomes more exposed, whereas the region containing class C becomes more internalized; upon Ca^{2+}-binding to the γ-sites class B becomes more internalized and class C becomes more exposed. It should be noted, however, that the thiol reactivity may be influenced by many other factors such as hydrophobicity, charge of the microenvironment and motion within the protein.

C. Conformational changes induced by nucleotide binding

The same type of experiment as that shown in Fig. 3 has been carried out in the presence of 5 mM ADP (Fig. 4) and 5 mM ATP (Fig. 5). The presence of ADP modifies the enzyme conformation in such a way that k_B and k_C decrease regardless of Ca^{2+}-controlled conformational states. There is not any change in the constant of class A. In the reaction scheme shown in Fig. 1, we can assume that the addition of ADP to the enzyme leads to the formation of only the ADP-enzyme complex in all the Ca^{2+}-controlled conformational states. Thus it is concluded that the decrease of both k_B and k_C reflects the conformational change induced by nucleotide binding. Con-

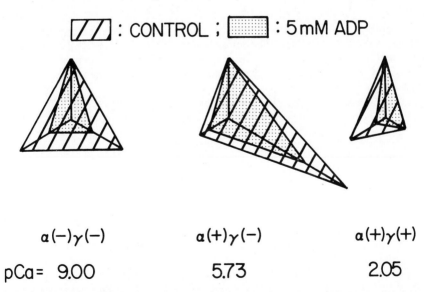

FIGURE 4. Effect of ADP on the kinetic constants of the three thiol classes. The constants were determined and expressed as described in the legend to Fig. 3, except that 5 mM ADP was included during the reaction of Dns.Cys.Hg with the enzyme.

sistent with this view, the addition of ATP to the enzyme in which both α- and γ-sites are empty and which would form only the ATP-enzyme complex (cf. Scheme 1, Fig. 1) results in the decrease of both k_B and k_C (Fig. 5).

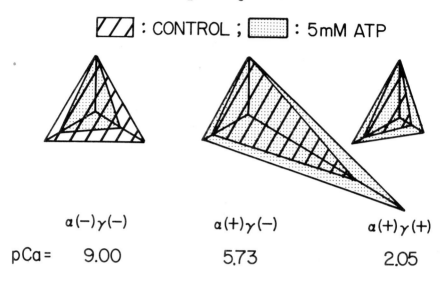

FIGURE 5. Effect of ATP on the kinetic constants of the three thiol classes. The constants were determined and expressed as described in the legend to Fig. 3, except that 5 mM ATP was included during the reaction of Dns.Cys.Hg with the enzyme.

D. Conformational changes induced by the formation of EP

If ATP is added to the enzyme in the α(+)γ(+) state both k_B and k_C increase, again with no change in k_A (Fig. 5). Since in the α(+)γ(+) state the formation, but not the decomposition, of EP occurs (cf. Fig. 1), the increase of both k_B and k_C may be ascribed to the new conformation produced by the formation of EP.

The addition of ATP to the enzyme in the α(+)γ(-) state also results in the increase of both k_B and k_C (Fig. 5). Although in this state both the decomposition and the formation of EP are activated, the increase of k_B and k_C seems to be produced chiefly by the EP formation for the following reasons.

Appropriate concentrations of propranolol (e.g. 1 mM) produce rather selective inhibition of the reaction steps in which EP-decomposition takes place (21,38). As shown in Table I, the addition of ATP to the enzyme in the α(+)γ(-)

state in the presence of 1 mM propranolol results in the increase of the k_2 or k_B value.

TABLE I. Effect of propranolol (PPL) and ATP on the kinetic constant of class B (k_2) at the $\alpha(+)\gamma(-)$ state. Dns.Cys.Hg reacted with the purified enzyme as described in the legend to Fig. 3 at $[Ca^{2+}] = 1.5 \times 10^{-6}$ M with various additions as indicated.

addition	$k_2 \times 10^{-4}$
1 mM PPL	2.1
1 mM PPL, 5 mM ATP	4.6

IV. CONCLUSION

Stopped flow fluorometry of the rapid reaction of free thiols of the purified ATPase of sarcoplasmic reticulum with the non-covalently reacting reagent, S-mercuric N-dansyl cysteine has permitted us to group the 12 free thiols into three classes and to determine the rate constants of individual classes. The kinetic constants of two (B and C) of the three classes vary with Ca^{2+}-binding to the high affinity α- and to the low affinity γ-sites. The changes of k_B and k_C take place always in opposite directions. In contrast, formation of the nucleotide-enzyme complex induces the decrease of both k_B and k_C, and the EP-formation the increase of both k_B and k_C. These results suggest that Ca^{2+}-binding makes different effects on the enzyme conformation in different domains, corresponding to class B and class C thiols, whereas nucleotide binding and EP formation produce changes in a rather broad domain of the molecule. Furthermore, it appears that the changes produced by nucleotide-binding and by EP-formation are relatively small compared with the changes produced by Ca^{2+}-binding. Thus various conformations determined by the Ca^{2+}-binding pattern are maintained throughout the enzyme reaction.

According to our recent results, as the $[Ca^{2+}]$ increases in the range of $8 \geq pCa \geq 5$, a particular SH group which presumably is a member of class B increases its reactivity with N-ethylmaleimide (39); at the same time other thiols which are in class C decrease their reactivity with iodoacetamide. These results strengthen the above conclusion concerning a differential effect of Ca^{2+}. This differential effect might be useful in future studies to increase the specificity of the labeling of a particular type of thiol or thiols. This would eventually permit one to follow dynamic changes of enzyme conformation during the Ca^{2+}-transport reaction with the use of the conformation reporters covalently attached to the well-defined sites of the enzyme molecule.

REFERENCES

1. Ikemoto, N. (1974). J. Biol. Chem. 249:649.
2. Ikemoto, N. (1975). J. Biol. Chem. 250:7219.
3. Hasselbach, W., and Makinose, M. (1961). Biochem. Z. 333:518.
4. Hasselbach, W., and Makinose, M. (1963). Biochem. Z. 339:94.
5. Martonosi, A. (1969). J. Biol. Chem. 244:613.
6. Makinose, M. (1969). Eur. J. Biochem. 10:74.
7. Inesi, G., Marling, E., Murphy, A.J., and McFarland, B.H. (1970). Arch. Biochem. Biophys. 138:285.
8. Kanazawa, T., Yamada, S., Yamamoto, T. and Tonomura, Y. (1971). J. Biochem. 70:95.
9. Weber, A. (1971). J. Gen. Physiol. 59:50.
10. deMeis, L. (1972). Biochemistry 11:2460.
11. Ikemoto, N. (1976). J. Biol. Chem. 251:7275.
12. Yamada, S., and Tonomura, Y. (1972). J. Biochem. 72:417.
13. Shigekawa, M., and Dougherty, J.P. (1978). J. Biol. Chem. 253:1458.
14. Kanazawa, T., and Boyer, P.D. (1973). J. Biol. Chem. 248: 3163.
15. deMeis, L., and Tume, R.K. (1977). Biochemistry 16: 4455.
16. Meissner, G. (1973). Biochim. Biophys. Acta 298:906.
17. Landgraf, W.C., and Inesi, G. (1969). Arch. Biochem. Biophys. 130:111.
18. Seelig, J., and Hasselbach, W. (1971). Eur. J. Biochem. 21:17.
19. Eletr, S., and Inesi, G. (1972). Biochim. Biophys. Acta 290:178.
20. Nakamura, H., Hori, H., and Mitsui, T. (1972). J. Biochem. 72:635.
21. Pang, D., Briggs, R.N., and Rogowski, R.S. (1974). Arch. Biochem. Biophys. 164:332.
22. Tonomura, Y., and Morales, M.F. (1974). Proc. Natl. Acad. Sci. U.S. 71:3687.
23. Yoshida, H., and Tonomura, Y. (1976). J. Biochem. 79: 649.
24. Coan, C.R., and Inesi, G. (1977). J. Biol. Chem. 252: 3044.

25. Yu, B.P., Masoro, E.J., Downs, J., and Wharton, D. (1977). J. Biol. Chem. 252:5262.
26. Vanderkooi, J.M., Ierokomas, A., Nakamura, H., and Martonosi, A. (1977). Biochemistry 16:1262.
27. Champeil, P., Büschlen-Boucly, S., Bastide, F., and Gary-Bobo, C. (1978). J. Biol. Chem. 253:1179.
28. Dupont, Y. (1976). Biochem. Biophys. Res. Commun. 71:544.
29. Dupont, Y. (1978). Biochem. Biophys. Res. Commun. 82:893.
30. Anderson, J.P. and Moller, J.V. (1977). Biochim. Biophys. Acta 485:188.
31. Murphy, A.J. (1976). Biochemistry 15:4492.
32. Murphy, A.J. (1978). J. Biol. Chem. 253:385.
33. Thorley-Lawson, D.A., and Green, N.M. (1977). Biochem. J. 167:739.
34. Ikemoto, N., Morgan, J.F., and Yamada, S. (1978). J. Biol. Chem. (in press).
35. Leavis, P.C., and Lehrer, S. (1974). Biochemistry 13:3042.
36. Ellman, G.L. (1959). Arch. Biochem. Biophys. 82:70.
37. Fiske, C.H., and SubbaRow, Y. (1925). J. Biol. Chem. 66:375.
38. Noack, E., Kurzmack, M., Verjovski-Almeida, S. and Inesi, G. (1978). J. Pharmacol. Exptl. Therap. (in press).
39. Yamada, S., and Ikemoto, N. (1978) J. Biol. Chem. (in press).

THE ROLE OF MAGNESIUM ON THE SARCOPLASMIC CALCIUM PUMP

Madoka Makinose
Werner Boll

Max-Planck-Institut für medizinische Forschung
Abteilung Physiologie
Heidelberg, West Germany

I. INTRODUCTION

It is well known that, in order to display the activity of the sarcoplasmic calcium pump, magnesium is one of the indispensable factors (Hasselbach and Makinose, 1961; Ebashi and Lipmann, 1962). That is, the rate of the calcium accumulation, of the extra ATP-splitting, of the ATP-ADP exchange reaction (Hasselbach and Makinose, 1962) and that of the phosphoprotein formation (Kanazawa et al. 1971) can essentially be activated only in the presence of magnesium in the testing mixtures. On the virtual role of the magnesium on this transport system, however, rather poor information is published, most of which is not discussed accurately enough.

Firstly, the questions are raised: 1. whether the magnesium in the solution activates the enzyme protein in binding directly with it and 2. which species, MgATP or free ATP, is the real substrate for the forward reaction, and which MgADP or free ADP, for the backward reaction.

In the first part of this paper, it will be shown theoretically that the alternatives asked can be differentiated from the kinetic pattern of the ATP-ADP exchange reaction qualitatively, and, in the second part, experimental data are presented which give clear answers to the questions raised.

II THEORETICAL PART

A. Symbols and abbreviations

K_T and K_D: apparent equilibrium (association) constant of Mg-ATP and Mg-ADP complex, respectively, at given pH.

K_1, K_2 and K_3: those of Mg-enzyme complexes in varied state of the enzyme (see the reaction schemes).

K_A and K_B: those for the ATP-enzyme complex (see the reaction schemes).

k_1, k_2, k_3 and k_4: rate constants (see the reaction schemes).

V_A and V_m: the rate of ATP-ADP-exchange reaction and its theoretical maximum, respectively.

The following symbols express the species of the reactants and, simultaneously, their concentrations, when they are used in the mathematical treatments.

E^o: enzyme without Mg actyvation.
E^*: activated enzyme by binding with Mg^{++}.
$E{\sim}P$ and $E^*{\sim}P$: phosphorylated enzymes.
A^o and A^*: enzyme substrate complex without and with Mg activation, respectively.
E_t, ATP_t and ADP_t: total concentration of enzyme, ATP and ADP, respectively.
ATP_f and ADP_f: ATP and ADP anions free from Mg.

B. Reaction schemes and differentiation

The general reaction scheme for the ATP-ADP exchange reaction by the SR vesicles can be written;

$$E^o + Mg \xrightleftharpoons{K_1} E^*, \quad A^o + Mg \xrightleftharpoons{K_2} A^*, \quad E^o{\sim}P + Mg \xrightleftharpoons{K_3} E^*{\sim}P.$$

$$E^* + T \underset{k_2}{\overset{k_1}{\rightleftharpoons}} A^* \underset{k_4}{\overset{k_3}{\rightleftharpoons}} E^*{\sim}P + D \qquad \text{(Scheme 1)}$$

where T and D represent ATP ions (MgATP or ATP_f) and ADP ions (MgADP or ADP_f), respectively. The rate of ATP-ADP exchange (V_A) is then expressed;

$$\frac{V_m}{V_A} = \left(\frac{1}{K_1 Mg} + 1\right)\frac{1}{K_A} \times \frac{1}{T} + \left(\frac{1}{K_2 Mg} + 1\right) + \left(\frac{1}{K_3 Mg} + 1\right)\frac{1}{K_B} \times \frac{1}{D} \qquad \text{(Eq.1)}$$

where $V_m = \dfrac{k_2 k_3}{k_2 + k_3} E_t$, $K_A = k_1/k_2$ and $K_B = k_4/k_3$.

It should be noticed that the slope of the Lineweaver-Burk plot of V_A versus 1/T (or 1/D) is independent of the species of D (or T) and its concentration. If it is assumed that Mg

The Role of Magnesium

Table I. Theoretical pattern of Lineweaver-Burk plot of the rate of the ATP-ADP exchange reaction (V_A).

Two series of assays are prescribed for different Mg^{++} concentrations (Mg1 > Mg2). In each series, T or D concentration is changed systematically while the substrate concentration for the counter process is kept constant. $\alpha1$ and $\alpha2$ are the slope of the plot for the series with Mg1 and Mg2, respectively.

Substrate assumed	Abscissa	Slope	Mode of $\alpha1$ and $\alpha2$	Pattern in Fig.1.
Eq.1: Mg^{++} is enzyme activator.				
MgATP	1/MgATP	$(\frac{1}{K_1 Mg} + 1)\frac{1}{K_A}$	$\alpha1 < \alpha2$	c
ATP_f	1/MgATP	$(\frac{1}{K_1} + Mg)\frac{K_T}{K_A}$	$\alpha1 > \alpha2$	b
MgADP	1/MgADP	$(\frac{1}{K_3 Mg} + 1)\frac{1}{K_B}$	$\alpha1 < \alpha2$	c
ADP_f	1/MgADP	$(\frac{1}{K_3} + Mg)\frac{K_D}{K_B}$	$\alpha1 > \alpha2$	b
Eq.2: Mg^{++} is no enzyme activator.				
MgATP	1/MgATP	$\frac{1}{K_A}$	$\alpha1 = \alpha2$	a
ATP_f	1/MgATP	$\frac{K_T}{K_A} Mg$	$\alpha1 > \alpha2$	b
MgADP	1/MgADP	$\frac{1}{K_B}$	$\alpha1 = \alpha2$	a
ADP_f	1/MgADP	$\frac{K_D}{K_B} Mg$	$\alpha1 > \alpha2$	b

does not activate the enzyme protein directly, the scheme is simplified as:

$$E^O + T \underset{}{\overset{K_A}{\rightleftharpoons}} A^O \underset{}{\overset{K_B}{\rightleftharpoons}} E^O{\sim}P + D \qquad \text{(Scheme 2)}$$

then

$$\frac{V_m}{V_A} = \frac{1}{K_A} \times \frac{1}{T} + 1 + \frac{1}{K_B} \times \frac{1}{D} \qquad \text{(Eq. 2)}$$

From these schemes and equations, one can expect characteristic patterns of Lineweaver-Burk plot of V_A for variable choice of T and D species as the virtual substrate for the forward and backward reaction process, respectively. In Table I and Fig. 1, the differential patterns for the possible choice of the substrates were summarised.

If a random binding sequence of Mg and substrates to the enzyme is assumed, the kinetic equation becomes more complicated. But, with increasing of the concentration of Mg^{++} ions, the equations approach to the one mentioned above and their differential characters stay unchanged.

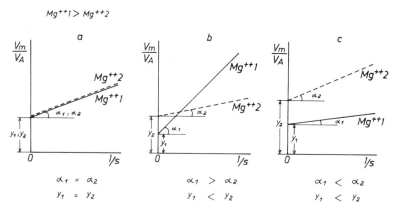

Fig. 1: Theoretical pattern of the Lineweaver-Burk plot of the rate of ATP-ADP exchange reaction (V_A).

The mode of Y_1 and Y_2 is valid only when the species of the substrate for the counter process is chosen correctly and its concentration is kept constant through the experiment.

III EXPERIMENTAL PART

The Role of Magnesium

A. Materials and Methods

1. ATP-ADP exchange reaction was carried out in the reaction mixture containing 20mM histidine or tris-maleate buffer (pH 7.0), 0.1mM $CaCl_2$ or 1mM Ca-EGTA. Accurate amount of ATP, (β^{32}P) ADP and $MgCl_2$ were added to the mixture to obtain desired concentration of the Mg^{++} ions, free and Mg bound substrates. ADP and ATP contamination in the commercial ATP and ADP preparation (Boehringer/Mannheim), respectively, were taken into account. The calculation of the concentration of each component was based on the equilibrium constants given by Martel and Schwarzenbach (1956). Ionic strength of the assay was adjusted to $\mu = 0.12$ by adding KCl. Temperature was kept constant at $20°C$. After the final pH control, the reaction was started by addition of the SR vesicles (0.02~0.05 mg prot./ml). After desired reaction time, 1ml of the mixture was filtrated through membrane filter. The filtrate was chromatographed on PEI DC plate. The ATP and ADP spots were localized, cut out and counted with liquid scintillation counter. The rate of the reaction is calculated according to Makinose (1969).

2. The measurement of the phosphoprotein formation and the preparation of the sarcoplamic vesicles, of (γ-^{32}P) ATP and of (β-^{32}P)ADP were carried out according to Makinose (1969).

B. Results

Fig. 2 shows the time course of the ATP-ADP exchange reaction under higher (1mM, Fig. 2a) and lower (0.2mM, Fig. 2b) Mg^{++} concentrations with varied Mg-ATP concentrations. The concentration of free ADP was kept constant in each series.

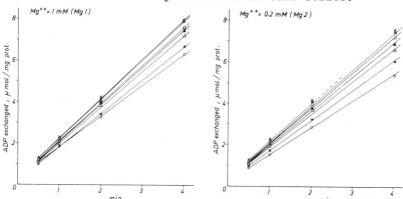

Fig. 2: Time course of the ATP-ADP exchange reaction by SR vesicles.
Two series of assays contain 1.0 or 0.2mM Mg^{++} (Mg1 and Mg2).

MgATP is varied from bottom to the top curve 1, 1.5, 2, 2.5, 3, 3.5 and 4mM. ADP_f is 0.49mM for Mg1 and 0.83mM for Mg2. 0.02Mg vesicles protein/ml assay. μ = 0.12. pH 7.0. 20°C.

Fig. 3 shows the Lineweaver-Burk plot of the ATP-ADP exchange activity obtained in Fig. 2 against reciprocal of MgATP concentration. The plot for the higher Mg concentration has lower slope and smaller intercept than the other. It is clearly recognizable that this pattern belongs to the scheme Fig. 1c, so that one has to accept the assumption i. e. Mg^{++} ions activate the enzyme protein directly and the real substrate for the enzyme phosphorylation is MgATP. The possibility for free ATP as phosphorylation substrate is ruled out.

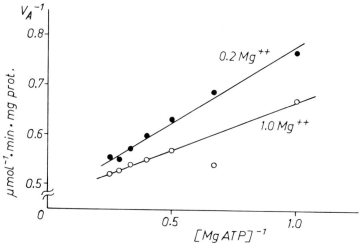

Fig. 3: Lineweaver-Burk plot of the V_A in Fig. 2 versus 1/MgATP. p = α1/α2 = 0.67. The mode of Y_1 and Y_2 is not valid because the ADP_f concentration is not the same for two curves.

For analysis of the backwards process, a similar kind of the experiment was prepared. In this case the concentration of free ADP (and of MgADP consequently also) was changed in the series of two different Mg^{++} concentrations (0.2 and 1mM). The concentration of MgATP was kept throughout constant (1.82mM). Lineweaver-Burk plot of this experiment is shown in Fig. 4. Two straight lines referring to the different Mg^{++} concentrations has higher slope and smaller intercept than the other corresponding to the pattern of Fig. 1b. The phosphorylated enzyme should accept free ADP as the substrate. From this result, it can also be concluded that the phosphorylated enzyme is activated by Mg^{++} ions in calculating the Mg-affinity of the active intermediate from the slopes of the plottings (see discussion).

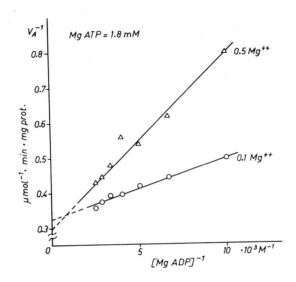

Fig. 4: Lineweaver-Burk plot of V_A versus reciprocal of MgADP concentration.
Two series of assays contain 0.5 and 0.1mM Mg^{++} (Mg1 and Mg2). ADP_f is varied from 0.19 to 0.76mM for Mg1 series and from 0.96 to 3.8mM for Mg2 series. p = 2.75. MgATP concentration is kept constant at 1.8mM for both series. The mode of Y_1 and Y_2 is valid. 0.1mg vesicles protein/ml assay. μ = 0.12. pH 7.0. 20°C.

Fig. 5 shows the phosphoprotein formation with varied total Mg concentration under the constand ATP and ADP (Curve I). The E~P level raises with increasing Mg^{++} concentration and approaches to a definite value. This simple result confirms the conclusions mentioned above exclusively, because such an upwards convex curve for the E~P formation can be expected only assuming the combination of the true substrate as MgATP and free ADP (see discussion).

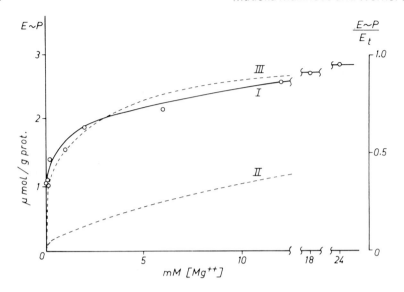

Fig. 5: Phosphoprotein formation with varied concentration of Mg^{++} ions.
Curve I: E~P measured. ATP_t = 5mM, ADP_t = 1mM. 20mM histidine (pH 7), 0.1mM $CaCl_2$, 0.5mg vesicles protein per ml assay, μ = 0.12. 5°C. Reaction time 15 sec.
Curve II: E~P calculated with Eq. 5 and K values in Table II.
Curve III. do: $K_A = K_B = 2mM^{-1}$.

IV Discussion

All the experiments presented in this study were carried out in the presence of 0.1mM $CaCl_2$ or 1mM CaEGTA to keep the ionic concentration of calcium higher than $5 \cdot 10^{-6}$ M throhghout the incubation time. The sites for the calcium activation of the sarcoplasmic enzyme is considered to be always closely near the saturation. So that the process of the Ca-activation of the enzyme is settled out of the scope of this paper.

It has been suggested since serveral years that MgATP is the substrate for the SR transport ATPase (Kanazawa et al. 1971, Meissner 1973, Vianna 1975), giving agreement tacitly with Michaelis-Menten mechanism. The results presented in Fig. 3 and 4 predict strictly 1. that the virtual substrates for the forward and the backward reaction of the sarcoplasmic Ca transport system are MgATP and Mg free ADP and 2. that the non-phosphorylated enzyme is activated by Mg^{++} ions directly. The question, whether the phosphorylated enzyme is also activated with Mg^{++} or not, can not be answered from the mode of the slope $\alpha 1$ and $\alpha 2$. If the phosphorylated enzyme does not need Mg ions to be activated, the value of K_3 should be infinite.

The Role of Magnesium

Table II Equilibrium constants in the ATP-ADP exchange reaction and conditions for the rate measurements.

All values are expressed in mM.

Mg^{++} Mg1 / Mg2	MgATP	ADP	$1/K_1$	$1/K_A$	$1/K_3$	$1/K_B$
1.0 / 0.2	1.0-4.0	0.49 / 0.83	0.140	0.175		
1.0 / 0.2	2.0-5.0	1.5	0.184	0.133		
0.5 / 0.1	0.3-1.3	0.60	0.052	0.179		
0.3 / 0.06	0.3-1.0	0.3	0.085	0.077		
0.5 / 0.1	2.5	0.50-2.0			0.112	0.063
0.5 / 0.1	2.5	0.50-2.0			0.138	0.043
0.5 / 0.1	1.8	0.19-0.76 / 0.96-3.8			0.077	0.048
0.5 / 0.1	1.8	0.19-0.76 / 0.96-3.8			0.129	0.040
		Average	0.115	0.141	0.114	0.048

The value of K_1 and K_3 can be calculated simply from the proportion of $\alpha 1$ and $\alpha 2$ ($p = \alpha 1/\alpha 2$) in Fig. 3 and 4 respectively.

$$K_1 = \left(\frac{1}{Mg1} - \frac{p}{Mg2}\right)/(p-1)$$

$$K_3 = (1-p)/(pMg2 - Mg1) \qquad \text{(Eq. 3)}$$

The calculated K_1 and K_3 values are summarized in Table II obtained in assays under varied conditions.

As was shown, K_3 calculated takes a finite value which is not much different from K_1. Apparently, the sarcoplasmic enzyme does not change its Mg affinity during the phosphorylation process. So, the reaction scheme for the ATP-ADP exchange reaction can be written therefore;

$$E^o + Mg \underset{}{\overset{K_1}{\rightleftharpoons}} E^*, \quad E^o MgATP + Mg \underset{}{\overset{K_2}{\rightleftharpoons}} E^* MgATP, \quad E^o Mg{\sim}P + Mg \underset{}{\overset{K_3}{\rightleftharpoons}} E^* Mg{\sim}P$$

$$E^* + MgATP \underset{}{\overset{K_A}{\rightleftharpoons}} E^* MgATP \underset{}{\overset{K_B}{\rightleftharpoons}} E^* Mg{\sim}P + ADP \qquad \text{(Scheme 3)}$$

Since now the values of K_1 and K_3 are known, the affinities of the enzyme to the substrates (K_A and K_B) can also be calculated (Table II). The affinity of the non-phosphorylated enzyme to the substrate in active form (K_A) is about 3 times higher than that of the phosphoprotein (K_B) in the activated state by Mg^{++}. It means that the phosphorylation reaction is coupled with small change of the standard free energy.

The simple experiment of the phosphoprotein formation shown in Fig. 5 supports in an independent way the conclusion mentioned above. As the first approximation, the phosphoprotein formation reaction is formulated

$$E + T \underset{}{\overset{K_o}{\rightleftharpoons}} E{\sim}P + D \qquad \text{(Scheme 4)}$$

It leads to the equation

$$\frac{E{\sim}P}{E_t} = \frac{T}{T + D/K_o} \qquad \text{(Eq. 4)}$$

In Fig. 6 the simulation curves of the E~P formation calculated by Eq. 4 were presented with possible combination of the substrates. The constant K_o is chosen as 1 conveniently. An upward convex curve of the E~P formation versus Mg concentration can only be expected by placing MgATP and free ADP as T and D respectively. When the value of K_o is changed, the curves are shifted, but they do not change the sign of the curvature. From the Scheme 3, the E~P formation is formulated:

$$\frac{\Sigma E{\sim}P}{E_t} = \frac{MgATP}{MgATP + K_B \cdot MgATP \cdot ADP + (K_B/K_A)ADP} \qquad \text{(Eq. 5)}$$

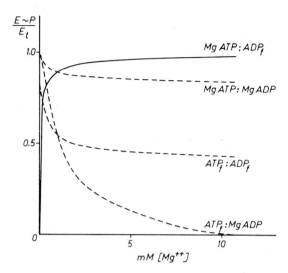

Fig. 6: Theoretical curve for the E~P formation.
Ordinate: $\dfrac{E\sim P}{E_t} = \dfrac{T}{T + D/K_o}$

Assumption for the calculation; 5mM ATP_t and 1mM ADP_t. K_o is chosen conveniently to be 1. Concentration of each components (ATP_f, MgATP, ADP_f) were calculated by using the stability constants given by Martel and Schwarzenbach (1956).

assuming that the enzyme does not change the Mg affinity in varied state, i. e. $K_1 = K_2 = K_3$. Using the average values of K_A and K_B in Table II, $\Sigma E\sim P/E_t$ calculated by Eq. 5 was shown in Fig. 5 (Curve II). The quantitative fitness of the curve is not satisfactory, but the general tendency of the real and calculated curve is just similar. Better fitness can be obtained by setting K_A and K_B equal to $2mM^{-1}$ (Curve III). The study on the significance of this quantitative difference between these values of the equilibrium constants belongs to the future.

The sarcoplasmic Ca transport ATPase behaves essentially differently from the classic Michaelis-Menten manner, because its intermediate level does not depend only on the Pi-donating substrate (MgATP) but also on the Pi-accepting substrate (ADP_f). From this angle, the kinetic data of the activity of the sarcoplasmic transport ATPase uptodate should be criticized again.

Garrahan (1976) has reported, that the equilibrium level of the E~P formation of the SR protein is independent of the

Mg added. This experiment was carried out, certainly, without ADP added. From the Eq. 4 and 5, it is clear that, in the absence of ADP, E~P formation in steady state should be always maximal, if a trace of Mg ions and, consequently, a finite amount of MgATP are present.

The Mg ions play double roles on the sarcoplasmic transport ATPase, namely, directly as an enzyme activator and indirectly by forming the virtual P-donating substrate (MgATP). Presumably, the active intermediate for the Pi-liberation is the Mg activated phosphoprotein (E*Mg~P), which contains apparently two Mg - one is the direct activator and the other is transferred together with energy rich phosphate to the enzyme.

Summary

From the qualitative nature of the analysis of the ATP-ADP exchange reaction and of phosphoprotein formation, it can be concluded that the virtual substrates for the forward and backward reaction of the sarcoplasmic calcium transport are MgATP and Mg free ADP. The participation of Mg free ATP and MgADP are definitively excluded. Mg activates the transport enzyme directly in binding with it in varied enzymatic states. The active intermediate for the calcium transport is Mg activated phosphoprotein. The affinities of the enzyme to Mg ions and substrates were calculated and discussed.

Acknowledgement

We wish to thank Miss A. Hommel and Miss G. Zörkler for excellent assistance in the work.

References

Ebashi, S. and Lipmann, F. (1962). J. Cell Biol. 14:389
Garrahan, P. J., Rega, A. F. and Alonzo, G. L. (1976) Biochim. Biophys. Acta 448:121.
Hasselbach, W. and Makinose, M. (1961) Biochem. Z. 333:518.
Hasselbach, W. and Makinose, M. (1962) Biochem. Biophys. Res. Comm. 7:132.
Kanazawa, T., Yamada, S., Yamamoto, T. and Tonomura, Y. (1971) J. Biochem. 70:95.
Maissner, G. (1973) Biochim. Biophys. Acta 298:906.
Makinose, M. (1966) Proc. 2nd Intern. Biophys. Congr. No. 276
Martel, A. E. and Schwarzenbach, G. (1956) Helv. Chim. Acta 39:653
Vianna, A. L. (1975) Biochim. Biophys. Acta 410:389

TRANSIENT KINETICS OF Ca^{2+} TRANSPORT OF SARCOPLASMIC RETICULUM
A COMPARISON OF CARDIAC AND SKELETAL MUSCLE:[1]

Michihiro Sumida[2]
Taitzer Wang
Frederic Mandel
Arnold Schwartz

Department of Pharmacology and Cell Biophysics
University of Cincinnati College of Medicine
Cincinnati, Ohio, U.S.A.

Jeffrey P. Froehlich

Laboratory of Molecular Aging
National Institute of Aging
National Institutes of Health
Baltimore, Maryland, U.S.A.

Formation and decomposition of an acid stable phosphorylated intermediate (E∿P) of Ca^{2+} transport ATPase of sarcoplasmic reticulum (SR) from cardiac muscle were studied in transient state with a quench flow apparatus and the similar reaction mechanism to that from skeletal muscle (1,2) was proposed (FASEB 1978).

The reaction step of association and dissociation of Ca^{2+} with ATPase was further studied with the quench flow apparatus and following model was proposed:

$$E + 2Ca^{out} \underset{k_{-1}}{\overset{k_1}{\rightleftharpoons}} E \cdot Ca_2 + ATP \rightleftharpoons E \cdot Ca_2 \cdot ATP \rightleftharpoons Ca_2 \cdot E{\sim}P \rightleftharpoons E \cdot P + 2Ca^{in} \rightleftharpoons E + Pi^*$$

---

[1] This work was supported by National Institutes of Health, National Heart, Lung and Blood Institute Grants HL22039-01 (A.S.) T32 HL7382-01 (A.S.), HL22109-01 (A.S.) and P01 HL22619-01(A.S.)
[2] Present Address, Department of Medical Biochemistry, School of Medicine, Ehime University, Shigenobu, Ehime, Japan.

The rates of dissociation of calcium-ATPase complex ($k_{-1} = 10$ s^{-1} vs. 15 s^{-1}) and E∿P formation of Ca-free SR (pre-treated with EGTA prior to E∿P formation) were lower for cardiac than for skeletal SR. The dissociation rate was increased with solubilized ATPase with deoxycholate (25.7 s^{-1} for dog cardiac SR). By analog simulation, the apparent rate constants (k_1) associated with the reduced rates of the E∿P formation of Ca-free SR were 12 s^{-1} for cardiac and 63 s^{-1} for skeletal SR. The difference in these rates may partly contribute to the lower levels of Ca^{2+} transport activity of cardiac SR.

* The first step ($E + 2Ca^{out} \rightleftharpoons E \cdot Ca_2$) was assumed to be an elementary without taking account of Mg^{2+} dependency (3,4).

REFERENCES

1. Sumida, M. and Tonomura, Y., J. Biochem., 75:283 (1974).
2. Froehlich, J. P. and Taylor, E. W., J. Biol. Chem., 251: 2301 (1976)
3. Kanazawa, T. and Boyer, P. D., *ibid.*, 248:3163 (1976).
4. Froehlich, J. P. and Schenk, F. J., Biophys. J., 17:162a (1977).

OCCURRENCE OF TWO TYPES OF ACID-STABLE
PHOSPHOENZYME INTERMEDIATE DURING ATP
HYDROLYSIS BY SARCOPLASMIC RETICULUM[1]

Munekazu Shigekawa[2]
Alfred A. Akowitz

Department of Medicine
University of Connecticut
Farmington, Connecticut USA

Two types of acid-stable phosphoenzyme were shown to occur in an ATPase preparation of sarcoplasmic reticulum during ATP hydrolysis in low Mg^{2+} and Ca^{2+} at 0° in the absence of added alkali metal salts (1). One of these phosphoenzymes is ADP-sensitive (E_1P) as it could donate its phosphate group to added ADP to form ATP while the other is ADP-insensitive (E_2P) as it did not donate its phosphate group to added ADP. During ATP hydrolysis under these conditions, E_1P was formed first and then converted to E_2P (1).

The phosphoenzyme formed at the steady state in high Mg^{2+} and low Ca^{2+} at 0° in the presence of 0 to 30 mM KCl could also be resolved into E_1P and E_2P. As shown in Table I, the steady state levels of E_1P and E_2P were markedly dependent on KCl concentration whereas KCl decreased the steady state level of total phosphoenzyme (EP_t) only slightly. E_2P which was 80-90% of EP_t in the absence of KCl, decreased to 25-30% of EP_t in 30 mM KCl. As the rate of ATP hydrolysis was stimulated by KCl, the ratio (V/E_2P) between the steady state rate of ATP hydrolysis (V) and the steady state level of E_2P increased markedly with increasing KCl concentrations. The value of V/E_2P obtained at each KCl concentration was in good agreement with

---

[1] Supported by NIH grants HL-22135 and HL-21812, and grant from Connecticut Heart Association.
[2] Present address: Department of Biochemistry, Asahikawa Medical College, Asahikawa, 078-11 Japan.

TABLE I. KCl dependence of Rate of ATP Hydrolysis, Phosphoenzyme Levels and Rate of E_2P Hydrolysis[a]

KCl	V[b]	Phosphoenzyme levels			V/E_2P	k_d
		EP_t	E_1P	E_2P		
mM			nmol/mg		min^{-1}	min^{-1}
0	5.91	5.39	0.58	4.81	1.23	1.09
6	10.2	4.42	1.40	3.02	3.36	4.08
10	12.5	4.05	1.67	2.38	5.26	5.85
30	13.3	4.40	3.13	1.27	10.4	9.03

[a]Conditions were 0.25 mg/ml of partially purified ATPase protein, 15 mM imidazole/HCl(pH 7.0), 2.0 mM $MgCl_2$, 20 μM $CaCl_2$ and 20 μM [γ-^{32}P]ATP at 0°. 30 s after the start of the reaction, excess EGTA and ADP were added to the reaction medium to induce phosphoenzyme decomposition, the time course of which exhibited a rapid phase followed by a slow phase. The amount of E_2P was estimated by extrapolating the time course of the slow phase to the time of addition of EGTA and ADP. The amount of E_1P was estimated by subtracting the amount of E_2P from that of EP_t. [b]Unit is nmol/mg/min.

the hydrolysis rate (k_d) of E_2P that was measured directly at each KCl concentration after E_2P was formed in the absence of KCl and isolated by addition of excess EGTA.

Thus, these and our previous (1-4) findings are consistent with a reaction scheme for ATP hydrolysis in which E_1P and E_2P occur sequentially, and P_i is derived only from the latter.

ACKNOWLEDGMENTS

We are grateful to Dr. Arnold M. Katz for his encouragement and discussion.

REFERENCES

1. Shigekawa,M.,and Dougherty,J.(1978). J. Biol. Chem. 253:1458.
2. Shigekawa,M.,and Pearl,L.(1976). J. Biol. Chem. 251:6947.
3. Shigekawa,M.,Dougherty,J.,and Katz,A.(1978). J. Biol. Chem. 253:1442.
4. Shigekawa,M.,and Dougherty,J.(1978). J. Biol. Chem. 253:1451.

EFFECTS OF TEMPERATURE, NUCLEOTIDES AND METAL CATIONS ON THE STATE OF Ca^{2+},Mg^{2+}-DEPENDENT ATPASE OF SARCOPLASMIC RETICULUM MEMBRANES AS STUDIED BY HYDROGEN-DEUTERIUM EXCHANGE REACTION KINETICS AND SATURATION TRANSFER ELECTRON SPIN RESONANCE

Yutaka Kirino
Kazunori Anzai
Taka'aki Ohkuma
Hiroshi Shimizu

Faculty of Pharmaceutical Sciences
The University of Tokyo
Bunkyo-ku, Tokyo

I. INTRODUCTION

In recent years there have been many reports claiming that Arrhenius plots of the Ca^{2+}-dependent ATPase (the Ca^{2+}-ATPase) activity and of the Ca^{2+} uptake activity of fragmented sarcoplasmic reticulum (SR) show a break at about 18°C (1-7). These phenomena have been discussed in relation to the state of membrane lipids, and some of the authors ascribed the break to a phase transition of the lipids from a gel to a liquid crystalline state. However, results of X-ray diffraction (6,8) and differential scanning calorimetry (9) have shown that at ordinary temperatures lipids of SR membranes are in a liquid crystalline state having fluidity and do not undergo a gel-liquid crystalline transition at around 18°C. The relation between the state of the lipids and the break at 18°C in the Arrhenius plot of the Ca^{2+}-ATPase activity is not yet clear.

The present work is an attempt to measure the physical state of Ca^{2+}-ATPase molecules of SR membranes isolated from rabbit skeletal muscle and correlate it with the activity of the membrane-bound enzyme. The techniques we have used are hydrogen-deuterium (H-D) exchange reaction kinetics (10,11) and saturation transfer electron spin resonance (ST-ESR) (12).

These are two of the few which are useful in obtaining the information about the physical state of membrane-bound proteins. The results obtained by both the methods in this study have revealed the thermotropic change in the physical state of the Ca^{2+}-ATPase at about 18°C, which may be responsible for the break at the same temperature in the Arrhenius plots of the Ca^{2+}-ATPase and Ca^{2+} uptake activity of SR.

II. EXPERIMENTAL

MEMBRANE PREPARATIONS AND ASSAYS OF ATPASE ACTIVITY. Fragmented SR was isolated from rabbit white skeletal muscle as described previously (10-11). Preparation of MacLennan's enzyme was according to MacLennan (13). It was dialyzed for 3 days at 4°C to remove ammonium acetate and deoxycholate that were used in the preparation. Replacement of endogenous membrane lipids by dioleoyllecithin (DOL) or egg yolk lecithin (egg PC) was according to Warren et al. (14,15). DOL was synthesized by the method of Robles and van den Berg (16). Egg PC was extracted and purified according to Rhodes and Lea (17). Protein concentrations, phospholipid content in the membranes and fatty acid composition of the membrane lipids were determined as described previously (11). To measure the "total" ATPase activity, the reaction mixture contained finally 0.1 M KCl, 5 mM $MgCl_2$, 1.08 mM $CaCl_2$, 1 mM ethyleneglycol-bis(2-aminoethylether)-N,N,N',N'-tetraacetic acid (EGTA), 0.1 mM [γ-^{32}P]ATP, and 0.02-0.1 mg/ml protein. $CaCl_2$ was omitted from this mixture to measure the Ca^{2+}-independent ATPase activity ("basic" activity). Determination of inorganic phosphate liberation and phosphoenzyme intermediate was carried out as described previously (10,11).

HYDROGEN-DEUTERIUM EXCHANGE REACTION. The samples were washed by centrifugation with a solution containing 0.1 M KCl and 10 mM N,N-bis(2-hydroxyethyl)-2-aminomethanesulfonic acid (BES), pH 7.0, and resuspended in a solution containing 0.1 M KCl, 10 mM BES (pH 7.0), and 0.1 M sucrose. Each sample was divided into small portions, placed in microtubes, lyophylized, and stored in a desiccator in vacuo at 4°C. The exchange experiments were carried out in a way similar to that reported elsewhere (10,11). The reaction was started by addition of deuterium oxide to freeze-dried samples and followed by IR absorption spectroscopy.

ESR EXPERIMENTS. An SR preparation (4 mg protein/ml) was first incubated at 0°C for 12 hr in a solution of pH 7.0 containing 0.05-0.4 mM N-ethylmaleimide (NEM), 80 mM KCl, and 50 mM Tris-maleate. After washing by repeated centrifugation and resuspension, SR (4 mg protein/ml) was allowed to react with 0.15 mM 4-maleimido-2,2,6,6-tetramethyl-1-piperidinooxyl (MSL) in a solution of 0.1 M KCl and Tris-HCl at pH 8.2, 0°C for 24 hr. The Ca^{2+}-ATPase activity of the labeled SR was about the same as that of the control which was incubated at pH 8.2 while the apparent Ca^{2+} uptake ability vanished after the labeling.

The ST-ESR measurements were performed in the absorption mode with field modulation at 50 kHz and phase-sensitive detection at 100 kHz, 90° out-of-phase, the resulting spectra having been designated V_2' (18). The modification of a JEOL JES-PE-1 X band spectrometer to permit these measurements was described elsewhere (12). In the later experiments the phase-sensitive detection of signals was performed using a commercial lock-in amplifier such as model 124A or 5203 of Princeton Applied Research (Princeton, New Jersey). The amplifier was operated in the second harmonic (2f) mode and the 50 kHz from the field modulation coil was used as the reference. This enabled one to minimize phase drift during the measurements. Unless otherwise noted, all spectra were obtained from SR membranes (20-40 mg proteins/ml) in 30% sucrose, 80 mM KCl, and 50 mM Tris-maleate, pH 7.0 placed in a JEOL LC-01 quartz capillary cell. A deoxygenated solution of Fremy's salt was used to calibrate the microwave field strength, H_1, (18) and the modulation amplitude, H_m, (19) received by the sample.

III. RESULTS AND DISCUSSION

THE TEMPERATURE DEPENDENCE OF Ca^{2+}-ATPase ACTIVITY. The temperature dependence of the Ca^{2+}-ATPase activity has been examined for intact isolated SR vesicles, MacLennan's enzyme, DOL-ATPase, and egg PC-ATPase. In the latter three preparations most proteins other than the ATPase are removed. The SDS-polyacrylamide gel electrophoretic profiles indicate that the ATPase protein comprises over 90% of the total proteins. In these preparations lipid contents are greatly reduced. Assuming the molecular weight of the ATPase is 100,000 and its content in total proteins is 70% for intact SR and 100% for the others, the phospholipid content per mole of ATPase is calculated to be 139, 82, 67, and 74 for intact SR, MacLennan's enzyme, DOL-ATPase, and egg PC-ATPase, respectively.

Table I shows the fatty acid composition of the lipids in intact SR, MacLennan's enzyme, and DOL-ATPase. The former two preparations have similar fatty acid compositions. In the DOL-ATPase, the content of DOL amounts to about 80% of the total lipids, which is in agreement with the value calculated by assuming equilibrium between the endogenous lipids and the exogenous DOL. Although the existence of lipids tightly bound to protein molecules has been questioned (20), the concept of boundary and bulk lipids in SR membranes is popular. If the

TABLE I. Fatty Acid Composition of Various ATPase Preparations

Samples	Composition (%)					
	14:0	16:0	18:0	18:1	18:2	Others
Intact SR	1.5	36.1	8.4	14.9	19.7	19.4
MacLennan's enzyme	0.5	36.0	10.5	13.3	18.2	21.5
DOL-ATPase	0.6	6.3	0.8	78.6	0.5	13.2

boundary lipids should be replaced only after all bulk lipids are replaced, the replacement of the boundary lipids is calculated to be about 60% assuming the number of boundary lipid molecules around an ATPase molecule is thirty (21). It seems, however, more reasonable to consider that the replacement occurred homogeneously since the comparison of the data for intact SR and MacLennan's enzyme shown in Table I suggest that the boundary lipids, if any, and the bulk lipids may be readily exchangeable with each other.

The Ca^{2+}-ATPase activity of SR is obtained by subtracting the "basic" activity from the "total" activity (see "EXPERIMENTAL"). This Ca^{2+}-ATPase activity is considered to couple with Ca^{2+} uptake. The other three kinds of preparations do not have the "basic" activity so that the "total" activity is equal to the Ca^{2+}-ATPase activity. In the reaction scheme proposed by Yamamoto and Tonomura (22), the rate-limiting step of the Ca^{2+}-ATPase reaction is the decomposition of the phosphorylated intermediate (EP) to free enzyme and inorganic phosphate. Therefore, in the steady state the Ca^{2+}-ATPase activity (v) can be written as $v = k_d \cdot [EP]$, where k_d is the rate constant of EP decomposition. The steady state level of EP was found to be almost constant between 0 and 37°C for all the preparations studied. Therefore, the temperature dependence of v represents the temperature dependence of k_d. Figure 1 shows Arrhenius plots of the Ca^{2+}-ATPase activity for the four kinds of preparations.

For all the preparations the plots do not show a straight line, but can be approximated with two straight lines intersecting at about 18°C. Since the gel-liquid crystalline phase transition temperature of DOL and egg PC is well below 0°C, these lipids are in the fluid liquid-crystalline state in the temperature range under study. The observation that the Arrhenius plots for both DOL-ATPase and egg PC-ATPase definitely show a break indicates that the lipid phase transition from a gel to a liquid crystalline state is not the cause of the break.

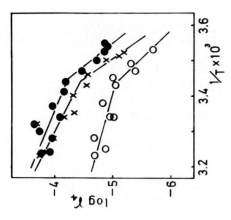

FIGURE 3. Plots of log γ_4 against reciprocal of absolute temperature for intact SR (○), MacLennan's enzyme (✗), and DOL-ATPase (●).

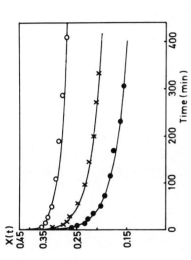

FIGURE 2. Semilogarithmic presentation of the time course of the H-D exchange reaction for MacLennan's enzyme at 11.0 (○), 25.7 (✗), and 35.1°C (●). Points are experimental data and curves are calculated according to Eq. 2 (see text) using the best sets of k_n values: $k_1=\infty$, $k_2=0.34$, $k_3=0.015$, $k_4=0.0016$, $k_5=0.00015$, $k_6-k_9=0$ at 11°; $k_1-k_2=\infty$, $k_3=0.28$, $k_4=0.033$, $k_5=0.0063$, $k_6=0.00061$, $k_7-k_9=0$ at 25.7°; $k_1-k_3=\infty$, $k_4=0.18$, $k_5=0.033$, $k_6=0.0059$, $k_7=0.00034$, $k_8-k_9=0$ at 35.1°.

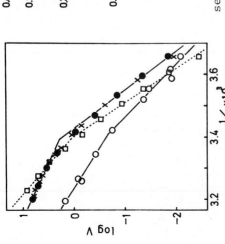

FIGURE 1. Arrhenius plot of the Ca^{2+}-ATPase activity for various ATPase preparations: ○, intact SR; ✗, MacLennan's enzyme; ●, DOL-ATPase; □, egg PC-ATPase. The unit of the activity (v) is μmol P_i liberated/mg protein/min.

STRUCTURAL FLUCTUATION OF Ca^{2+}-ATPASE MOLECULE AS STUDIED BY HYDROGEN-DEUTERIUM EXCHANGE REACTION KINETICS. The conformation of protein molecules is considered to fluctuate constantly under ordinary temperature conditions as

$$N_i \rightleftarrows D_i \quad \text{for } i = 1, 2, \cdots \qquad (1)$$

where N_i and D_i stand for conformations where the ith one of the peptide NH groups is folded and unfolded, respectively (23). The thermodynamic state of the protein molecule is, therefore, related with the structural fluctuation through the fluctuation amplitude, $\gamma_i = [D_i]/([D_i] + [N_i])$, which can be conveniently measured by the H-D exchange reaction kinetics technique.

The exchange reaction of the peptide NH protons of the ATPase protein for deuterium ions of the solvent D_2O was followed by IR spectroscopy. The absorbance of the amide II band at 1545 cm^{-1} decreases with deuteration while the absorbance of the amide I band at 1645 cm^{-1} remains constant and is proportional to the total number of peptide groups. The ratio of the absorbance of the amide II band to that of the amide I band, which is written as $X(t)$, is proportional to the amount of undeuterated peptide NH groups at time t. At t = 0 the ratio is assumed to be 0.45, in accord with the values reported for many kinds of protein molecules (24).

As a typical example, time courses of H-D exchange for MacLennan's enzyme obtained at three temperatures are shown in Fig. 2. Similar plots were obtained for intact SR and DOL-ATPase. The absorbance ratio as a function of time, $X(t)$, can be expressed well by

$$X(t) = 0.05 \sum_{n=1}^{9} \exp(-k_n \cdot t) \qquad (2)$$

where $k_n > k_{n+1}$. In this equation we tentatively assumed that the peptide groups should be classified effectively into several categories with different kinetic behavior from each other. The rate constants k_n were determined by computer simulation using the least-squares method to get the best fit between the simulated and the observed curves. Most reliable values were obtained for k_3-k_5 throughout the temperature range studied.

The rate constant k_n may be written as $k_n = \gamma_n \cdot k_e$, where

$$k_e = 50 \times (10^{-pH} + 10^{pH-6.0}) \times 10^{0.05 \times (\theta - 20)}$$, the latter being the

empirical equation for the rate constant for D form given by Hvidt and Nielsen (23) where θ represents the temperature (°C). Figure 3 shows van't Hoff plots of γ_4 for intact SR, MacLennan's enzyme, and DOL-ATPase. They are well represented by two straight lines intersecting at about 18°C. These are most reasonably interpreted as indicating that there is a phase change at about 18°C. The enthalpy change for the fluctuation is larger at temperatures below 18°C than above. This result resembles the change in the temperature dependence of the Ca^{2+}-ATPase activity around 18°C.

Several authors (7,25) have reported that the replacement of endogenous lipids of SR by exogenous dipalmitoyllecithin (DPL) greatly reduced the Ca^{2+}-ATPase activity below the gel-liquid crystalline transition temperature of DPL. The very low activity has been ascribed to the crystalline state of DPL. Based on the results with DPL-ATPase, some of the above authors ascribed the break at about 18°C in the Arrhenius plot for intact SR to phase transition of the endogenous lipids. However, the extrapolation from DPL-ATPase to intact SR is not appropriate, since the endogenous lipids of SR do not undergo phase transition in the range of about 5-40°C. The present study shows that both the Ca^{2+}-ATPase activity and the conformation of the ATPase protein have a discontinuity in their temperature dependence profile even when the lipid is in a fluid state.

Effects of nucleotide binding on the structural fluctuation was also examined. The H-D exchange reaction was followed for intact SR at 29.3°C in the presence or absence of 5 mM ADP or AMPPNP. However, significant difference in the exchange rate was not detected with the addition of these nucleotides.

SATURATION TRANSFER ESR STUDY. A new technique referred to as saturation transfer ESR (ST-ESR) has recently been developed (18,26,27). The technique involves nitroxide spin labeling and is sensitive to the rotational motion of the label with correlation times (τ_2) in the range $10^{-7} < \tau_2 < 10^{-3}$ sec. Therefore, ST-ESR should be very suitable for the study of slow molecular dynamics occurring in biomembranes. Selective and tight binding of the nitroxide group to a specific protein under study is essential in the application of ST-ESR.

A conventional first harmonic in-phase absorption spectrum (V_1) of SR vesicles labeled with MSL was recorded at 20°C and is shown in Fig. 4a. Since the ATPase protein contains about 95% of all the SH groups of SR proteins (28) and since the pretreatment with NEM serves to prevent MSL from binding to the other SH-containing proteins according to the recent study of Hidalgo and Thomas (29), the spin label can be regarded as bound solely of the ATPase protein. The spectrum is predominantly due to the signal of immobilized labels with a slight contribution from weakly immobilized labels. The strongly immobilized label, we believe, binds rigidly to the ATPase molecule and hence is likely to represent the motion of the

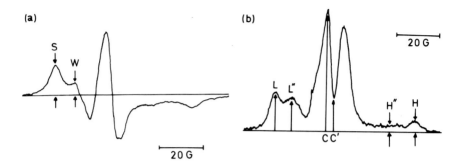

FIGURE 4. ESR spectra of spin-labeled SR membranes measured at 20°C. SR (4 mg protein/ml) was pretreated with 0.1 mM NEM at pH 7.0, 0°C, for 12 hr and labeled with 0.15 mM MSL at pH 8.2, 0°C for 24 hr. (a) V_1 spectrum: 10 mW power, 100 kHz modulation, H_m = 1.3 G; (b) V_2' spectrum: 60 mW power (H_1 = 0.25 G), 50 kHz modulation, H_m = 4.7 G, 90° out-of-phase.

molecule. The ratio of the two peak intensities in the low field, designated as W and S in Fig. 4a, was varied depending on the concentration of NEM used in the pretreatment and also depends on the measurement temperature. The value of W/S at 20°C was smallest (0.33) when 0.1 mM NEM was used. Therefore, the pretreatment with 0.1 mM NEM was used in the subsequent experiments.

Figure 1b is a reproduction of an out-of-phase second harmonic spectrum (V_2'). Weakly immobilized nitroxide affects the spectrum slightly in the low field region and considerably in the central region while no effect can be seen in the high field region. The ratio of peak intensities, H"/H, C'/C, and L"/L in Fig. 1b, is an important parameter, from which the rotational correlation time (τ_2) for the strongly immobilized nitroxide can be deduced (18). The values (with standard deviations) obtained from the repeated recordings seven times of the V_2' spectrum are H"/H = 0.67 ± 0.11, C'/C = 0.24 ± 0.12, and L"/L = 0.90 ± 0.06. If the calibration curves obtained from the measurements for MSL-hemoglobin (18), which undergoes isotropic Brownian rotation, are applied, these values lead, respectively, to the correlation times, $\tau_2(H)$ = (2.3 ± 1.4) × 10^{-4}, $\tau_2(C)$ = (1.9 ± 1.1) × 10^{-5}, and $\tau_2(L)$ = (2.4 ± 1.0) × 10^{-4} sec. The value deduced from C'/C, $\tau_2(C)$, is much smaller than the other two. This probably results from an anisotropy of the rotational motion under measurement and also from the superposition of signals from a small amount of the weakly immobilized probes. The same reasoning may hold for the finding that the disagreement between $\tau_2(H)$ and $\tau_2(L)$ was also observed when the temperature was changed. In the subsequent part of this paper the apparent correlation time deduced from H''/H, which is free from the interference by the weakly immobilized labels, will be used as a measure of the rotational motion of the ATPase molecule.

The view that the motion measured here is not the isotropic tumbling of the entire vesicle is supported by the finding that the τ_2 value is unaffected by sucrose addition (up to 30%) and the increase in the protein concentration, which affects the effective viscosity of the medium. Translational diffusion of the ATPase on the sphere of the vesicle cannot be fast enough to account for the τ_2 value determined above. A translational diffusion coefficient $D_t = 10^{-9}$ cm^2/sec leads to $\tau_2 = 2.5$ sec through the equation $r^2 = 4D_t \cdot \tau_2$, where the radius of the vesicle is assumed to be $r = 100$ nm.

Based on these considerations, the motion observed in the experiment is ascribable to the rotation of the ATPase molecule in the membrane. If the allowed rotation is assumed to be an axial one about an axis perpendicular to the plane of the membrane, saturation transfer is proportional to $\omega \cdot \sin\alpha$, in the first approximation, where ω denotes the mean angular velocity and α denotes the angle between the rotation axis and the nitrogen $2p_z$ orbital of the nitroxide. Therefore, an apparent correlation time experimentally obtained from an ST-ESR spectrum may be larger by a factor of $1/\sin\alpha$ than the true correlation time for the axial rotation. However, the difference may not be very serious unless $\alpha < 15°$. Recently, Manuck and Sykes (30) reported an unusually small τ_2 value (3×10^{-7} sec as an upper limit) for the axial rotation of the ATPase molecule, which was obtained from the NMR line-width measurements of UTP solutions in the presence and the absence of SR. Their value is much smaller than the calculated value (about 1×10^{-5} sec) using the equation of Saffman and Delbrück (31) and probable values for parameters (3). The motion measured by the NMR method seems to be some rapid motion of a part of the protein molecule.

The tempeature dependence of the apparent rotational correlation time was examined. Below 17°C, τ_2 decreases with temperature, ranging from 1×10^{-3} sec at 2°C to 2×10^{-4} sec at 17°C, while τ_2 increases with temperature above 17°C. An Arrhenius plot of the mean rotational angular velocity, i.e., $1/\tau_2$ plotted against $1/T$, is shown in Fig. 5. The plot shows a break at 17°C. The apparent activation energy is about 16 kcal/mol below and -6 kcal/mol above the transition temperature. This result is significant irrespective of some uncertainty in the absolute value for τ_2. This finding suggests that increase of the temperature brings about a decrease in the membrane viscosity, while some change in the ATPase mole-

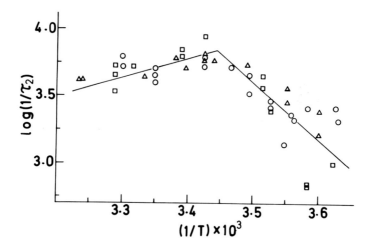

FIGURE 5. Arrhenius plot of the rotational motion expressed as the inverse of the correlation time (τ_2). Different symbols represent different SR preparations.

cule begins at the transition temperature to reduce the apparent rotational velocity. Although there are many possible candidates for the change at 17°C (for example, oligomer formation), the most probable one may be a conformational change in the ATPase molecule which results in a slight decrease in the angle α between the rotation axis and the nitrogen $2p_z$ orbital of the nitroxide.

In an attempt to detect spectroscopically the intermediary states of the enzyme defined by kinetic studies (32) of the ATPase reaction, effects of Mg^{2+}, Ca^{2+}, and AMPPNP were examined. Results are shown in Table II. The presence or absence of Ca^{2+} or Mg^{2+} does not make a difference in the τ_2 value, while the decrease in τ_2 with the addition of AMPPNP seems to be significant. This is in contrast to the result of the H-D exchange study where no effect was detected with the addition of AMPPNP or ADP. Effects of nucleotides and divalent cations to the physical states to be observed by the present technique are rather subtle, if any, compared with the effect of the temperature. Obviously a more refined measurement will be needed to detect the change in the states of the ATPase molecule during the ATPase reaction cycles.

TABLE II. Effects of Divalent Cations and AMPPNP to the Apparent Rotational Correlation Time (τ_2) at 24°C.

Additives	$\tau_2 \times 10^4$ sec
None	1.7-2.2
2 mM EGTA	1.4-2.0
2 mM EGTA, 20 mM MgCl$_2$	1.3-2.3
2 mM EGTA, 20 mM MgCl$_2$, 8 mM AMPPNP	0.7-1.1
5-20 mM CaCl$_2$	1.5-1.9
5-20 mM CaCl$_2$, 8 mM AMPPNP	0.8-1.2

IV. CONCLUSION

Our results both from the H-D exchange experiment and ST-ESR measurment show that the ATPase molecule of SR changes its physical state, probably its conformation, at around 18°C. We conclude this thermotropic conformational change is the main cause of the break in Arrhenius plots of the Ca^{2+}-ATPase activity. It may also be related to the result of Sumida and Tonomura (33) that the ratio of the rate of Ca^{2+} transport and ATP hydrolysis is 2 at 22°C and 1 at 0°C, and to the result of Ikemoto (34) that the number of α sites changes from 1 per ATPase molecule at 0°C to 2 at 22°C.

A similar transition in protein physical states at about 18°C has been recently found for cytochrome oxidase, too (35). Thermotropic transition in states around 18°C could be a common phenomenon for a wide variety of membrane-bound proteins.

ACKNOWLEDGMENT

We thank Prof. Y. Tonomura and Dr. T. Yamamoto of Osaka University for their kind guidance in SR preparation and measurement of the ATPase activity and Dr. M. Nakanishi and Prof. M. Tsuboi of the University of Tokyo for valuable advice and support on the H-D exchange experiments. We also thank Mr. T. Kaizu for his assistance in carrying out a part of the ESR experiments.

REFERENCES

1. Eletr, S. and Inesi, G. (1972). *Biochim. Biophys. Acta* 290:178.
2. Inesi, G., Millman, M., and Eletr, S. (1973). *J. Mol. Biol.* 81:483.
3. Madeira, V.M.C., Antunes-Madeira, M.C., and Carvalho, A.P. (1974). *Biochem. Biophys. Res. Commun.* 58: 897.
4. Madeira, V.M.C. and Antunes-Madeira, M.C. (1975). *Biochem. Biophys. Res. Commun.* 65:997.
5. Lee, A.G., Birdsall, N.J.M., Metcalfe, J.C., Toon, P.A. and Warren, G.B. (1974). *Biochem.* 13:3699.
6. Davis, D.G., Inesi, G., and Gulik-Krzywcki, T. (1976). *Biochem.* 15:1271.
7. Hidalgo, C., Ikemoto, N., and Gergely, J. (1976). *J. Biol. Chem.* 251:4224.
8. Mitsui, T. (1978). *Advan. Biophys.* 10:97.
9. Martonosi, A. (1974). *FEBS Lett.* 47:327.
10. Kirino, Y., Anzai, K., Shimizu, H., Ohta, S., Nakanishi, M., and Tsuboi, M. (1977). *J. Biochem.* 82:1181.
11. Anzai, K., Kirino, Y., and Shimizu, H. (1978). *J. Biochem.* 84:815.
12. Kirino, Y., Ohkuma, T., and Shimizu, H. (1978). *J. Biochem.* 84:111.
13. MacLennan, D.H. (1970). *J. Biol. Chem.* 245:4508.
14. Warren, G.B., Toon, P.A., Birdsall, N.J.M., Lee, A.G., and Metcalfe, J.C. (1974). *FEBS Lett.* 41:122.
15. Warren, G.B., Toon, P.A., Birdsall, N.J.M., Lee, A.G., and Metcalfe, J.C. (1974). *Proc. Natl. Acad. Sci. USA* 71:622.
16. Robles, E.C. and van den Berg, D. (1969). *Biochim. Biophys. Acta* 187:520.
17. Rhodes, D.N. and Lea, C.H. (1975). *Biochem. J.* 65:526.
18. Thomas, D.D., Dalton, L.R., and Hyde, J.S. (1976). *J. Chem. Phys.* 65:3006.
19. Wahlquist, H. (1961). *J. Chem. Phys.* 35:1708.
20. Davoust, J. and Devaux, P.P. (1978). *Sixth International Biophysics Congress, September 3-9, 1978, Kyoto, Japan.*
21. Warren, G.B., Houslay, M.D., Metcalfe, J.C., and Birdsall, N.J.M. (1975). *Nature* 255:684.
22. Yamamoto, T. and Tonomura, Y. (1968). *J. Biochem.* 64:137.
23. Hvidt, A. and Nielsen, A.O. (1966). *Advan. Protein Chem.* 21:287.
24. Nakanishi, M. and Tsuboi, M. (1976). *Biochim. Biophys. Acta* 434:365 and references cited therein.

25. Warren, G.B., Toon, P.A., Birdsall, N.J.M., Lee, A.G., and Metcalfe, J.C. (1974). *Biochem. 13*:5501; Hesketh, T.R., Smith, G.A., Houslay, M.D., McGill, K.A., Birdsall, N.J.M., Metcalfe, J.C., and Warren, G.B. (1976). *Biochem. 15*:4145; Nakamura, H., Jilka, R.L., Boland, R., and Martonosi, A.N. (1976). *J. Biol. Chem. 251*:5414.
26. Dalton, L.R., Robinson, B.H., Dalton, L.A., and Coffey, P. (1976). In *"Advances in Magnetic Resonance, Vol. 8"* (J.S. Waugh, ed.), p.149. Academic Press, New York.
27. Hyde, J.S. (1978). In *"Methods in Enzymology, 49G, No.19"* (C.H.W. Hirs and S.N. Timasheff, eds.), p.480. Academic Press, New York.
28. Murphy, A.J. (1976). *Biochem. 15*:4492.
29. Hidalgo, C. and Thomas, D.D. (1977). *Biochem. Biophys. Res. Commun. 78*:1175.
30. Manuck, B.A. and Sykes, B.D. (1977). *Can. J. Biochem. 55*:587.
31. Saffman, P.G. and Delbrück, M. (1975). *Proc. Natl. Acad. Sci. USA 72*:3111.
32. Kanazawa, T., Yamada, S., Yamamoto, T., and Tonomura, Y. (1971). *J. Biochem. 70*:95.
33. Sumida, M. and Tonomura, Y. (1974). *J. Biochem. 75*:283.
34. Ikemoto, N. (1975). *J. Biol. Chem. 250*:7219.
35. Kawato, S., Ikegami, A., Yoshida, S., and Orii, Y. (1978). *Sixth International Biophysics Congress, September 3-9, 1978, Kyoto, Japan.*

EFFECT OF Ca^{2+} IONS ON THE REACTIVITY
OF THE NUCLEOTIDE BINDING SITE OF THE SARCOPLASMIC
RETICULUM Ca^{2+},Mg^{2+}-ADENOSINE TRIPHOSPHATASE

Masao Kawakita, Kimiko Yasuoka and Yoshito Kaziro

The Institute of Medical Science
The University of Tokyo, Tokyo, Japan

I. INTRODUCTION

In recent years, much attention has been focused to the understanding of the molecular mechanism of the Ca^{2+}-pump of sarcoplasmic reticulum(SR)[1] (for a recent review, see Tada et al., 1978). The accumulating evidence suggests that the process of Ca^{2+} transport could be defined by sequential conformational transitions of the Ca^{2+},Mg^{2+}-ATPase protein, induced through interactions with substrates. Several observations on the Ca^{2+}- or ATP-induced conformational changes of the ATPase have recently been reported(Champeil et al., 1976 and 1978; Dupont, 1976; Ikemoto, 1978; Coan and Inesi, 1977)

In this paper, we describe effects of Ca^{2+} on the reactivity of ATP-binding site of the enzyme. Reactivity of this site was assessed by the susceptibility of the enzyme to the modification with pyridoxal-5'-phosphate(PLP)(Murphy, 1977) and also by the protection with ATP analogs, AMP-P(CH_2)P and AMP-P(NH)P, against PLP modification. The results indicate that the reactivity of ATP binding site of the enzyme is modulated through interaction with Ca^{2+} at the high affinity site.

---

[1]The abbreviations used are: SR, sarcoplasmic reticulum; AMP-P(CH_2)P, 5'-adenylyl methylene diphosphonate; AMP-P(NH)P, adenyl-5'-yl imidodiphosphate; PLP, pyridoxal-5'-phosphate; HEPES, N-2-hydroxyethyl piperazine-N'-2-ethanesulfonic acid; EGTA, ethylene glycol bis(β-aminoethyl ether)-N,N,N',N'-tetraacetic acid.

II. MATERIALS AND METHODS

SR was prepared from rabbit skeletal muscles by the method of Kanazawa et al.(1971).

Modification by PLP was carried out, unless otherwise specified, at 30°C for 10 min at an SR protein concentration of 0.3-1.0 mg/ml in 40 mM K-HEPES(pH 7.0), 0.1 M KCl, 5 mM $MgCl_2$, 0.05 mM $CaCl_2$, 0.05 mM EGTA(7 μM free Ca^{2+}), and 1 mM PLP. The reactions were started by adding PLP and stopped by the addition of $NaBH_4$ to 10 mM.

Activities were measured at 30°C in 40 mM Tris-maleate (pH 6.5), 0.1 M KCl, 5 mM $MgCl_2$(buffer A) supplemented with 0.05 mM $CaCl_2$, 0.05 mM EGTA(16 μM free Ca^{2+}) and 0.1 mM ATP. ATPase activity was determined after solubilizing PLP-treated SR by Triton X-100(2 mg/mg of SR protein) by measuring ^{32}Pi released from [γ-^{32}P]ATP(see Kawakita et al., 1974). For $^{45}Ca^{2+}$ transport assay, 5 mM K-oxalate was also included in the reaction mixture. After stopping the reaction by adding 1 ml of buffer A, SR was collected on a nitrocellulose membrane filter and $^{45}Ca^{2+}$-radioactivity retained on the filter was measured.

Binding of $^{45}Ca^{2+}$ and [^{3}H]AMP-P(NH)P to SR membrane was measured by a modification of the method of Penefsky(1977), using a small Sephadex column to separate protein-bound ligand. In the modified procedure, a 1-ml Sephadex G-25. (coarse) column was equilibrated with a solution containing a radioactive ligand, namely buffer A containing 0.05 mM EGTA, 0.05 mM $CaCl_2$ and 0.02 mM AMP-P(NH)P, either of Ca^{2+} or AMP-P(NH)P being labeled. After forced draining of external liquid by centrifugation, 30 μl of an SR sample in the same solution was placed on top of the column and effluent recovered by centrifugation(1,100 rpm, 2 min). The column was washed by another 30-μl aliquot of the above solution in the same way as above. The radioactivity of the combined effluent(60 μl) in excess of that in the same volume of equilibration solution was determined to give the amount of bound ligand. Further details will be published elsewhere.

III. RESULTS AND DISCUSSION

1. <u>Effect of PLP Treatment on Ca^{2+} Transport and ATPase Activities of SR Membranes</u>. PLP treatment followed by reduction with $NaBH_4$ lead to an irreversible loss of both ATPase and Ca^{2+} transport activities as shown in Fig. 1. Effect of the modification on ATPase was similar to that reported by Murphy(1977). In the present experiment, it

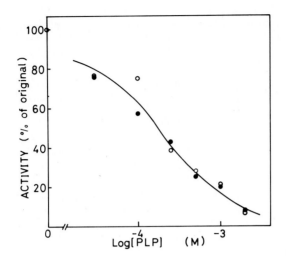

Fig. 1. Effect of PLP treatment on Ca^{2+} transport(o) and ATPase(●) activities of SR membrane.

should be noted that both ATPase and Ca^{2+} transport activities showed the same PLP concentration dependence of inactivation. Although the actual size of the functioning molecule of Ca^{2+} transport system and of ATPase is still a matter of debate (Vanderkooi et al., 1977; Dean and Tanford, 1978), the parallel inactivation of both activities suggests that it may be identical. It may also be noted that there is no indication of cooperativity in this inactivation curve.

2. <u>Impaired AMP-P(NH)P Binding to PLP-Treated SR</u>. PLP-treated SR membranes were recovered by centrifugation and were tested for the ability to bind Ca^{2+} and AMP-P(NH)P by centrifugal gel filtration. As shown in Table I there was a marked decrease in the amount of AMP-P(NH)P bound, which paralleled with the extent of inactivation, whereas the binding of Ca^{2+} was not affected by PLP modification.

TABLE I. Effect of PLP Modification on Substrate Binding to SR Membranes[a]

	AMP-P(NH)P bound	Ca^{2+} bound
Control SR	1.27	3.19
PLP-treated SR	0.26	3.27

[a]Units are in nmol/mg SR protein.

Impairment of AMP-P(NH)P binding is not unexpected in view of the suggestion that PLP is directed toward the phosphate binding sites of enzymes(Rippa et al., 1967), since phosphate binding site in this system may be very close to or even a part of the nucleotide binding site.

3. **Protection Against PLP-Inactivation by ATP Analogs.**
Fig. 2 shows a time course of PLP-inactivation of Ca^{2+} transport activity. AMP-P(CH_2)P was found to protect the activity from inactivation. AMP-P(NH)P had the same effect and ATPase activity was also protected similarly(data not shown). This is also compatible with the interpretation that the PLP-reactive site on the ATPase molecule lies close to the nucleotide binding site. A similar and even better protection of ATPase activity by ATP was reported(Murphy, 1977). In our hand, however, the protection by nucleotides was only partial. Neither the use of ATP instead of its analogs nor nucleotides at higher concentrations improved the extent of protection significantly(data not shown). Apparently, there are at least two classes of PLP-reactive groups essential for both Ca^{2+} transport and ATPase, and ATP and its analogs seem to protect only one of them.

4. **A Ca^{2+}-Dependent Change in the Reactivity of Nucleotide Binding Site.** Although evidence suggesting Ca^{2+}-induced conformational changes of ATPase has accumulated considerably in recent years(Champeil et al., 1976 and 1978; Dupont, 1976; Ikemoto, 1978; Murphy, 1978), a change occurring

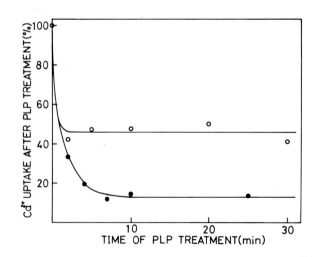

Fig. 2. Time course of PLP-inactivation of Ca^{2+} transport activity with(o) and without(●) 1.75 mM AMP-P(CH_2)P.

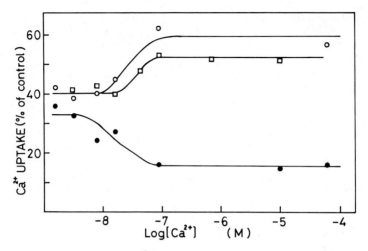

Fig. 3. Effect of Ca^{2+} concentration on PLP-inactivation of Ca^{2+} transport activity and on the effectiveness of protection by ATP analogs. Incubation was carried out in the presence of 0.2 mM AMP-P(CH$_2$)P (○), 0.05 mM AMP-P(NH)P (□), or of no added nucleotide (●).

in functionally important sites of the ATPase has not been sorted out. We have attempted to investigate the effect of Ca^{2+}-induced conformational change on the reactivity of ATP binding site by measuring the change of susceptibility of the enzyme to PLP at various concentrations of Ca^{2+}.

When free Ca^{2+} concentration during PLP treatment was decreased by the addition of EGTA, Ca^{2+} transport became definitely less prone to PLP-inactivation as shown in Fig. 3. The transition to a state more accessible to PLP occurs at around 10^{-7} M Ca^{2+}, suggesting an involvement of a high affinity Ca^{2+}-binding site. Fig. 3 further demonstrates that the protection by ATP analogs described in the previous section was also dependent on Ca^{2+} concentration. Thus, at 7 μM free Ca^{2+} where standard PLP treatment was carried out, we observed an extensive inactivation in the absence of ATP analogs, but a good protection if either AMP-P(CH$_2$)P or AMP-P(NH)P was added. On the other hand, little protection was afforded by ATP analogs at sufficiently low free Ca^{2+}. The free Ca^{2+} concentration at which the ATP analogs become effective was again around 10^{-7} M. It is remarkable that a similar concentration dependence on Ca^{2+} was observed for both the susceptibility to PLP and the protective effect of ATP analogs against PLP modification.

TABLE II. Ca^{2+} Dependence of ATP Analog Binding to SR

Nucleotide	Nucleotide bound[a]	
	0.03 µM free Ca^{2+}	12 µM free Ca^{2+}
AMP-P(CH$_2$)P (10 µM)	0.28	0.49
AMP-P(NH)P (5 µM)	0.20	0.33

[a] Units are in nmol/mg SR protein.

The Ca^{2+}-dependent protection seems to be related with an increased affinity of the ATP analogs to the ATPase. This is born out as shown in Table II where definitely more ATP analogs are bound to SR membrane in the presence of 12 µM Ca^{2+} and at sub-saturating nucleotide concentrations.

The observed effects of Ca^{2+} on the reactivity of the nucleotide binding site can be well understood if we assume a conformational change of the ATPase caused by binding of Ca^{2+} at the high affinity binding site. In this altered conformation the nucleotide binding site may become more accessible to ATP.

REFERENCES

Champeil, P., Bastide, F., Taupin, C., and Gary-Bobo, C. M. (1976). *FEBS Lett.* 63: 270.
Champeil, P., Buschlen-Boucly, S., Bastide, F., and Gary-Bobo, C., (1978). *J. Biol. Chem.* 253: 1179.
Coan, C. R., and Inesi, G. (1977). *J. Biol. Chem.* 252: 3044.
Dean, W. L., and Tanford, C. (1978). *Biochem.* 17: 1683.
Dupont, Y. (1976). *Biochem. Biophys. Res. Commun.* 71: 544.
Ikemoto, N. (1978). *Ann. N. Y. Acad. Sci.* 307: 221.
Kanazawa, T., Yamada, S., Yamamoto, T., and Tonomura, Y. (1971). *J. Biochem.* 70: 95.
Kawakita, M., Arai, K., and Kaziro, Y. (1974). *J. Biochem.* 76: 801.
Murphy, A. J. (1977). *Arch. Biochem. Biophys.* 180: 114.
Murphy, A. J. (1978). *J. Biol. Chem.* 253: 385.
Penefsky, H. S. (1977). *J. Biol. Chem.* 252: 2891.
Rippa, M., Spanio, L., and Pontremoli, S. (1967). *Arch. Biochem. Biophys.* 118: 48.
Tada, M., Yamamoto, T., and Tonomura, Y. (1978). *Physiol. Rev.* 58: 1.
Vanderkooi, J. M., Ierokomas, A., Nakamura, H., and Martonosi, A. (1977). *Biochem.* 16: 1262.

ATP-INDUCED Ca BINDING OF Ca-ATPASE IN THE ABSENCE OF ADDED Mg ION

Jun Nakamura

Biological Institute
Faculty of Science
Tohoku University
Sendai, Japan

It has been reported that the affinity of Ca-ATPase for Ca ion is much greater than that for Mg ion in the absence of ATP, while in the presence of ATP the affinity for Ca ion is nearly the same as that for Mg ion (1). The present study deals with the effects of ATP on the Ca binding of Ca-ATPase in the absence of added Mg ion.

Sarcoplasmic reticulum was solubilized with DOC (1/3=DOC/prot.) in a solution containing 0.5 M KCl, 0.3 M sucrose, 10 µM $CaCl_2$, 0.1 % β-mercaptoethanol and 20 mM Tris-maleate (pH 8.0). The solubilized material was passed through a column of Sephadex G-100. The fractions eluted at void volume were used as Ca-ATPase. In the binding assay, the membranous fraction was removed by filtration with two pieces of Whatmann glass filter (0.7 µm diameter) and by centrifugation at 109,000xg for 60 min.

The amount of bound Ca increased from 2.0 to 2.6 n moles/mg prot. of Ca-ATPase at 22 µM of free Ca ion in the absence of added Mg ion by ATP, and the bound Ca was released by the addition of Mg ion (Fig. 1). The amount of ATP-induced Ca binding was decreased by increasing the concentration of Mg ion above 100 µM, while the Ca binding in the absence of ATP was not affected by this ion untill 1 mM. The ATP-induced Ca binding,was not observed at the concentration of ATP more than 1 mM. The amount of ATP-induced Ca binding at acidic pH was greater than that at alkaline pH. On the other hand, Mg ion decreased the amount of ATP-induced Ca binding at alkaline pH more than at acidic pH. The association constant of the Ca-ATPase-

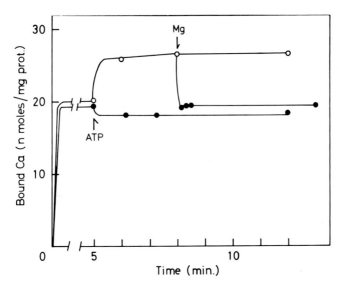

FIGURE 1. ATP-induced Ca binding in the absence of added Mg ion and release of the Ca by the addition of Mg ion. The reaction was carried out in a solution containing 1.1 mg prot./ml Ca-ATPase, 0.12 M KCl, 22 µM $CaCl_2$, 1 mM creatin phosphate, 0.3 mg prot./ml creatin kinase, 20 mM Tris-maleate (pH 6.8), 1 mM $MgCl_2$ (●) and/or not (○) at 0 C. ATP (final conc. 20 µM) and/or $MgCl_2$ (final conc. 1 mM) was added to the solution at the time indicated in Figure.

calcium complex in the presence of ATP was approx. 0.4 $µM^{-1}$, same as that in the absence of ATP. ATP increased the number of the Ca binding sites from 11 to 16 n moles/mg prot.

No Ca binding proteins except for the high affinity Ca binding protein (mw. 55,000., 1 mole of Ca/mole) and Ca-ATPase (mw. 105,000., 2 moles of Ca/mole) were presented in this preparation. The amount of high affinity Ca binding protein in the preparation was at maximum 4 % of Ca-ATPase. Therefore, the binding by the high affinity Ca binding protein is estimated about 4 % of the total amount of Ca binding.

These observstions suggest that a step of the ATP-induced Ca binding is involed in the process of Ca transport in sarcoplasmic reticulum.

I am grateful to professor Kazuhiko Konishi for his encouragement and advice during the course of this work.

1. Tonomura, Y. (1972). "Muscle Protein, Muscle Contraction and Cation Transport" p. 305. University of Tokyo Press, Tokyo.

ENTROPY-DRIVEN PHOSPHORYLATION WITH Pi
OF THE TRANSPORT ATPase
OF SARCOPLASMIC RETICULUM

Tohru Kanazawa
Yuichi Takakuwa
Fumio Katabami

Department of Biochemistry
Asahikawa Medical College
Asahikawa, Japan

We previously demonstrated that the transport ATPase of sarcoplasmic reticulum solubilized with Triton X-100 was phosphorylated with Pi (1). The phosphorylation was quite likely to result from reversal of late steps in the ATPase reaction (1,2). In the present study, thermodynamic analysis of the phosphorylation with Pi was made in the absence of detergent over a wide range of temperature (6-38°).

Sarcoplasmic reticulum vesicles isolated from rabbit skeletal muscle were pretreated with excess EGTA in alkaline medium (pH 9.0) to remove endogenous internal Ca^{2+}, and then the pH of the medium was readjusted to 7.0. The transport ATPase was phosphorylated by incubating the vesicles with ^{32}Pi in the presence of saturating concentration of $MgCl_2$.

The equilibrium constant of the phosphorylation reaction, $K = [Mg \cdot EP]/[Mg \cdot E][Pi]$, was measured at different temperatures. As shown in FIGURE 1, the linear plot of log K against $1/T$ showed a remarkable break at 18°. The thermodynamic analysis of the plot showed that, in the whole range of temperature tested, the phosphorylation reaction was driven by a marked increase in entropy, which compensated for the concurrent unfavourable increase in enthalpy. The apparent standard enthalpy change, $\Delta H^{\circ\prime}$, the apparent standard entropy change, $\Delta S^{\circ\prime}$, and the apparent standard free energy change, $\Delta G^{\circ\prime}$, in the phosphorylation reaction ($Mg \cdot E + Pi \rightleftharpoons Mg \cdot EP$) in the lower temperature range (6-18°) were found to be, respec-

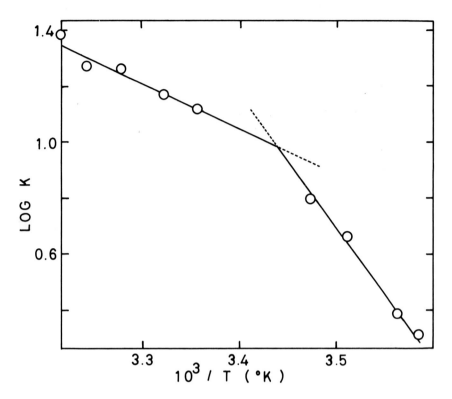

FIGURE 1. Temperature dependence of the equilibrium constant in the phosphorylation reaction with Pi of the transport ATPase of sarcoplasmic reticulum. K represents the equilibrium constant, [Mg·EP] / [Mg·E] [Pi], in the phosphorylation reaction, Mg·E + Pi ⇌ Mg·EP, which was measured in the presence of 100 mM $MgCl_2$, 100 mM KCl and 5 mM EGTA at pH 7.0. The true substrate (Pi) for this reaction was assumed to be $MgHPO_4$.

tively, 21.7 kcal/mol, 79.0 e.u. and -0.8 kcal/mol (at 12°), while those in the higher temperature range (18-38°) were, respectively, 7.5 kcal/mol, 30.1 e.u. and -1.9 kcal/mol (at 37°). The results strongly suggest an essential role of entropy in the energy transduction of this system.

References

1. Kanazawa, T. (1975) J. Biol. Chem. 250:113.
2. Kanazawa, T., and Boyer, P.D. (1973) J. Biol. Chem. 248: 3136.

KINETIC PROPERTIES OF Ca^{2+}, Mg^{2+}-DEPENDENT ATPase
OF DETERGENT-TREATED SARCOPLASMIC RETICULUM

Haruhiko Takisawa

Department of Biology
Osaka University
Toyonaka, Osaka, Japan

I. INTRODUCTION

The accumulation of Ca^{2+} inside the vesicle of fragmented sarcoplasmic reticulum (FSR) is an energy driven process. The energy comes from the hydrolysis of ATP by Ca^{2+}, Mg^{2+}-dependent ATPase of FSR. It is now known that the ATPase molecule itself is a carrier protein for Ca^{2+} transport. Several investigators have studied the reaction mechanism of Ca^{2+}, Mg^{2+}-dependent ATPase using FSR preparations. Although an acid stable phosphoprotein (EP) has been identified as a true reaction intermediate, many problems remain to be solved concerning the reaction mechanism of FSR ATPase (1). For example, the overshoot in the phosphorylation reaction and two kinds of P_i burst in the presteady state of the ATPase reaction cannot be explained by a simple reaction mechanism as proposed by various workers (2,3,4). Furthermore, when the membrane structure is destroyed by detergent, the ATPase does not show any complex phenomena in the presteady state (3,4), and the reaction itself can be studied in the absence of Ca^{2+} gradient.

This paper deals with the kinetic properties of SR ATPase treated with detergent. The ATPase reaction was studied both in the steady and presteady states and could be explained by a simple reaction mechanism;

$$E + S \rightleftharpoons ES \rightarrow EP \rightarrow E + Pr.$$

Recently, Shigekawa et al. (5) reported the existence of two

kinds of EP in the reaction of SR ATPase in the absence of
KCl. One kind could react with ADP to form ATP (ADP-sensitive
EP) and the other could not (ADP-insensitive EP). However,
it is not known whether these two types of EP exist in the
presence of KCl or not. In the present work, both were found
in the reaction of detergent-treated SR ATPase in the presence
of a sufficient amount of KCl. Furthermore, the amount of
ADP-sensitive EP increased with an increase in $CaCl_2$ concentration and decreased with an increase in $MgCl_2$ concentration
and the overall reaction could be explained by assuming the
sequential formation of the two kinds of EP; ADP-sensitive
EP is formed first and then converted into ADP-insensitive EP.

II. EXPERIMENTAL PROCEDURE

FSR was prepared from rabbit skeletal muscle as described
previously (6). Detergent-treated SR ATPase was prepared as
follows. 200 mg of FSR protein was suspended in 100 ml of
sucrose-Tris buffer containing 0.25 M sucrose, 50 mM KCl, 0.1
mM $CaCl_2$, 0.1 mg/ml Tween 80, and 20 mM Tris-HCl, pH 8.0 at
10°C, then solubilized with 400 mg of dodecyl octaoxyethyleneglycol monoether ($C_{12}E_8$). The solubilized suspension was
centrifuged at 100,000 x g for 40 min to remove insoluble
materials. The clear supernatant was put on DEAE-cellulose
column (5 x 10 cm), equilibrated with the sucrose-Tris buffer.
The column was thoroughly washed with the buffer to remove
$C_{12}E_8$. ATPase was eluted with sucrose-Tris buffer containing
0.35 M KCl.

The amount of EP and the ATPase activity in the steady
state were measured as described previously (4). A Durrum
D-133 multi-mixing apparatus was used to follow the rapid
reaction. Details of the experimental procedures have been
described elsewhere (4). The amounts of ADP-sensitive and
-insensitive EP were estimated as follows. The phosphorylation of ATPase was started by addition of 1 µM $AT^{32}P$ to the
reaction mixture containing SR, 0.1 M KCl, 0.1 M sucrose,
$MgCl_2$, $CaCl_2$, 0.1 mg/ml Tween 80 and 100 mM Tris-HCl, pH 9.0
at 10°C. After an appropriate interval, a solution containing
ADP and ATP (final concentrations were 1 mM) was added to the
reaction mixture. The reaction was terminated 5 seconds later
by the addition of 4% TCA. The amounts of EP remaining and
P_i liberated after the addition of ATP + ADP were considered
to represent the amount of ADP-insensitive EP. The amount of
ADP-sensitive EP was obtained by subtracting the amount of
ADP-insensitive EP from the amount of EP just before the
addition of ADP + ATP or by measuring the amount of $AT^{32}P$

formed after the addition of ADP + ATP. Both methods gave similar values.

III. RESULTS AND DISCUSSION

A. Steady State

The dependence on the ATP concentration of the ATPase activity and the amount of EP in the steady state was measured over a wide range of ATP concentrations (0.33 μM - 20 μM) in the presence of 0.1 M KCl, 0.1 M sucrose, 1 mM $MgCl_2$, 0.1 mM $CaCl_2$, 0.1 mg/ml Tween 80 and 50 mM Tris-maleate, pH 7.0 at 10°C. The double reciprocal plots of the ATPase activity and

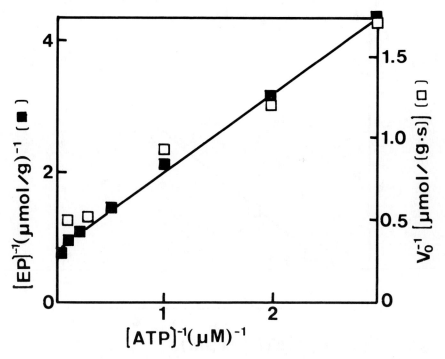

Fig. 1. Double reciprocal plots of the rate of the ATPase reaction and the amount of EP of the detergent-treated ATPase in the steady state against the ATP concentration. The rate of ATPase (□) and the amount of EP (■) were measured in the presence of 10 (□) or 100 (■) μg/ml SR protein in 0.1 M KCl, 0.1 M sucrose, 1 mM $MgCl_2$, 0.1 mM $CaCl_2$, 0.1 mg/ml Tween 80, and 50 mM Tris-maleate, pH 7.0 at 10°C.

the amount of EP formed against the ATP concentration gave straight lines. The Michaelis constants for both ATPase activity and EP formation were 1.4 µM (Fig. 1).

B. Presteady State

The initial velocity of EP formation was measured over a wide range of ATP concentrations under the conditions described in Fig. 2. As shown in Fig. 2, the double reciprocal plot of the initial velocity of EP formation versus the ATP concentration gave a straight line, and the Michaelis constant (K_S) was 16 µM.

The amount of ADP bound to the enzyme during the ATPase

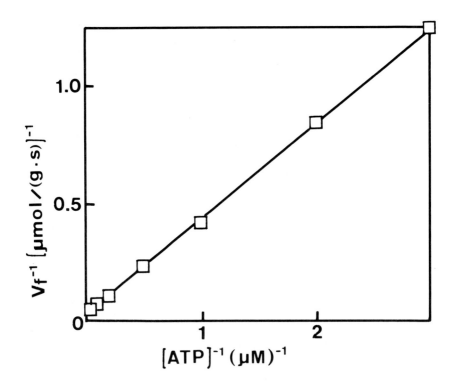

Fig. 2. Double reciprocal plot of the initial velocity of EP formation of detergent-treated ATPase against the ATP concentration. The initial velocity of EP formation was measured in the presence of 100 µg/ml SR protein in 0.1 M KCl, 0.1 M sucrose, 1 mM $MgCl_2$, 0.1 mM $CaCl_2$, 0.1 mg/ml Tween 80, and 50 mM Tris-maleate, pH 7.0 at 10°C.

reaction is negligibly small (4). Therefore, the reaction of detergent-treated ATPase can be explained by a simple mechanism:

$$E + ATP \xrightleftharpoons{K_s} E \cdot ATP \xrightarrow{k_f} EP \xrightarrow{k_d} E + P_i$$

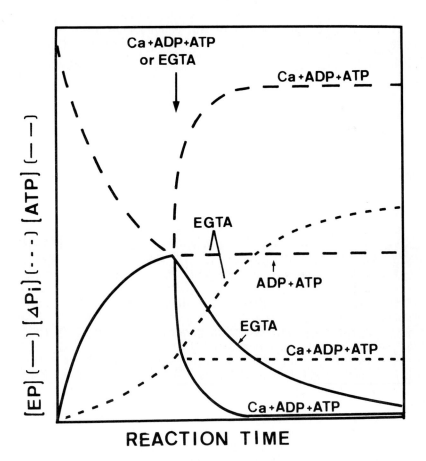

Fig. 3. ATP-formation from EP and added ADP in the presence or absence of Ca^{2+}. SR was phosphorylated with 1 μM $AT^{32}P$ in 0.1 M KCl, 0.1 M sucrose, 1 mM $MgCl_2$, 0.1 mM $CaCl_2$, 0.1 mg/ml Tween 80 and 100 mM Tris-HCl, pH 9.0 at 10°C. After 5 sec, the phosphorylation reaction was stopped by addition of a sufficient amount of EGTA or unlabeled ATP. ATP-formation was tested in the presence or absence of Ca^{2+}. For details see text.

The value of k_d was obtained from the rate of the ATPase reaction per unit of EP concentration at steady state and was about 2.5 s^{-1}. The value of k_f was obtained as the maximum rate of EP formation per unit of active site concentration and was about 28 s^{-1}. The relationship between K_s and K_m was:

$$K_m = K_s \frac{k_d}{k_f + k_d}.$$

The value of K_m calculated from the above equation was about 1.3 μM, which was consistent with the experimental value 1.4 μM.

C. Partial Reactions

After the SR had been phosphorylated with ATP, further phosphorylation was stopped by the addition of an excess amount of unlabeled ATP or EGTA. When Ca^{2+} and ADP were added simultaneously with ATP, almost all the EP formed reacted with ADP to form ATP i.g. in the presence of a sufficient amount of Ca^{2+}, almost all the EP was ADP sensitive (see Fig. 3). On the other hand, 5 sec after the addition of EGTA, addition of ADP did not cause the formation of ATP from EP and ADP i.g. in the absence of Ca^{2+}, all the EP formed was ADP insensitive (see also Fig. 3).

Kinetic studies on the reaction of FSR ATPase showed that the phosphorylation by ATP requires Ca^{2+} and the decomposition of EP requires Mg^{2+}. The decomposition of EP was inhibited by Ca^{2+} which is known to compete with Mg^{2+}. Thus, it is proposed that the reaction proceeds through EPCa to EPMg, which in turn is decomposed into E + P_i (cf. 1). To investigate the relationship between ADP sensitive or -insensitive EP and their bound divalent cations, the ratio of ADP-sensitive EP to total EP was studied with various concentrations of Mg^{2+} or Ca^{2+} at steady state. Fig. 4A shows the dependence of the ratio of ADP-sensitive EP to total EP on Ca^{2+} concentration. The ratio increases with an increase in Ca^{2+} concentration. On the other hand, the amount of ADP-sensitive EP decreases when Mg^{2+} concentration increases (Fig. 4B).

In the presence of a sufficient amount of Ca^{2+}, all the EP formed was ADP-sensitive, while in the absence of Ca^{2+} or in the presence of a high concentration of Mg^{2+}, all the EP formed was ADP insensitive. These results indicate that in the reaction of SR ATPase, at least two kinds of EP intermediates exist: ADP-sensitive and -insensitive EP. ADP-sen-

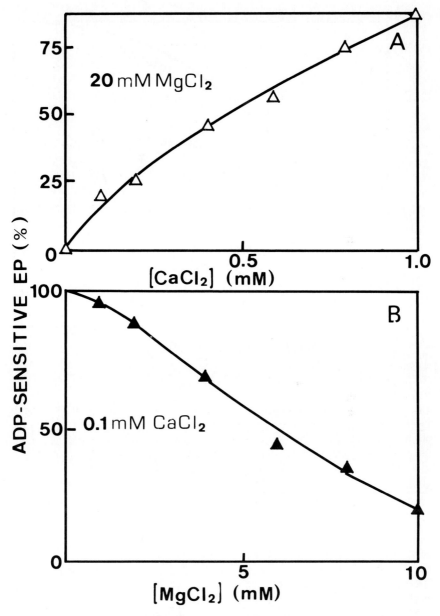

Fig. 4. Concentration dependence of the fraction of ADP-sensitive EP on $CaCl_2$ or $MgCl_2$. The amounts of ADP-sensitive EP were measured as described in "Experimental procedures." Concentration dependence on $CaCl_2$ or $MgCl_2$ was measured in the presence of 20 mM $MgCl_2$ or 0.1 mM $CaCl_2$.

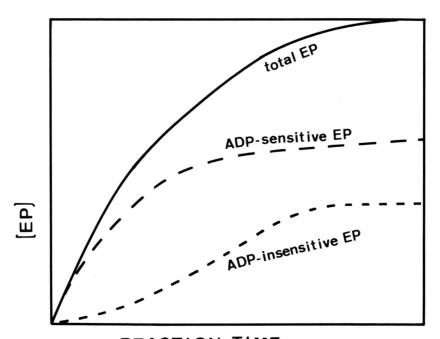

Fig. 5. Time courses of ADP-sensitive EP, ADP-insensitive EP and total EP during the initial phase of the reaction. The reaction was performed in 0.1 M KCl, 0.1 M sucrose, 10 mM $MgCl_2$, 0.1 mM $CaCl_2$, 0.1 mg/ml Tween 80, and 100 mM Tris-HCl, pH 7.0 at 10°C. The amounts of total EP (———), ADP-sensitive EP (— —) and ADP-insensitive EP (- - -) were measured.

sitive EP exists as EPCa and ADP-insensitive EP as EP without bound Ca^{2+} or EPMg.

Finally, the reaction sequence of the two kinds of EP was studied. The amount of ADP-insensitive EP was estimated as described in the experimental procedures. Fig. 5 clearly indicates that ADP-sensitive EP is formed first then converted into ADP-insensitive EP. All of these results are consistent with Kanazawa's proposal that EPCa is formed first then bound Ca^{2+} is replaced by Mg^{2+} thus forming EPMg (7).

In summary, the reaction of Ca^{2+}, Mg^{2+}-dependent ATPase of SR treated with detergent could be explained reasonably by the sequential formation of two kinds of EP, and ADP-sensitive EP was assumed as being EPCa and ADP-insensitive EP as EPMg in the presence of a sufficient amount of Mg^{2+}. The concentration of Ca^{2+} required to retain 50% of ADP-sensitive EP was about 0.5 mM, which was about 1,000 times greater than the K_{Ca} for the formation of EP from ATP (1). Therefore, the change in affinity for Ca^{2+} occurs during the phosphorylation reaction which probably arises from a change in the conformational state of the enzyme. Studies of the UV spectrum of detergent-treated SR ATPase have shown that the spectrum changes when Ca^{2+} is removed from the enzyme solution (8). This also indicates that the conformational state of ATPase changes when Ca^{2+} is removed from the enzyme. The change in the conformational state of the SR ATPase is undoubtedly an important step in the active transport of Ca^{2+} by FSR.

ACKNOWLEDGMENTS

The author wishes to thank Professor Y. Tonomura for his valuable suggestions and encouragement. The author also wishes to thank Mr. Y. Nakamura for his technical assistance, and Dr. S. K. Srivastava for critically reading the manuscript.

REFERENCES

1. Yamamoto, Y., Takisawa, H., and Tonomura, Y. (1978). In "Current Topics in Bioenergetics" (R. Sanadi, ed.) Vol. 9. Academic Press, New York, in press.
2. Froehlich, J. P., and Taylor, E. W. (1976). J. Biol. Chem. 250, 2013-2021.
3. Yamada, S., Yamamoto, T., Kanazawa, T., and Tonomura, Y. (1971). J. Biochem. 70, 279-291.
4. Takisawa, H., and Tonomura, Y. (1978). J. Biochem. 83, 1275-1284.
5. Shigekawa, M., and Dougherty, J. P. (1978). J. Biol. Chem. 253, 1458-1464.
6. Yamada, S., Yamamoto, T., and Tonomura, Y. (1970). J. Biochem. 67, 789-794.
7. Kanazawa, T., Yamada, S., Yamamoto, T., and Tonomura, Y. (1971). J. Biochem. 70, 95-123.
8. Nakamura, Y., and Tonomura, Y. (1978). Unpublished observations.

PROTON RELAXATION STUDIES OF THE INTERACTION BETWEEN MANGANESE(II) AND ATPASE OF SARCOPLASMIC RETICULUM

Ryoichi Kataoka
Toshiyuki Shibata
Akira Ikegami

The Institute of Physical and Chemical Research
Wako-shi, Saitama, Japan

The binding of manganese(II) to sarcoplasmic reticulum of rabbit skeletal muscle has been studied by pulse nuclear magnetic resonance spectroscopy.

Sarcoplasmic reticulum controls intracellular calcium concentration in the excitation-contraction coupling in skeletal muscle(1). With the excitation of muscular cell membrane, calcium is released from sarcoplasmic reticulum to cause contraction in Actin-Myosin filaments. And then, relaxation occurs when calcium is re-uptaken by Ca-ATPase into sarcoplasmic reticulum coupled with ATP hydrolysis.

Manganese can play the role of magnesium in the ATPase activity of both sarcoplasmic reticulum and of the purified enzyme(2). Manganese also binds to cooperative sites for calcium, which can be related to the inhibition of the calcium transport at high concentration of manganese or calcium(3).

The substrate analogue, adenyl-5'-yl imidodiphosphate (AMPPNP) is not cleaved, and binds competitively to the ATP hydrolytic site(4). Caffeine releases calcium from sarcoplasmic reticulum(5).

The hydration number, q, or the number of water molecules fast exchanging between the bound manganese and bulk solvents, is calculated from net longitudinal and transverse proton relaxation times, T_{1p} and T_{2p}, respectively, due to manganese-enzyme complex, when the concentration of bound manganese, N, is known(6).

TABLE I. Effect of AMPPNP and Caffeine

	Ca-ATPase		Sarcoplasmic reticulum	
Addition	N (μM)	q	N (μM)	q
None	20	0.71	13	0.76
AMPPNP	22	0.40	21	0.50
Caffeine	12	1.10	14	0.63
AMPPNP and caffeine	20	0.40	20	0.48

[a] 2 mM AMPPNP and/or 5 mM caffeine were added to solutions containing 60 μM $MnCl_2$, 3 mg protein/ml of Ca-ATPase or sarcoplasmic reticulum, 5 mM $MgCl_2$, 20 mM Tris-maleate, pH 6.8, and 50 mM KCl.

The binding of manganese was depressed by the addition of potassium. The hydration number was, however, increased from about two-third to one when potassium was added. In the absence of potassium, nonspecific ionic binding of manganese may increase. Therefore, the hydration number of manganese nonspecifically bound at low ionic strength may be less than those bound at high ionic strength.

A part of bound manganese is replaced by calcium or magnesium, but the hydration number of remaining manganese was the same with initial one.

The differences between the effect of AMPPNP and that of caffeine are clearly shown in TABLE I. The bound manganese was concealed by the enzyme interacting with Mg-AMPPNP but exposed to water with caffeine, and the effect of caffeine was cancelled out by Mg-AMPPNP. In the absence of magnesium, the different effect of AMPPNP on the bound manganese was observed, that is, all the bound manganese were released from the enzyme with AMPPNP.

REFERENCES

(1) Ebashi, S., and Endo, M. (1968) Progr. Biophys. Mol. Biol. 10:123.
(2) MacLennan, D. H. (1970) J. Biol. Chem. 245:4508.
(3) Kalbitzer, H. R., Stehlik, D., and Hasselbach, W. (1978) Eur. J. Biochem. 82:245.
(4) Dupont, Y. (1977) Eur. J. Biochem. 72:185.
(5) Ogawa, Y. (1970) J. Biochem. 67:667.
(6) Navon, G. (1970) Chem. Phys. Lett. 7:390.

ELECTROGENIC CALCIUM TRANSPORT IN THE SARCOPLASMIC RETICULUM MEMBRANE

Yves DUPONT

Laboratoire de Biophysique
Moléculaire et Cellulaire (1)
D.R.F. - C.E.N.G.
Grenoble, France

I. INTRODUCTION

Intact Sarcoplasmic Reticulum (SR) in muscle or fragmented in vesicles is much too small to permit measurements with micro-electrodes. Therefore data concerning the electric properties of this membrane system are very rare.

Optical probes offer a unique mean of detecting potential changes across membranes of small organelles. They have been used successfully in nerve, muscle and vesicular suspensions [1-3]. We describe here a study of some electrical properties of the Sarcoplasmic Reticulum membrane and of the calcium pump using these dyes. The experiments we will describe have been designed to answer mainly two questions :
- Is calcium transport electrogenic ? or, in other words : are the charges of calcium fully balanced by the transport of a counter cation ? (K^+, Mg^{++}...)
- What are the effects of the membrane potential on the activity of the calcium pump ?

These two points are often ignored in the study of the calcium pump and of the ATPase activity using Sarcoplasmic Reticulum vesicles. It is not possible to propose any satisfactory transport model without answering these questions.

---

1 Equipe de Recherche CNRS : E.R. 199.
Supported by grants from the Délégation Générale à la Recherche Scientifique.

Measurement of the electric properties of the Sarcoplasmic Reticulum membrane is rendered difficult by the high permeability of the Membrane to Cl^-, Na^+, K^+ and other small ions. This was measured earlier with radioactive ions [4], fluorescent dyes [5] or light scattering [6]. Zimniak and Racker [7] reported recently a study on the electrogenicity of the calcium pump in reconstituted phospholipid vesicles. This system was preferred because the relative impermeability of the reconstituted membranes facilitates studies of the electrical properties of the calcium pump. We will demonstrate that also in intact SR vesicles, it is possible to create and maintain a relatively stable electrical potential. The main conclusions of this work are that calcium transport is electrogenic and that the electric potential modifies the activity of the ATPase.

II. MEASUREMENT OF POTENTIALS

A. Calibration of the Dye Response

Numerous studies of membrane potential with optical probes have been reported. (For a review see : Ref. : [1-3]). Cyanine dyes have been used mostly with vesicular suspensions and the mechanism of the response was studied on red blood cells and phosphatidylcholine vesicles by Sims et al. [3]. They have shown that the incorporation of the dye into the vesicles is a potential-dependent process and that the fluorescence quenching in the membrane is due to the formation of dye aggregates.

The figure 1 shows the emission spectrum of the dye DiS-C_3- (5) for increasing concentrations of S.R. vesicles. The results are compatible with the interpretation of Sims et al [3]. The fluorescence quenching with increasing vesicle concentration can be explained by the formation of low fluorescence aggregates in the membrane. If the concentration of vesicles is raised further, almost all of the dye is incorporated in the vesicles and the high concentration of lipids produces a dilution of the aggregates, the fluorescence is then enhanced with an emission maximum corresponding to the dye dissolved in a non aqueous solvent.

These measurements are necessary to optimize the optical response of the dye to small changes in partition between the lipid and water phase. We have usually used a dye concentration of 2 to $10.10^{-6}M$ with 15 to 100 µg/ml proteins.

The only way to calibrate the response of the dye to membrane potentials is to artificialy establish a diffusion

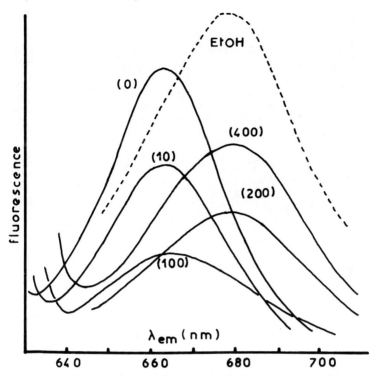

Fig. 1. Emission spectrum of diS-C_3-(5) (3,3'-dipropyl-thiodicarbocyanine iodide : 10 µM) in the presence of various amounts of vesicles (µg/ml) or in ethanol (dashed line). Details of the S.R. preparation are given in the Ref. : 17. λ_{exc} = 622 nm.

potential with mixtures of permeant and less permeant ions in various concentrations. From previous studies [4-6] the following sequence of permeabilities has been obtained : Sucrose, Ca^{++}, Mn^{++} < gluconate$^-$ < TRIS$^+$, choline$^+$ < methane sufonate$^-$ < Na$^+$, K$^+$, glycerol, urea < Li$^+$, Cl$^-$ [4]. We decided to use K$^+$ or Cl$^-$ diffusion potentials with two distinct systems : (TRIS$^+$ + K$^+$), gluconate$^-$ or TRIS$^+$, (Cl$^-$ + gluconate$^-$). The conditions are summarized in the table I.

The vesicles loaded with the internal medium by overnight dialysis are diluted in the external buffer in the presence of the dye. Due to the high permeability of the membrane to K$^+$ or Cl$^-$ the potential is supposed to be established just after dilution of the vesicles in the external medium. The experiments shown in figure 2 demonstrate that the potential created by the diffusion of K$^+$ or Cl$^-$ can be readily detected by the

TABLE I.

	Internal medium	External medium	Calculated internal potential
I	K^+-$gluc^-$ = 200mM	K^++$TRIS^+$ = 200mM $gluc^-$ = 200 mM	-130mV to 0mV
II	$TRIS^+$-$gluc^-$ =200mM K^+-$gluc^-$ = 1 mM	"	0 to +130mV
III	$TRIS^+$-$gluc^-$ =200mM $TRIS^+$-Cl^- = 1mM	$TRIS^+$ = 200mM Cl^-+$gluc^-$ = 200mM	-130mV to 0mV
IV	$TRIS^+$-Cl^- = 200mM	"	0 to +130mV

dye fluorescence. As expected from previous studies a negative potential inside is detected by a greater incorporation of dye and a quenching of fluorescence, but it is important to note that the dye is also sensitive to positive potentials, although with a lower sensitivity.

During the few minutes following the dilution of the vesicles the fluorescence returns slowly to the signal obtained without potential. This can be explained by a slow dissipation of the potential due to the diffusion of the less permeant ions. The rate of decay depends on the temperature and on the ionic species present.

Using the ionic conditions described in the table I we have tried to obtain a calibration of the response of the dye expressed in relative fluorescence change as a function of the diffusion potential calculated with the Nerst equation for the permeant ions (Fig. 3).

The most prominent feature of the calibration curves are the discrepancy between the measurement with Cl^- and K^+. The slope of the curves are identical for high voltage but curves II and III cannot be fitted with a simple straight line down to the origin. Although the explanation might not be completely satisfactory we think that curves II and III are not correct. Both are obtained in conditions where the permeant ion is in low concentration inside : 1mM K^+ or Cl^- in the dialysis medium. It is probable that this does not represent the true activity of ion inside and that any residual Cl^- in K^+-$gluconate^-$ or K^+ in $TRIS^+$-Cl^- can easily perturb the measurement. An other effect, which is difficult to quantify, is the incidence of the incorporation of the charged dye into the membrane. The effect might not be negligible if low

Electrogenic Calcium Transport

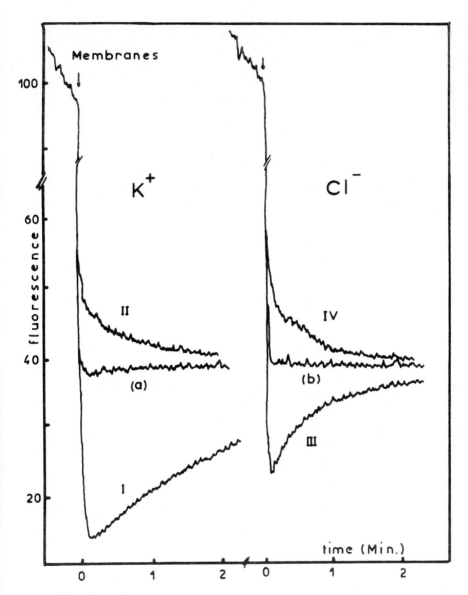

Fig. 2. Dye fluorescence at 670nm (λ_{exc} = 622nm). Conditions I to IV are described in the table I. The permeant ions concentrations are : I : $\frac{K^+ext}{K^+int} = \frac{1}{200}$, II : $\frac{K^+ext}{K^+int} = \frac{200}{1}$,

(a) $\frac{K^+ext}{K^+int} = 1$, III : $\frac{Cl^-ext}{Cl^-int} = \frac{200}{1}$, IV : $\frac{Cl^-ext}{Cl^-int} = \frac{1}{200}$ and

(b) $\frac{Cl^-ext}{Cl^-int} = 1$. Dye=10μM, S.R. proteins=75μg/mℓ, t=2°C.

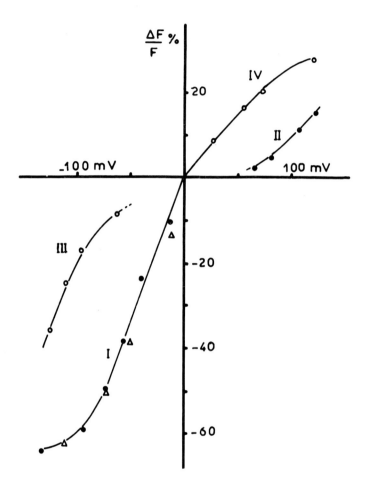

Fig. 3. *Calibration curves obtained with the conditions described in table 1. Two different internal K^+ concentration have been used for the calibration 1 : full circles K^+_{in} = 200mM, triangles K^+_{in} = 100mM.*

internal concentration of ions are used for the calculation of potentials. We think that the measurements in conditions I and IV (Table I), with 200mM permeant ion inside, are less subject to these limitations and in the rest of this paper we have used the calibration obtained in these conditions.

The figure 4 illustrates the response of the dye to transient potential changes. Hyperpolarization and depolarization

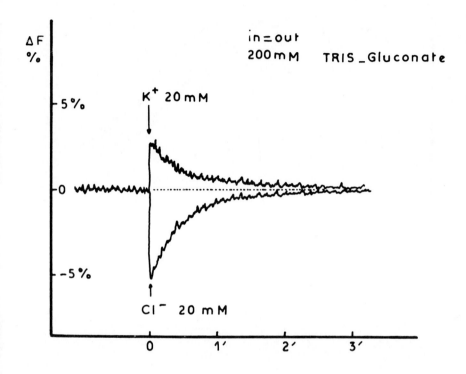

Fig. 4. *Transient potential changes produced by the injection of K^+-gluconate$^-$ or $TRIS^+$-Cl^-. Temperature is 22°.*

are induced by the addition of K^+ or Cl^- to vesicles loaded by $TRIS^+$-gluconate$^-$ and diluted in the same medium.

B. Charge Movement Associated with Calcium Transport

1. <u>Measurement of the Electrogenic Activity of the Calcium Pump</u>. K^+ ions being necessary for the activity of the pump, the measurements are made in K^+-gluconate buffered with 20mM TRIS. The experiment shown in figure 5 indicates that a large fluorescence change is observed during the uptake of Ca^{++} by the S.R. vesicle (phase-1).

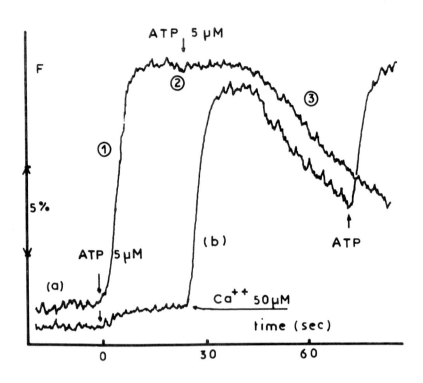

Fig. 5. Record of the dye response to an injection of ATP. Dye concentration is 10μM and protein 75μg/ml = 0,25μM of high affinity ATP sites. Conditions are : int. = ext., K^+-gluconate$^-$ = 200mM, $TRIS^+$-gluconate$^-$ = 20mM, Mg^{++} = 5mM, pH 7.2 and 22°. a) Ca^{++}_{ext} = 50μM. b) EGTA = 50μM. (Ca^{++}_{ext} < 10^{-8} before Ca^{++} injection).

When no oxalate is present the fluorescence remains high as long as ATP is present in the solution (phase-2). A slow decay is observed when ATP is completly exhausted (phase-3), this can be checked by varying the concentration of ATP. The figure 5 shows that both ATP and Ca^{++} are necessary to obtain an important fluorescence enhancement. The effect observed is then directly related to the ATPase activity, but before concluding that the calcium pump is electrogenic one must preclude all possible artefacts. Effect of light scattering can be eliminated from the spectrum shown in figure 1. Some other controls have been made without dye and the contribution of the scattering at 670nm is evaluated at 2 to 3 % of the intensity ($\lambda_{exc.}$ = 622 nm). A record of the light scattering at

670 nm, with or without oxalate, shows no effect of ATP, while internal precipitation of calcium oxalate during the transport produces a significant increase in intensity. This increase however is too small to perturb the fluorescence measurement.

A direct effect of calcium ions inside the vesicles on the dye incorporation or fluorescence is a more serious artefact which has to be examined. Little is known on the location and on the state of calcium in the inner space of the vesicles, local charge effects might affect significantly the incorporation of the dye. We have therefore designed an experiment to test the effect of internal calcium on the dye response. The experiment shown on the figure 5 was repeated with vesicles previously loaded with 5 to 10mM Ca^{++} by dialysis. The results are shown on the figure 6. The internal calcium concentration is completly different in both cases, nevertheless we have found little or no effect of the internal

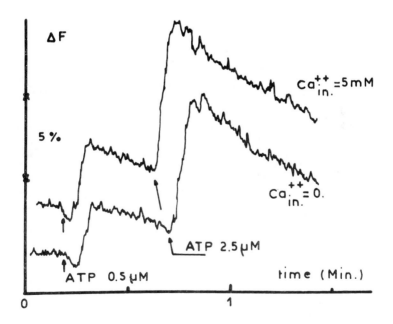

Fig. 6. Effect of internal calcium on the dye response. Conditions are described in Fig. 5. Ca^{++}_{ext} = 50μM. The traces are vertically displaced for clarity.

load either on the fluorescence level or on the fluorescence change associated with calcium transport. From these experiments we can safely reject an effect of the internal calcium on the fluorescence of the dye and the data shown in figures 5 and 6 are in favor of an electrogenic activity of the calcium pump.

2. Effect of Permeant Anions.

a. Oxalate.

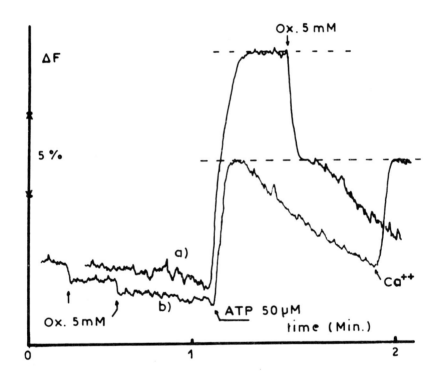

Fig. 7. Response of the fluorescence to an injection of oxalate before (trace b) or after ATP addition (trace a).

In K^+-gluconate buffer and if enough ATP is added, the fluorescence reaches rapidly a plateau corresponding to a potential increase of about 60mV; in this state the addition of oxalate produces a fluorescence drop corresponding to the incorporation of negative charges into the vesicles (fig.7).

The fluorescence returns to the reference level when all the calcium has been incorporated inside the vesicles, addition of Ca^{++} reactivates the system. An injection of oxalate has little effect on the fluorescence if no ATP is present even if the vesicles are previously loaded with 5mM Ca^{++}.

It is usualy assumed that the incorporation of oxalate during calcium transport is produced by a passive diffusion sustained by the precipitation of Ca^{++}-oxalate inside the vesicles [8-11]. This process reduces the activity of the calcium ions thus removing the inhibition of the calcium pump by internal calcium. Our experiments tend to support an other view where, at least in a first stage, the inward movement of oxalate is driven by the membrane potential. This is summarized in the table II.

Oxalate produces a fluorescence drop only if the potential is positive inside, independently of the internal calcium concentration. Indeed the precipitation of Ca^{++}-oxalate can still be involved in a latter stage in the reduction of the concentration of calcium and in the reactivation of the calcium pump.

b. Cl^-. If chlorine is added to the external medium the saturation level in high ATP is reduced as shown on the figure 8.

A burst of Cl^- added at any time during the experiment produces a transient depolarisation which is greater with vesicles polarized by ATP (fig. 9).

TABLE II. Effect of oxalate on the fluorescence

conditions					
inside	outside	ATP	estimated pot. ins.	estimated Ca^{++} ins.	ΔF OX^{--} (5mM)
K^+gluc^-	K^+gluc^-	0	0	< 1mM	≃ 0
$\begin{cases} K^+gluc^- \\ + 5mM\ Ca^{++} \end{cases}$	"	0	0	5mM	≃ 0
K^+gluc^-	"	10μM	+ 60mV	2-5mM	-7 to -8%
$\begin{cases} K^+gluc^- \\ + 5mM\ Ca^{++} \end{cases}$	"	10μM	+ 60mV	7-10mM	

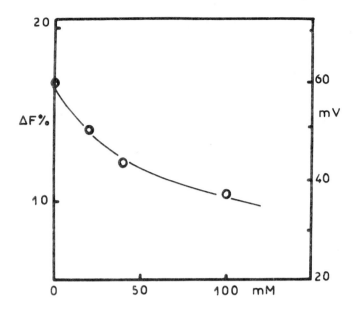

Fig. 8. Level of saturation in the presence of excess ATP and various concentrations of Cl^-.

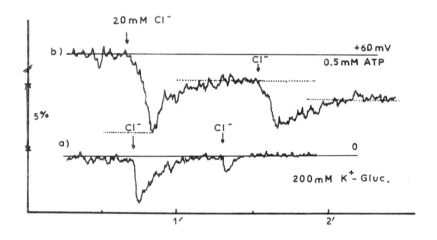

Fig. 9. Transient potential changes induced by an injection of $Tris^+$-Cl^- on membranes at rest (a) or polarized by an excess of ATP (b). Temperature = 22°. The two traces are placed arbitrarily on the vertical scale.

3. **Effect of** ____ **Potential on the Activity of the Calcium Pump.**

a. <u>Negative potential</u>. A negative potential can be created in condition I (see Table I). The half-life of the potential is around 40 sec. and ATP can be added during that time. The experiment is shown on the figure 10.

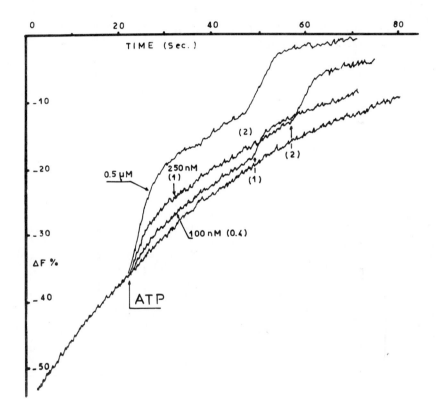

Fig. 10. Effect of ATP on membranes polarized negatively (condition I of Table I). Numbers in parenthesis indicate the ratio of ATP injected to the concentration of high affinity ATP sites. Initial K^+ concentrations are : In : 200mM, out : 5mM. Temperature 22°. Note that the decay of the potential is much faster than at 2° (compare with trace I, figure 2).

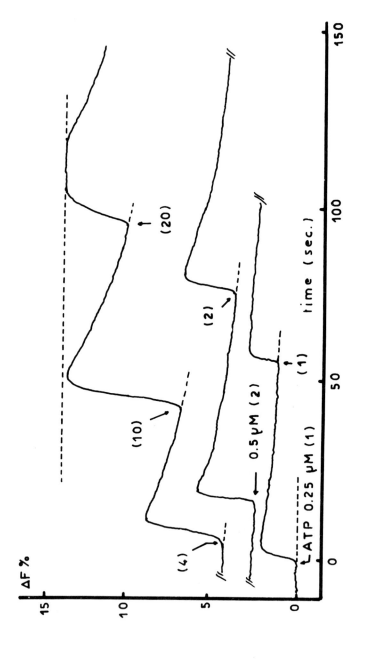

Fig. 11. Polarization of the S.R. membrane by multiple injection of small quantities of ATP. Conditions are those described in figure 5.

Electrogenic Calcium Transport

At any negative potential the fluorescence increase is proportional to the amount of ATP injected and hydrolysed. This change corresponds to 7 ± 1 mV per turn-over of the enzyme. Oxalate ions have no effect in these conditions.

b. <u>Positive potentials</u>. ATP is added in various concentrations on S.R. vesicles thereby generating an increase in fluorescence (Fig. 11). When ATP is completly hydrolysed the fluorescence decreases slowly. It is remarkable that this reversal of the signal is not observed in the negative potential conditions where the activity of the pump facilitates simply the neutralization of the potential. As the potential increases the fluorescence change produced by the hydrolysis of a given amount of ATP declines and finaly a plateau is reached where ATP produces no further effects (Figure 5 and 11). This cannot be attributed to an effect of the dye response since the calibration curve is linear in these conditions. The more likely explanation is a potential dependent reduction of the apparent electrogenicity of the pump. This can be obtained from experiments like those of the figure (11) and the results are shown below (fig. 12).

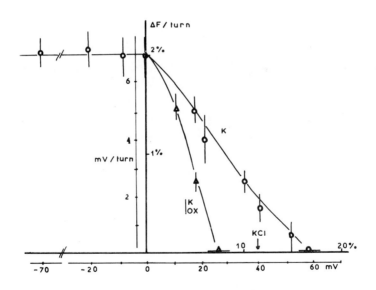

Fig. 12. *Apparent electrogenicity of the calcium pump as a function of the membrane polarization, in K^+-gluconate (circles) or in K^+-gluconate + 5mM oxalate (triangles). The saturation in 100mM KCl is indicated.*

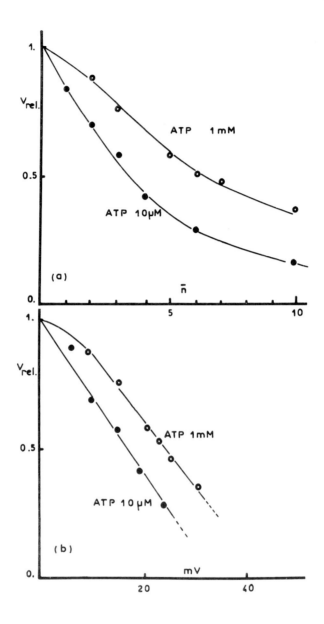

Fig. 13. Relative ATPase rate as a function: a) of the mean number of turn-overs after ATP injection, or b) of the membrane potential. The experiments are made in 100m KCl, 5mM Mg^{++} and 4°. The ATPase rate and the number of ATP cleaved are calculated from the H^+ production associated with the ATP cleavage.

This curve is specific to a particular medium. In the presence of Cl^- or oxalate the saturation potential is reduced as described in the precedent paragraphs.

4. $\underline{\Delta pH}$. This study was initiated by a recent report [13] indicating an H^+ transport preceding the Ca^{++} translocation. We have used 9-aminoacridine as a probe of pH gradients [14-16]. By preincubation in Tris-maleate adjusted at pH 6 to 8.5, a pH gradient $\Delta pH = 0$ to -2.2 pH unit was imposed to the SR membrane ($\Delta pH = pH.in - pH.out$). The response of the acridine dye is well observed by its fluorescence at 460nm and calibrated. At 0° the pH gradient is stable for minutes and depends on the concentration of buffer. As described in ref. 14-16 the dye does not respond if the pH inside is higher than that of the medium. We have therefore imposed a negative pH gradient to sensitize the dye to ΔpH change in any direction during the calcium transport. In even very weakly buffered media we have failed to observe any ΔpH changes in the conditions where an intense signal is observed with the potential sensitive dyes.

C. Membrane Potential and ATPase Activity

In the absence of oxalate the rates of calcium uptake and of ATP hydrolysis decline very rapidly few seconds after tion of ATP [11,12]. This was noted by various authors and attributed to an inhibitory effect of the internal calcium. This assumption was made basically on the observations that oxalate or detergent solubilization are able to reactivate the ATPase activity. The inhibitory effect of internal calcium can be reproduced by preloading the vesicles by high concentrations of calcium [11]. If, as suggested by the present experiments, the calcium transport is electrogenic, one might expect that the membrane potential build up during the first few seconds could also have an effect on the ATPase activity.

We have measured the ATPase rate on vesicles at 4° by adding controlled amount of ATP and expressed the rate as a function of the mean turn-over number of the enzyme $\bar{n}$ = (number of ATP cleaved/number of ATP sites). The figure 13a shows that the ATPase rate is reduced to a low value after 5 to 10 turn-overs. We have shown in the preceding discussion that the membrane potential was saturated at its maximum value ($\simeq 60mV$) after about the same number of turn-overs (Fig. 11). It is then tempting to associate the two observations and to establish a relation between the rate of the ATPase activity and the membrane potential. This gives the result shown on   figure 13b which represents an important characteristic of the calcium pump.

III. DISCUSSION AND SUMMARY

The first conclusion of the present study is that it is possible to create an ion diffusion potential accross the membrane of S.R. vesicles and that this potential can be evaluated with the cyanine dye diS-C3-(5). In spite of a lower sensitivity for positive potentials it is feasible to calibrate the response of the dye from - 100mV to + 100mV. It is also clear that the membrane potential is very rapidly dissipated by the diffusion of the counter ions.

Zimnyak and Racker [7], using reconstituted vesicles and ANS as a probe, have found an electrogenic activity of the S.R. calcium pump. Our results provide evidence that this phenomenon is basically the same in intact vesicles. Having tried to reject as far as possible other explanations or artefacts our present conclusion would be that we have observed an important electrogenic activity of the calcium pump during active transport in SR vesicles.

The apparent electrogenicity of the pump is constant around 7 ± 1mV per turn-over as long as a negative potential is maintained inside the vesicles, but it is gradualy reduced with increasing positive polarization. Indeed the electrogenic ratio is a complex function of the intrinsic electrogenicity of the pump, the rate of calcium uptake and the rate of diffusion of other ions through the membrane. Most of these parameters are potential dependent and may account for the apparent reduction of the electrogenicity of the pump. Present data are unsufficient to support the hypothesis that the variable electrogenicity is an intrinsic property of the protein. This is indeed a very attractive hypothesis and could be produced by a variable, potential dependent gearing between Ca^{++} and an hypothetic counter ion actively transported by the pump. (K^+ was the only cation present inside the vesicles in most experiments).

The calcium uptake being electrogenic one should expect that the activity of the pump can be influenced by the membrane polarization. This fact,practically ignored in earlier studies on vesicles,can be very important and in agreement with Zimniak and Racker [7] our results suggest an inhibition of the activity of the pump for positive potentials
After few turn-over cycles the membrane potential stabilizes around + 60mV   , at this stage the ATPase is inhibited and

the low electrogenic activity just compensates the leakage of other ions through the membrane. This is demonstrated by the effect of permeant anions and the leakage can be evaluated when the activity of the pump ceases by ATP exhaustion. The response of the dye and the saturation effect is the same if the vesicles are previously loaded with a high concentration of calcium by dialysis. It is then unlikely that, in the conditions described in this paper, the saturation of the ATPase is caused by the binding of calcium to the internal sites of the protein.

The consequences of these observations are important ; in contradiction with the current view we conclude that, in the absence of oxalate, the building up of a positive potential inside the vesicles causes the early inhibition of the calcium transport and of the ATPase activity. We do not exclude a late effect of the calcium binding which can become important as the inside concentration increases. In the presence of oxalate the negative charges of oxalate neutralize the charges of calcium and the subsequent depolarization reactivates the transport. It is then interesting to note that, in this case, the steady state rate of the ATPase measured on vesicles can be strongly limited by the rate of oxalate influx.

ACKNOWLEDGMENTS

The author expresses his thanks to Dr Alan Waggoner for the gift of the dye used in this work.

REFERENCES

1. Waggoner, A.S. (1976). J. Membrane Biol. 27, 317-334.
2. Waggoner, A.S. and Grinvald, A. (1977). Annals for N.Y. acad. Sciences, 303, 217-241.
3. Sims, P.J. Waggoner, A.S. Chao-Huei, Wang and Hoffman, J.F. (1974). Biochemistry, 13, 3315-3330.
4. Meissner, G. and Mc. Kinley, D. (1976). J. Membrane Biol. 30, 79-98.
5. Mc Kinley, D. and Meissner, G. (1977). FEBS Letters, 82, 47-50.
6. Kometani, T. and Kasai, M. (1978). J. Membrane Biol., 41, 295-308.
7. Zimniav, P. and Racker, E. (1978). J. Biol. Chem. 253, 4631-4637.

8. Hasselbach, W. and Maninose, M. (1963). Biochem. Z., 339, 94-111.
9. Hasselbach, W. and Weber, H.H. (1974). in Membrane Proteins in transport and phosphorylation (Azzone et al. eds) p. 103-111.
10. Weber, A., Herz, R. and Reisse, I. (1966). Biochem. Z., 345, 329-369.
11. Weber, A. (1971). J. Gen. Physiol. 57, 50-63.
12. Yamada, S., Yamamoto, T. and Tonomura, Y. (1970). J. Biochem., 67, 789-794.
13. Madeira, V.M.C. (1978). Arch. Biochem. Biophys., 185, 316-325.
14. Deamer, D.W., Prince, R.C. and Crofts, A.R. (1972). Biochim. Biophys. Acta, 274, 323-335.
15. Casadio, R., Baccarini-Melandri, A. and Melandri, B.A. (1974). Eur. J. Biochem. 47, 121-128.
16. Casadio, R. and Melandri, B.A. (1977). J. of Bioenergetics and Biomembranes, 9, 17-29.
17. Dupont, Y. (1977). Eur. J. Biochem. 72, 185-190.

INITIAL RATE OF Ca UPTAKE BY FRAGMENTED SARCOPLASMIC RETICULUM FROM BULLFROG SKELETAL MUSCLE[1]

Yasuo Ogawa
Nagomi Kurebayashi
Takao Kodama

Department of Pharmacology
Juntendo University School of Medicine
Tokyo, Japan

The transient kinetic studies of Ca uptake by the sarcoplasmic reticulum should be a great help for the elucidation of the mechanism of Ca uptake as well as the explanation of the relaxation of the skeletal muscle. It is reasonable to take the advantage of the fact that many physiological findings have been obtained with frog skeletal muscle (1).

Fragmented sarcoplasmic reticulum was prepared from bullfrog skeletal muscle according to the method described previously (2). Transient change in Ca^{2+} was determined in the presence of tetramethylmurexide, a Ca indicator, by Aminco dualwavelength spectrophotometer equipped with a stopped-flow apparatus. The reaction was started by adding ATP. The dependence of the initial rate of Ca uptake on the Ca^{2+} concentration changed according to pH of the medium. The optimum pH was 6.6-6.8 at the initial concentration of 40μM Ca^{2+}, and 6.8-7.2 at 15μM Ca^{2+}. In both cases the maximum rates were almost of the same value, respectively. Therefore, the rate of Ca uptake was determined at pH 6.80. The initial rate of Ca uptake increased as the concentration of ATP increased, and reached the maximum around 0.5mM ATP.

---

[1]Supported in part by research grants from the Ministry of Education, Science and Culture of Japan, and Naioto Foundation.

Initial burst of the rate of 90-190 nmoles/mg/sec of Ca uptake was observed during 100 msec or less in the presence of 0.5 mM ATP at 15°C, which was followed by a slower uptake, e.g., at the rate of about 50 nmoles/mg/sec. This burst was hardly observed at lower concentrations of ATP. The burst rate could explain the twitch relaxation of bullfrog skeletal muscle at 15°C which took about 0.2 sec, and the slower rate could explain the tetanus relaxation which was prolonged by a factor of about 2.5 times the twitch relaxation(3)(see also Ref.(1)).

Pi liberation and EP formation were examined in the presence of ATP-γ-^{32}P by a rapid quenching apparatus under the same conditions. After about 50 msec, the rapid phase of Pi liberation was followed by the phase of a slower rate, while the level of EP was constant from 40 to 100 msec; Ca uptake/Pi ratio being near 1 at 100 msec. Further experiments should be required to explain the discrepancy between the rapid phases of Ca-uptake and Pi liberation.

The calorimetric determination showed that the process was thermally neutral as a whole so far as the initial transient Ca uptake was concerned. It should be mentioned that extra H^+ liberation of 0.2 mole per mole of Ca sequestered was observed in addition to H^+ release due to ATP hydrolysis. Therefore, the Ca-uptake process itself is slightly endothermic after the correction for the heat of ionization of the buffer, Tris-maleate.

REFERENCES

1. Endo, M., Physiol. Rev., 57:71 (1977).
2. Ogawa, Y., J. Biochem., 67:667 (1970).
3. Fraser, A. and Carlson, R. D., J. Gen. Physiol., 62:271 (1973).

MECHANISM OF CALCIUM RELEASE FROM FRAGMENTED SARCOPLASMIC RETICULUM: DEVELOPMENT OF AN OPTICAL METHOD FOR THE RAPID KINETIC STUDIES

S. Tsuyoshi Ohnishi[1]

Biophysics Laboratory
Dept. of Anesthesiology and Dept. of Biological Chemistry
Hahnemann Medical College
Philadelphia, PA U.S.A.

An optical method was developed for the study of the mechanism of Ca-triggered Ca release (1) and depolarization induced Ca release (1,2) from fragmented sarcoplasmic reticulum (SR). The concentration of ionized Ca was measured by metallochromic indicator method (3-5). Ca-triggered Ca release: Fig. 1(A) shows an example of this release. Upon addition of ATP, Ca concentration fell to 0. Then, 20μM Ca was added repeatedly as indicated by the bottom arrows. After the second addition, the level of Ca reached to 40μM indicating the release of 20μM from the SR. The amount of loaded Ca necessary to initiate the Ca-triggered Ca release was 80 n moles/mg. protein. Local anesthetics or high concentration of Mg (5mM) inhibited the release. In the presence of clinically-used concentrations of halothane and enflurance, the release was increased by 100-150%. Ethanol (100mM) also increased the release by 50%. Halothane and enflurance by themselves triggered the release, but ethanol did not. Increase of Ca release by these drugs are linked to the decrease of Ca-loading of SR and may therefore be the cause of the negative inotropic effect of the drugs. Caffeine-induced Ca release: As shown in Fig. 1(B), 4mM caffeine added instead of the 2nd addition of 20μM Ca induced a Ca release. Depolarization-induced Ca release: After Ca concentration fell to 0 (Fig. 2(A)), the gluconate solution was rapidly diluted with 9 volumes of KCl solution (Fig. 2(B)). In addition to the turbidity change (indicated

[1]Supported in part by NIH grant HL 15799.

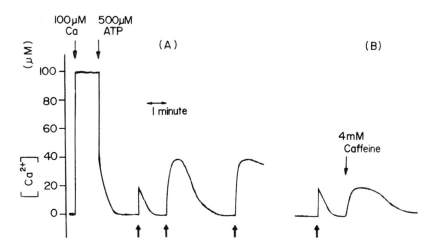

Fig. 1. 1.6 mg protein/ml, 75mM KCl, 0.5mM MgCl, pH 7, 25 C°.

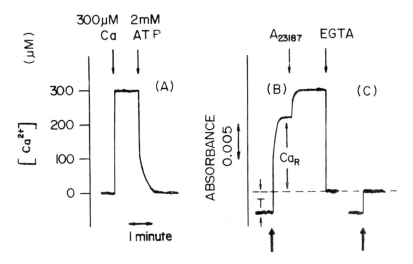

Fig. 2. (A) 1.9mg protein/ml, 150mM K-gluconate, 1mM Mg, pH 7, 25°. Ten times dilution with 150 mM KCl(B) and 150mM K-gluconate (C).

by T), Ca-release was observed as shown by Ca_R. No release was observed after dilution with gluconate solution (Fig. 2(C)). The release was insensitive to local anesthetics and high Mg, but was suppressed by sucrose. The membrane potential as measured by diO-C_2 (5) was 0, 25, 110 and 125 mV (negative inside) for the dilution with gluconate, chloride, Tris and choline, respectively. With Tris or choline dilution, Ca was not released unless some amount of K or Na was added to the outside of SR, suggesting the requirement of permeable cations for the release. <u>Prostaglandin-induced Ca release</u>: Prostaglandins were found to cause the release of Ca from SR with the order of activity of $Bx > B_2, A_1 > E_2 > F_1 \alpha=0$ (Ohnishi, S.T. & Devlin, T.M. (submitted for publication)).

<u>References</u>:
(1) Endo, M. Physiol. Rev. <u>57</u>:71 (1977).
(2) Kasai, M. & Miyamoto, H. J. Biochem. <u>79</u>:1053 (1976).
(3) Ohnishi, T. & Ebashi, S. J. Biochem. <u>54</u>:506 (1963).
(4) Ohnishi, S.T. Anal. Biochem. <u>85</u>:165 (1978).
(5) Dipolo, R. et al. J. Gen. Physiol. <u>67</u>:433 (1976).

IONIC PERMEABILITY
OF SARCOPLASMIC RETICULUM MEMBRANE

Michiki Kasai
Tadaatsu Kometani

Department of Biophysical Engineering
Faculty of Engineering Science
Osaka University
Toyonaka, Osaka, Japan

I. INTRODUCTION

In the previous papers (1,2), we examined calcium release from sarcoplasmic reticulum (SR) vesicles and showed that SR responds to the anion replacement from methanesulfonate (MS) to chloride by releasing calcium ions. This anion exchange was considered to cause depolarization of the SR membrane. To clarify that such an anion replacement truely depolarized the SR membrane, the measurement of ionic permeability became important.

The measurement of permeability of various ions across the SR membrane has been carried out by several investigators mainly with tracer methods (2-5). However, various important ions such as Na^+, K^+, and Cl^- permeate very fast and their permeabilities could not be determined. New methods for the determination of fast permeation were needed.

Rapid flow of solute or water can be measured by following the time course of the osmotic volume change of membrane vesicles. The volume change of the vesicles or cells have been followed by measuring the light scattering or the turbidity (absorbance) of the membrane suspensions by various workers (6-9). However, such works mainly treated the problem of the permeability of non-electrolytes. There are few studies on the permeability of ions (9, 10).

In this paper, the osmotic volume change of SR vesicles was followed by measuring the change of the light scattering intensity by using a stopped flow apparatus. By analysing the time course of the turbidity change, permeabilities of various ions and neutral molecules were determined. On the basis of these data, a discussion was given about the mechanisms of ion transport and calcium release from sarcoplasmic reticulum. A part of this work has been published in the previous paper (11).

II. EXPERIMENTALS AND RESULTS

A. Osmotic Volume Change of SR Vesicles

Sarcoplasmic reticulum vesicles were prepared from rabbit skeletal muscle by the method of Weber et al (12) with slight modifications (1). SR vesicles incubated in a low osmotic solution were mixed with an equal volume of the salt solution of a high osmotic pressure using a stopped flow apparatus, and the change in the light scattering intensity was followed. As shown in Fig. 1, the scattering intensity increased rapidly and then decreased. The fast increase in the scattering intensity is caused by the decrease of the volume of SR vesicles due to the outflow of water through the membrane, which is driven by the osmotic pressure difference. From this rate, the permeation time of water can be calculated. The later decrease in the light scattering intensity is caused by the increase of the volume due to the inflow of ions and water which was accompanied with the inflow of salt ions driven by the chemical potential difference of these ions. From the later rate of change of light scattering, approximate permeation times of salts can be calculated. Although the change of the light scattering intensity was proportional to the volume change (11), the time course curve is expected to be complexed (13). Therefore, as a measure of the permeation time, half permeation time, τ, was defined as the time to reach the half value of the maximal increment of the scattering intensity, ΔI_o, which was estimated by an extraporation of the later phase to time zero (see Fig. 1). At 23°C, the permeation times of water and KCl were 0.1 and 10 seconds, respectively.

Contrary to this experiment, when SR vesicles incubated in a high osmotic salt solution was mixed with a low osmotic solution, only a slight decrease of the light scattering intensity was observed, which is also shown in Fig. 1. This shows only a slight increase in the volume from the incubated (resting) state. In the case of red blood cells, a symmetric change of the light scattering intensity has been reported (7). Probably, SR vesicles exist in the most swollen state under the incubated condition.

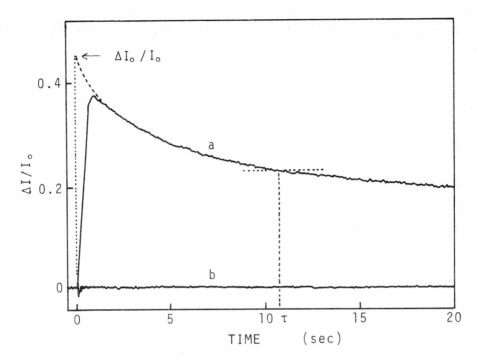

Fig. 1. Change in the light scattering intensity of SR vesicles caused by the volume change. (a). SR vesicles incubated in 2 mM KCl, 5 mM Tris-Maleate (pH 6.5) and 0.4 mg SR protein/ml was mixed with an equal volume of a solution containing 100 mM KCl and 5 mM Tris-Maleate using a stopped flow apparatus (UNION RA-401 and RA 450, Japan) and the change of the light scattering intensity was followed. The scattering was measured at 90° direction to the incident light using 400 nm at 23°C. Definition of τ and $\Delta I_o/I_o$ is also shown. (b). SR incubated in 100 mM KCl and 5 mM Tris-Maleate was mixed with an equal volume of a solution containing 2 mM KCl, 5 mM Tris-Maleate. The ordinate shows the change of the relative scattering intensity.

B. Permeabilities of Various Salts and Neutral Molecules

When different kinds of salts or neutral molecules were used in similar experiments to the above section, their permeation times were obtained. In the case of slowly permeable ions or molecules, two solutions were mixed and the change of the light scattering intensity was followed with a spectrofluorimeter (UNION FS-501, Japan).

In Table I, permeation times, τ, and maximal increments of

TABLE I. *Permeability Properties of the Sarcoplasmic Reticulum Vesicle Membrane*

Species	Permeation time (τ)	Permeability[a] (P)	Maximal increment of scattering intensity ($\Delta I_o/I_o$)
KCl	10 sec	1.2×10^{-7} cm/sec	0.45
K-acetate	10	1.2	0.45
K-propionate	10	1.2	0.45
K-butyrate	10	1.2	0.45
K_2-oxalate	90	0.13	0.51
KMS	50	0.23	0.48
NaMS	50	0.23	0.48
NaCl	13	0.89	0.45
LiCl	18	0.64	0.48
RbCl	8	1.41	0.45
Choline-Cl	360	0.032	0.69
Tris-Cl	600	0.019	0.73
$MgCl_2$	1800	0.0064	0.75
$CaCl_2$	6000	0.0019	0.75
K_2-EDTA	3000	0.0038	0.51
Tris-maleate	900	0.013	0.71
Thiourea	0.6	19	0.69
Urea	1.2	9.6	0.69
Ethyleneglycol	0.6	19	0.69
Glucose	1500	0.0077	0.69
Water	0.1	120	

[a] Permeability was calculated according to Eq. (5).

the scattering intensity, $\Delta I_o/I_o$, are summarized. In the case of neutral molecules, the permeation time shows the movement of the molecule itself. However, in the case of salt ions, since ions move as a pair, the permeation time does not necessarily correspond to the permeation of single ions. From this table, in the case of potassium salts, when the anion was changed from chloride to acetate, propionate, butyrate, etc, the permeation time did not change. On the contrary, in the case of chloride salts, when the cation was changed from potassium to lithium, sodium, rubidium, choline, etc, the permeation time changed. This result indicates that such anions permeate faster than the cations and the permeation time of salts depends on the permeation of the slower ion.

In the case of methanesulfonate salts, i.e., KMS or NaMS, the permeation times were larger than those of KCl or NaCl. This might be due to the slow permeation of MS ions.

In the case of slowly permeable ions or molecules, such as $CaCl_2$, $MgCl_2$, choline chloride, glucose, etc, $\Delta I_o/I_o$ were about 50 % larger than those of NaCl or KCl. Small difference less than 10 % could not be eliminated because the osmolarity of these solutions was not necessarily the same. At first, specific effect of such ions on the light scattering intensity could be considered. However, after a long time of the mixing, the scattering intensity reached the initial level and there was little difference between $CaCl_2$, and KCl. The effect of binding of divalent cations to the membrane seems not to contribute much to the scattering intensity. Therefore, the large change of the light scattering intensity also should be attributed to the volume change. A possible interpretation is following. In the case of KCl, because the permeation of KCl is not slow enough, a fairly amount of inflow of KCl occurred before the efflux of water had been completed. Accordingly, in the case of KCl, the maximal increment of the light scattering intensity becomes smaller than that in the case of $CaCl_2$. As a result, the permeation rate of KCl might be underestimated when compared with $CaCl_2$.

C. Effect of Valinomycin on the Permeability of Ions

It is well known that valinomycin increases the permeability of potassium ions. As described in the above section, if the permeation rate of potassium salts such as KCl is limited by the permeability of K^+, an increase in the permeation rate of KCl is expected in the presence of valinomycin. Experiments similar to those shown in Fig. 1 were carried out in the presence of valinomycin. As shown in Fig. 2, a tremendous increase in the permeation rate of KCl was observed. In this case the permeation rate of water was not changed.

To clarify the specificity of the effect of valinomycin on SR membrane, similar experiments were carried out using KMS, NaCl, LiCl, RbCl, choline chloride, $MgCl_2$, $CaCl_2$, and glucose. The effect of valinomycin was specific for K^+ and Rb^+.

In Fig. 2, the effect of valinomycin on the permeation of KCl was shown when the concentration of valinomycin was increased. In the presence of a large amount of valinomycin a decrease in the maximal increment of the scattering intensity was observed. This decrease is also explained by the concept that the permeability of KCl becomes close to that of water. For this reason, the time at which the increment of the scattering intensity reached a half value of the maximal increment of the control sample was taken as a measure of the permeation time. In Fig. 3, the increase of the permeation rate was plotted as a function of the valinomycin concentration. As a measure of the incease in the permeability,

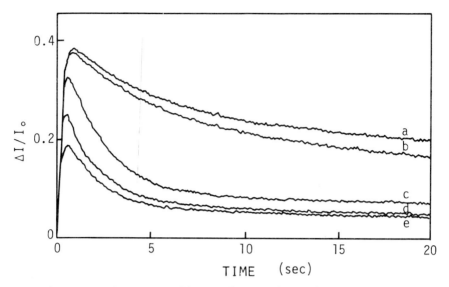

Fig. 2. Effect of valinomycin on the volume change of SR vesicles in KCl solution. An experiment similar to that in Fig. 1 was carried out in the presence of valinomycin. Valinomycin was added 30 min before the mixing the SR suspension. Concentration of valinomycin in g/ml was following: (a) 0; (b) 4 ×10^{-9}; (c) 5 ×10^{-8}; (d) 1.5 ×10^{-7}; (e) 1.5 ×10^{-6}. Final concentration of SR was 0.2 mg/ml.

$1/\tau - 1/\tau_o$ was used, where τ_o and τ are the permeation times in the absence and in the presence of valinomycin, respectively. The permeation rate initially increased proportionally to the concentration of valinomycin and then reached a constant value. From our assumption, the following interpretation can be derived. In the low concentration of valinomycin, the permeability of K^+ increases but it is still smaller than that of Cl^-. Thus, the permeation rate of KCl remains dependent on the permeation of K^+. In the higher concentration of valinomycin, the permeability of K^+ exceeds that of Cl^-, and the permeation of KCl becomes dependent on the permeability of Cl^-.

In the case of KMS, a similar experiment was carried out. The increase in the permeation rate of KMS was also observed in the presence of valinomycin, which is also shown in Fig. 3.

D. Determination of the Permeability of Single Ions

As mentioned in the above section, the net movement of salt ions depended on the movement of the slower ions. The

Ionic Permeability

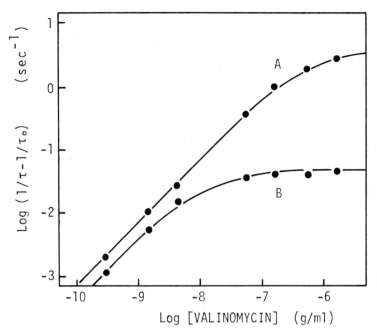

Fig. 3. Increase of the permeability of KCl and KMS as a function of valinomycin concentration. The data were taken from Fig. 2 in the case of KCl. In the case of KMS, similar experiment to Fig. 2 was made. The curves were calculated using Eq. (4).

movement of ions across the membrane seems to be described by the following Nernst's equation.

$$\frac{1}{D} = \frac{1}{2} \left(\frac{1}{D_+} + \frac{1}{D_-} \right) \quad (1)$$

where D is a diffusion constant of the salt, D_+ and D_- are those of the cation and the anion. Since the permeability is proportional to the diffusion constant, the apparent permeation time given in this paper is considered to be proportional to the inverse of the diffusion constant. That is,

$$\tau = \frac{1}{2} (\tau_+ + \tau_-) \quad (2)$$

If we assume that the permeability of K^+ increases linearly to the concentration of valinomycin, C, as follows,

$$P_K = P_{K_o} + \alpha C \quad (3)$$

TABLE II. *Permeability of ions. Experiments similar to those in Fig. 3 were carried out using various salts. Permeation time of ions, τ, was determined by using Eq. (4). Permeability, P, was calculated using Eq. (5).*

Ions	Permeation time (τ)	Permeability (P)	Relative permeability
K^+	20 sec	0.58×10^{-7} cm/sec	1
Na^+	26	0.44	0.77
Li^+	36	0.32	0.56
Rb^+	16	0.72	1.3
Cl^-	0.4	29	50
MS^-	20	0.58	1
Oxalate	40	0.29	0.5

where P_K and $P_{K\circ}$ are permeabilities of K^+ in the presence and in the absence of valinomycin, respectively, and α is a constant. Since the permeability is inversely proportional to the permeation time, the following relation can be obtained using Eq. (2).

$$\frac{\tau_0}{\tau} - 1 = \frac{\alpha P_- C}{P_K (P_- + P_{K\circ} + \alpha C)} \tag{4}$$

where P_- is the permeability of anion. The curves in Fig. 3 were calculated using this relation.

It is clearly shown that this relation is held in Fig. 3. From Fig. 3, the permeation times of ions were determined as follows, K^+ 20, Cl^- 0.4, MS^- 20 in second at room temperature. Thus, the permeability of Cl^- is about 50 times larger than that of K^+ or MS^-.

A similar anlaysis was carried out on some interesting anions such as oxalate. Their permeation times were determined and shown in Table II.

E. Estimation of the Absolute Values of Permeability

If it is assumed that τ is equal to the diffusional permeation time defined under the condition without volume change and the vesicles are homogeneous spheres of a constant radius, the absolute value of the permeability can be calculated from the permeation time by the following equation,

Ionic Permeability

$$P = 0.231 \frac{r}{\tau} \qquad (5)$$

where P is the permeability and r is the radius of the vesicles. However, because the first assumption is not valid and the size and the permeability of the vesicles are distributed heterogeneously, the calculated values are only approximate but give us important suggestion. Assuming the radius of SR to be 50 nm (14), the calculated values are shown in Tables I and II. The abosolute values of the permeability of ions and water in SR are about 10-100 times smaller than those reported for red blood cells (9, 15). However, the relative values are quite similar to those found in red blood cells.

F. Temperature Dependence of Permeability

Since we have succeeded in obtaining the permeabilities of single ions, temperature dependence of the permeability was examined in order to elucidate the mechanism of the ion permeation. Fig. 4 is an Arrhenius plot of the obtained data.

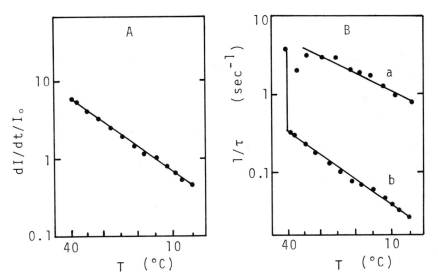

Fig. 4. Temperature dependence of the permeability. Similar experiment in Fig. 1 was carried out by changing temperature. (A). Permeability of water. As a measure of the permeability, the initial shrinking rate was used. (B) Permeability of K^+ (b) and Cl^- (a). As a measure of the permeability, $1/\tau$ was used.

In all cases, the permeabilities changed linearly and the activation heats were calculated as follows, water 11.7, K^+ 11.5, and Cl^- 8.3 in Kcal/mol. These values are somewhat smaller than that of the chloride permeation in red blood cells (16). These results suggest that cations permeate by the similar mechanism to that of water permeation. Probably, cations permeate through water filled pores or channels.

III. DISCUSSION AND CONCLUSION

It was demonstrated that the permeability of the ions which is too fast to be followed by the tracer method could be determined by the optical method. This should be an useful method. From the experimental results obtained using SR vesicles, the permeability of anions such as Cl^- is about 50 times larger than that of cation such as K^+. Then the SR membrane is an anion selective one.

This result suggests that two kinds of ion channels exist in SR membrane, i.e., cation channel and anion channel. Permeability of cations seems to be related to their hydrated radius from Table II. It might be a water filled pore without specificity. Anion channel might be a specific because of the high permeability. This suggests the existence of a similar channel in SR to that found in red blood cells. As a preliminary result, we found that the anion permeability was inhibited by SITS (4-acetoamido-4'-isothiocyanostilbene 2, 2'-disulfonate) which is a potent inhibitor for the anion channel of red blood cells (9).

As far as the absolute value of the permeability is concerned, another estimation is possible. From Fig. 3, the concentration of valinomycin which caused twice increase in the permeability of K^+, is about 10^{-8} g/ml. SR membrane consists mainly of phosphatidylcholine (PC) and phosphatidylethanolamine (17). To black lipid membrane made of PC 10^{-8} g/ml valinomycin causes the increase of K^+ conductance about 5×10^{-5} $ohm^{-1}cm^{-2}$ in 50 mM KCl (18). This value corresponds to the permeability of about 2.7×10^{-7} cm/sec. This value is only about 5 times larger than that obtained by Eq. (5) and shown in Table II. These two values are in good agreement. This small discrepency is probably due to the inadequacy of Eq. (5).

Finally, as a result of the experiment, we showed the permeability of Cl^- is about 50 times faster than that of MS^- and K^+. In the case of the previous experiments in which the ionic environment of the SR membrane was changed from KMS to KCl and calcium release was observed (1,2), it is probable that such an exchange of ions caused the depolarization of the

Ionic Permeability

vesicular membranes *in vitro*. Although we do not know the permeability properties of the SR membrane *in vivo*, it is reasonable to assume that they are the same as those of the isolated SR. A similar change of the ionic composition resulted in a calcium release from SR *in vivo* and this release might be caused by the depolarization or the change of membrane potential (19-21).

REFERENCES

1. Kasai, M., and Miyamoto, H. (1976). J. Biochem. 79:1053.
2. Kasai, M., and Miyamoto, H. (1976). J. Biochem. 79:1067.
3. Meissner, G., and McKinley, D. (1976). J. Membrane Biol. 30:79.
4. Duggan, P. F., and Martonosi, A. N. (1970). J. Gen. Physiol. 56:147.
5. Jilka, R. L., Martonosi, A. N., and Tillack, T. W. (1975). J. Biol. Chem. 250:7511.
6. Bangham, A. D., De Gier, J., and Greville, G. D. (1977). Chem. Phys. Lipids 1:225.
7. Rich, G. T., Sha'afi, R. I., Romualdez, A., and Solomon, A. K. (1968). J. Gen. Physiol. 62:941.
8. Sha'afi, R. I., Rich, G. T., Mikulecky, D. C., and Solomon, A. K. (1970). J. Gen. Physiol. 55:427.
9. Knauf, P. A., Fuhrmann, G. F., Rothstein, S. S., and Rothstein, A. (1977). J. Gen. Physiol. 69:363.
10. de Kruijff, B., Gerretsen, W. J., Oerlemans, A., Demel, R. A., and Van Deenen, L. L. (1974). Biochim. Biophys. Acta 339:30.
11. Kometani, T., and Kasai, M. (1978). J. Membrane Biol. 41:295.
12. Weber, A., Herz, R., and Reiss, I. (1966). Biochem. Z. 345:329.
13. Sha'afi, R. I., and Gary-Bobo, C. M. (1973). Prog. Biophys. Molec. Biol. 26:103.
14. Arrio, B., Chevallier, J., Jullien, M., Yon, J., and Calvayrac, R. (1974). J. Membrane Biol. 18:95.
15. Knauf, P. A., and Fuhrmann, G. F. (1974). Fed. Proc. 33:1591.
16. Hunter, M. J. (1977). J. Physiol. 268:35.
17. MacLennan, D. H., Seeman, P., Iles, G. H., and Yip, C. C. (1971). J. Biol. Chem. 246:2702.
18. Stark, G. and Benz, R. (1971). J. Membrane Biol. 5:133.
19. Endo, M. (1977). Physiol. Rev. 57:71.
20. Endo, M., and Nakajima, Y. (1973) Nature New Biol. 246:216.
21. Fabiato, A., and Fabiato, A. (1977). Circulation Res. 40:119.

MECHANISM OF CYCLIC AMP REGULATION
OF ACTIVE CALCIUM TRANSPORT
BY CARDIAC SARCOPLASMIC RETICULUM[1]

Michihiko Tada, Makoto Yamada, Fumio Ohmori,
Tsunehiko Kuzuya and Hiroshi Abe

First Department of Medicine
Cardiology Division
Osaka University School of Medicine
Osaka, Japan

I. INTRODUCTION

In the excitation-contraction coupling of the mammalian myocardium, as in that of skeletal muscle, the ATP-supported active calcium transport by sarcoplasmic reticulum is known to produce relaxation by lowering cytoplasmic Ca^{2+}. In this process, the 100,000-dalton ATPase enzyme within the membrane is assumed to serve as an energy-transducer as well as a translocator of Ca^{2+}, and the formation of an acyl phosphoprotein intermediate (EP) of ATPase is considered to play an essential role (Tada et al., 1978b). While the energy-dependent transport of Ca^{2+} across the membrane of sarcoplasmic reticulum from both cardiac and skeletal muscles exhibits common features in basic mechanisms, the cardiac membrane was shown to possess a control mechanism that is not seen in the membrane of fast-contracting skeletal muscle (Tada et al., 1974, 1975; Kirchberger et al., 1974; Kirchberger and Tada, 1976). Thus, both calcium uptake and Ca^{2+}-dependent ATPase of cardiac membranes were markedly enhanced when a cyclic AMP-dependent protein kinase catalyzes phosphorylation of a microsomal protein of 22,000-daltons. The latter phosphoprotein exhibited stability characteristics of a phosphoester in which the phosphate is largely incorporated in-

---

[1]Supported by Muscular Dystrophy Association of America, and by Ministry of Education, Science and Culture of Japan.

to serine and threonine. Based on these observations, a regulatory mechanism of calcium transport was proposed in which protein kinase-catalyzed phosphorylation of the 22,000-dalton protein, termed as "phospholamban" (Tada et al., 1973, 1975), serves as a modulator of Ca^{2+}-dependent ATPase of cardiac sarcoplasmic reticulum (Katz et al., 1975; Tada et al., 1978a). While these studies may raise the possibility that phospholamban functions as a regulatory component of the ATPase enzyme, none of these presented the direct evidence to explain the molecular mechanism by which Ca^{2+}-dependent ATPase enzyme is controlled by cyclic AMP-dependent protein kinase-catalyzed phosphorylation of phospholamban. The present paper reports studies undertaken to elucidate the relationship between ATPase and phospholamban that may explain, at the molecular level, the mechanism of cyclic AMP-mediated regulation of active calcium transport by cardiac sarcoplasmic reticulum.

II. MATERIALS AND METHODS

Cardiac microsomes, which consist mainly of fragmented sarcoplasmic reticulum, were prepared from dog heart ventricle by the previously described procedures (Harigaya and Schwartz, 1969; Kirchberger et al., 1974). Microsomes were stored on ice and used within several hours after preparation, unless otherwise indicated. Cyclic AMP-dependent protein kinase was partially purified from pooled supernatants of cardiac microsomal preparations, through DEAE-cellulose chromatography step by the method of Miyamoto et al. (1969). Proteins of cardiac microsomes were fractionated by gel filtration on a column (2.5 x 90 cm) of Ultrogel AcA 34 (LKB-Produktor) in 0.5% sodium dodecyl sulfate (SDS), 1% β-mercaptoethanol, 3 mM NaN_3, and 10 mM sodium phosphate buffer, pH 7. Purity of fractions were determined by SDS-polyacrylamide gel electrophoresis (Tada et al., 1975). For electron microscopic observations, a droplet of the suspension of cardiac microsomes in 0.1 M KCl and 10 mM histidine buffer, pH 7 (1 to 3 mg protein/ml) was placed on a collodion and carbon-coated grid and stained by 1% uranyl acetate, pH 4 to 5. They were examined in a JEM 100C electron microscope (acceleration voltage, 80 kV). Microsomal phosphorylation catalyzed by cyclic AMP-dependent protein kinase was performed as described previously (Tada et al., 1975). Cardiac microsomes incubated in the presence of 2 mM $MgCl_2$ (phosphorylated microsomes) and 2 mM EDTA (control microsomes) at pH 6.8 and 25°C were passed through a column (1.3 x 2 cm) of Dowex 1 x 8, and the eluted microsomes were subjected to assay for ATPase and formation of its intermediate. Assay for ATPase was carried out in the presence of pyruvate kinase and phosphoenolpyruvate, and the simultaneous measurements of EP and P_i were performed as described previously (Tada et al., 1979).

III. RESULTS

A. Characteristics of Cardiac Sarcoplasmic Reticulum

1. Electron Microscopy. Electron microscopic examination made within several hours after the preparation demonstrated that cardiac microsomal fraction consists of vesicles formed from fragmented sarcoplasmic reticulum, which are indistinguishable from those obtained from skeletal muscle (Fig. 1). When the microsomal fraction was aged by storing on ice, the vesicles underwent considerable alterations: in 20 hours the spherical configuration was lost and they became to show flattened or shrunken appearances. At the same time their abilities to take up calcium and to hydrolyze ATP were also diminished precipitously upon aging.

2. Formation of Two Classes of Phosphoproteins. Cardiac microsomes formed two classes of chemically distinct phosphoproteins. The phosphoprotein formed by the reaction with cyclic AMP and protein kinase was stable in hydroxylamine and in hot acid, whereas the other was hydroxylamine-labile acyl phosphoprotein that was formed as an intermediate of ATPase (EP) upon incubation with Ca^{2+}. SDS-polyacrylamide gel electrophoresis revealed that the acyl phosphoprotein had a molecular weight of about 100,000, while the hydroxylamine-stable phos-

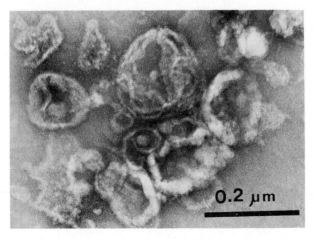

FIGURE 1. Electron microphotograph of cardiac microsomes.

phoester phosphoprotein (phospholamban) exhibited a molecular weight of about 22,000. When cardiac microsomes were kept on ice after preparation, the amount of the latter phosphoprotein remained almost unchanged, whereas the amount of the ATPase intermediate EP decreased precipitously, becoming less than half within 20 hours after preparation. Freezing the sample at $-80°C$ in the presence of sucrose prevented to a certain extent the rapid deterioration of EP formation.

B. Proteins of Cardiac Sarcoplasmic Reticulum

Cardiac microsomes that were solubilized in SDS showed several distinct protein bands after SDS-polyacrylamide gel electrophoresis. The 100,000-dalton ATPase protein was found to account for about 25% of the total protein when estimated by measuring the density profile of stained proteins after electrophoresis. Column chromatography on Ultrogel in SDS gave purification of the 100,000-dalton protein to near homogeneity (Fig. 2). Although the identification of this protein as the ATPase through activity measurement was not possible, it was quite likely that it represents the ATPase, since the immuno-precipitation test with anti-skeletal ATPase antibodies was positive, and its amino acid composition was quite similar to that of the skeletal enzyme (Tada et al., 1978a). SDS-polyacrylamide gel electrophoresis also demonstrated several other protein bands, which were found to appear at the positions corresponding to the molecular weight between 60,000 to 20,000

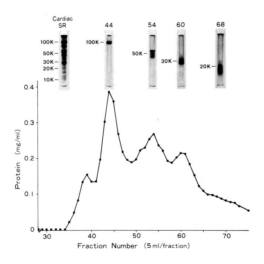

FIGURE 2. Fractionation of Solubilized Cardiac Microsomes.

TABLE I. Effect of Deoxycholate on Phospholamban.

Microsomes	Amount of ^{32}P bound[a]	
	Control	Deoxycholate[b]
Unphosphorylated	0.78 ± 0.03	0.43 ± 0.02
Phosphorylated	0.81 ± 0.04	0.79 ± 0.03

[a] Units are nmol/mg microsomal protein (Mean ± S.E.; n = 5).
[b] 0.1 mg/mg protein (performed by method of MacLennan (1970)).

(Fig. 2, inset). The 22,000-dalton protein phospholamban accounted for about 4% of the total protein. Gel filtration on Ultrogel in SDS did not clearly separate this protein from the neighboring components (Fig. 2).

Treatment of skeletal muscle microsomes with deoxycholate was reported by MacLennan (1970) to eliminate most of extrinsic and contaminating proteins, resulting in purification of the major protein, the ATPase enzyme. We attempted to examine whether the 22,000-dalton protein in cardiac microsomes can be eliminated by such procedures (Table I). When the treatment with deoxycholate was carried out after cardiac microsomes were phosphorylated by cyclic AMP-dependent protein kinase, virtually all of the ^{32}P label remained associated with the 22,000-dalton protein of the pellet after centrifugation. When cardiac microsomes were subjected to deoxycholate treatment prior to phosphorylation by protein kinase, about half of the ^{32}P label was recovered. These observations may suggest that unphosphorylated and phosphorylated states of phospholamban determine the extent to which this protein can be associated with the membrane: at phosphorylated state phospholamban may become much less dissociable from the membrane.

C. Enzyme Kinetics of Ca^{2+}-dependent ATPase: Effects of Cyclic AMP-dependent Protein Kinase

1. Dependence on ATP Concentration. Since the Ca^{2+}-dependent ATPase activity of cardiac microsomes, like that of its skeletal muscle counterpart (Tada et al., 1978b), was shown to exhibit a complex MgATP dependence (Shigekawa et al., 1976), we examined the effect of phospholamban phosphorylation on the ATP concentration dependence of cardiac microsomal ATPase in the presence of pyruvate kinase as the ATP regenerating system. Direct linear plots of Eisenthal and Cornish-Bowden (Cornish-

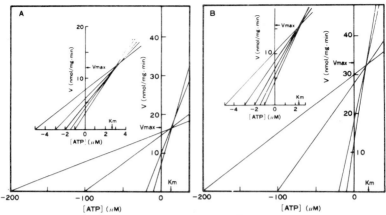

FIGURE 3. Direct Linear Plots of Ca^{2+}-dependent ATPase of unphosphorylated (A) and phosphorylated (B) cardiac microsomes at high ATP (10 to 200 μM) and low ATP (1 to 5 μM, insets).

Bowden, 1976) indicated that at both high and low ATP concentration ranges phosphorylation of phospholamban results in significant increases in the values of V_{max} with no appreciable change in K_m values (Fig. 3). The effect was more evident at the higher ATP concentration range (see also Table III).

2. Dependence on Ca^{2+} Concentration. The observed enhancement of the cardiac microsomal ATPase could accompany changes in the elementary steps of the ATPase reaction in view of the well-documented reaction scheme of the skeletal microsomal ATPase (Tada et al., 1978b). We therefore undertook a series of determination to examine effects of protein kinase on the intermediary steps of ATPase. It was found that within a range of Ca^{2+} between 0.1 to 10 μM the steady-state level of EP was significantly lower in phosphorylated microsomes, while the rate of P_i liberation (v) was markedly higher in phosphorylated microsomes. Thus, the ratio v/[EP], which was independent of Ca^{2+}, increased more than two-fold when microsomes were phosphorylated by protein kinase (Table II). It was noted that phospholamban phosphorylation resulted in profound reduction in EP levels at lower Ca^{2+}, whereas the effect was less evident at high Ca^{2+}. These results indicate that the rate at which the intermediate EP is decomposed is markedly enhanced by phospholamban phosphorylation. No significant alteration in the rate of EP formation was indicated when control and phosphorylated microsomes were incubated at various Ca^{2+} for 0.05 to 0.1 sec. However, in view of an extreme rapidity at which EP is formed (Kanazawa et al., 1971; Froehlich and Taylor, 1975), it still remains to be examined by high-performance apparatus if EP formation is altered by phospholamban phosphorylation.

TABLE II. Ca^{2+} Dependence of the Rates of P_i Liberation (v) and the Concentrations of EP ([EP]), Measured at 15°C and pH 7.

pCa	Control microsomes			Phosphorylated microsomes		
	v^a	$[EP]^b$	$v/[EP]^c$	v^a	$[EP]^b$	$v/[EP]^c$
7.0	3.8	0.11	0.58	4.1	0.06	1.14
6.3	4.8	0.23	0.35	6.8	0.11	1.03
6.0	7.2	0.36	0.33	11.3	0.17	1.11
5.3	15.0	0.44	0.57	21.0	0.35	1.00
5.0	16.8	0.44	0.64	22.5	0.38	0.99
Average	-	-	0.49	-	-	1.05

Units are: ^{a}nmol P_i/mg microsomal protein/min, ^{b}nmol P bound/mg microsomal protein, and ^{c}sec^{-1}.

3. Kinetics of Decay in the Amount of EP. In view of the remarkable augmentation of the ratio v/[EP] by phospholamban phosphorylation, we examined whether phospholamban phosphorylation produces an increase in the rate of decay in the amount of EP when further formation of EP is terminated by the addition of excess EGTA. The rate of EP decomposition of phosphorylated microsomes was about two-fold higher than that of control (Fig. 4, inset). The first-order rate constant, k_d, of EP decomposition of phosphorylated microsomes was markedly higher than that of control within a wide range of the temperature (Fig. 4). The Arrhenius plots for k_d of both microsomes indicated different slopes below and above 18°C: the activation energies of the phosphorylated microsomes were 11.4 and 19.2 kcal/mol at high and low temperatures, respectively, and were almost similar to those of control microsomes.

IV. DISCUSSION

A. Protein Kinase Effects on Intermediary Steps of ATPase

The present study demonstrated that the enzymatic parameters of Ca^{2+}-dependent ATPase of cardiac microsomes undergo profound alterations when cyclic AMP-dependent protein kinase catalyzes phosphorylation of the microsomal protein phospholamban: the values of V_{max}, v/[EP], and k_d increased by about two-fold over control (Table III). These results strongly suggest that the probable site of action of phospholamban is an

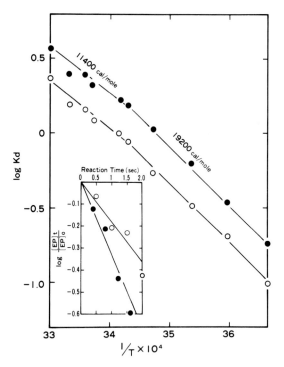

FIGURE 4. Arrhenius plots of the rate constant k_d of EP decomposition in control (○) and phosphorylated (●) microsomes. The <u>inset</u> shows the time courses of decay in EP at 15°C.

TABLE III. Comparison of enzymatic parameters of Ca^{2+}-dependent ATPase, measured at 15°C and pH 7, of unphosphorylated (control) and phosphorylated microsomes.

Microsomes	V_{max}[a]		v/[EP][b]	k_d[c]
	low ATP[d]	high ATP[e]		
Control	12.0	16.7	0.49	0.55
Phosphorylated	15.7	33.1	1.05	1.03

Units are: [a]nmol pyruvate/mg microsomal protein/min, [b]sec^{-1} and [c]sec^{-1}. ATP concentrations are: [d]1 - 5 μM and [e]10 - 100 μM.

intermediary step of the ATPase at which the reaction intermediate EP is decomposed.

The present demonstration that V_{max} of ATP hydrolysis is markedly enhanced by protein kinase-catalyzed phosphorylation (Fig. 3 and Table III) is consistent with the previous observations that the overall rates of P_i liberation and Ca^{2+} accumulation by cardiac microsomal vesicles are greatly enhanced by protein kinase with the stoichiometric coupling of 2 moles of calcium taken up per mole of ATP hydrolyzed (Tada et al., 1974; Kirchberger et al., 1974; LaRaia and Morkin, 1974). Enhancement of any of the intermediary steps of ATPase can account for the observed stimulation. No significant alteration in the rate of EP formation was indicated (see above). In contrast, the rate of EP decomposition, determined either by estimating the ratio v/[EP] or by directly measuring the rate constant k_d (Kanazawa et al., 1971; Tada et al., 1978b), was significantly augmented by pretreatment with protein kinase (Table III). Such enhancement is presumably more effective in reducing the steady-state levels of EP at lower Ca^{2+} concentrations (Table II), where the EP formation, rather than its decomposition, is the rate-determining step, whereas no effect on those is seen at higher Ca^{2+} (Table II), where the EP decomposition is the rate-determining step. The resultant increase in the turnover rate would account for about two-fold increases in both Ca^{2+}-dependent ATPase and the rate of calcium transport (Fig. 3; Table III; Tada et al., 1974; Schwartz et al., 1976; Will et al., 1976) when cardiac microsomes were subjected to phosphorylation by cyclic AMP-dependent protein kinase. Thus, the present report indicates that cyclic AMP-dependent protein kinase can control the rate of Ca^{2+}-dependent ATPase by regulating the rate of EP decomposition and that this process is highly associated with phosphorylation of phospholamban, a microsomal substrate for kinase. Such conclusion is in accord with the view (Tada et al., 1975, 1978a; Katz et al., 1975) that phospholamban may serve as a regulator controlling the active calcium transport (see below).

B. Interactions between ATPase and Phospholamban

Ca^{2+}-dependent ATPase complexed with phospholipids is considered to form an essential part of the active calcium transport system in skeletal muscle microsomes, and a similar mechanism is applicable to cardiac microsomes (Tada et al., 1978b). In these systems, the translocation of calcium ions from outside to inside the membrane is closely associated with the formation and decomposition of the reaction intermediate EP (Fig. 5). At the outer surface of the membrane two moles of Ca^{2+} and one mole of ATP bind one mole of the ATPase (E) to form the Michaelis complex EATP. This is immediately followed

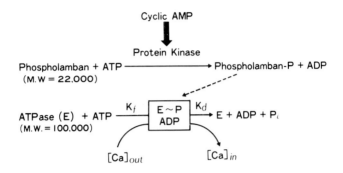

FIGURE 5. Diagram of Effect of Phospholamban on Ca^{2+}-dependent ATPase Enzyme of Cardiac Microsomes.

by the formation of the phosphorylated intermediate EP when calcium is translocated from outside to inside the membrane. Calcium is subsequently released from the enzyme to the interior of the vesicle, with the simultaneous decomposition of the intermediate into E and P_i. Among these steps, the one at which EP is decomposed (k_d in Fig. 5) is rate-limiting in the presence of saturating concentrations of ATP and Ca^{2+}, whereas under these conditions the EP formation step (k_f) is not rate-limiting. The present findings indicate that phosphorylation of phospholamban by cyclic AMP-dependent protein kinase is associated with marked increase in the rate constant k_d, resulting in increased turnover of the enzyme reaction and calcium transport.

The present observations thus lent support for our previous proposal that phospholamban functions as a regulatory factor of the active calcium transport (Tada et al., 1975, 1978a; Katz et al., 1975). However, the mechanism by which the enzymatic properties of ATPase is altered by the putative regulator remains to be elucidated. The observed alterations could be induced by the direct molecular interactions between phospholamban and ATPase. Alternatively, it could also be caused by an indirect effect such as that involving the lipid-protein interactions within the membrane. The present and previous results are suggestive of the former possibility, in that at phosphorylated state phospholamban appears to become tightly associated with the ATPase (Table I) and to be less affected by proteolytic digestion (Tada et al., 1975). It remains to be seen, however, in the reconstituting system whether there exists a direct molecular interaction between these proteins.

If, however, the hypothetical mechanism is functional, it is of interest to estimate the molecular stoichiometry between these key components within the microsomal membrane. Taking 25% and 4% for the contents of the 100,000-dalton ATPase and 22,000-

dalton phospholamban, respectively, in the total membrane protein (see above), the molecular stoichiometry can be estimated to be about 1:1. The apparent stoichiometry between protein kinase-catalyzed phosphate incorporation, which is 0.6 to 0.8 nmol P per mg of microsomal protein (Tada et al., 1979), and that for the incorporation of phosphate into the ATPase in the same membrane preparations, which is $\sim$ 0.8 nmol P per mg of microsomal protein (Tada et al., 1979), is approximately 1:1. Making the same assumption for protein contents and molecular weights for these proteins, up to 0.3-0.4 mole of phosphate each is estimated to be incorporated into a single mole of each protein. The possible functional relationship between these phosphorylatable proteins warrants further analysis.

C. Physiological Relevance

1. Inotropic and Relaxing Effects. Cyclic AMP regulation of the active calcium transport by cardiac microsomes documented in this communication may provide a biochemical basis for the mediation by cyclic AMP of the relaxing effects of catecholamines on cardiac muscle because of the increased rate at which calcium would be removed from troponin, as was mentioned in detail elsewhere (Tada et al., 1978a; Katz et al., 1975). Catecholamine-induced augmentation of myocardial contractility (positive inotropic effect) could also be interpreted by the related mechanisms.

2. Existence of Similar Phosphorylatable Proteins. Kirchberger and Tada (1976) were the first to report the possible occurrence of a phosphoprotein similar to cardiac microsomal phospholamban in the microsomal preparation obtained from slow contracting skeletal muscle of various species. While this possibility was challenged (Schwartz et al., 1976), possible existence of a similar mechanism has been considered in microsomal membranes from other tissues such as vascular (Bhalla et al., 1978) and visceral (Kimura et al., 1977) smooth muscles and blood platelets (Käser-Glanzmann et al., 1977; Haslam et al., 1978). Elucidation of the functional significance of protein kinase-catalyzed phosphorylation in producing regulation of motility in muscular and nonmuscular tissues remains a challenge for future investigation.

ACKNOWLEDGMENTS

We are greatly indebted to Professor Yuji Tonomura for his valuable comments during the course of kinetic studies of ATPase reactions. We also express our sincere gratitude to Dr. R. Nagai for her help in performing electron microscopic studies.

REFERENCES

Bhalla, R.C., Webb, R.C., and Singh, D. (1978). Am. J. Physiol. 234:H508-H514.
Cornish-Bowden, A. (1976). "Principles of Enzyme Kinetics," Butterworths, London.
Froehlich, J.P., and Taylor, E.W. (1975). J. Biol. Chem. 250: 2013-2021.
Harigaya, S., and Schwartz, A. (1969). Circ. Res. 25:781-794.
Haslam, R.J., Davidson, M.M.L., Davies, T., Lynham, J.A., and McClenaghan, M.D. (1978). Advan. Cycl. Nuc. Res. 9:533-552.
Kanazawa, T., Yamada, S., Yamamoto, T., and Tonomura, Y. (1971). J. Biochem. 70:95-123.
Käser-Glanzmann, R., Jákabová, M., George, J.N., and Lüscher, E.F. (1977). Biochim. Biophys. Acta 446:429-440.
Katz, A.M., Tada, M., and Kirchberger, M.A. (1975). Advan. Cycl. Nuc. Res. 5:453-472.
Kimura, M., Kimura, I., and Kobayashi, S. (1977). Biochem. Pharmacol. 26:994-996.
Kirchberger, M.A., and Tada, M. (1976). J. Biol. Chem. 251:725-729.
Kirchberger, M.A., Tada, M., and Katz, A.M. (1974). J. Biol. Chem. 249:6166-6173.
LaRaia, P.J., and Morkin, E. (1974). Circ. Res. 35:298-306.
MacLennan, D.H. (1970). J. Biol. Chem. 245:4508-4518.
Miyamoto, E., Kuo, J.F., and Greengard, P. (1969). J. Biol. Chem. 244:6395-6402.
Schwartz, A., Entman, M.L., Kaniike, K., Lane, L.K., van Winkle, W.B., and Bornet, E.P. (1976). Biochim. Biophys. Acta 426:57-72.
Shigekawa, M., Finegan, J.M., and Katz, A.M. (1976). J. Biol. Chem. 251:6894-6900.
Tada, M., Kirchberger, M.A., Iorio, J.M., and Katz, A.M. (1973). Circulation 48 [Suppl. 4], 25.
Tada, M., Kirchberger, M.A., and Katz, A.M. (1975). J. Biol. Chem. 250:2640-2647.
Tada, M., Kirchberger, M.A., Repke, D.I., and Katz, A.M. (1975). J. Biol. Chem. 249:6174-6180.
Tada, M., Ohmori, F., Kinoshita, N., and Abe, H. (1978a). Advan. Cycl. Nuc. Res. 9:355-369.
Tada, M., Ohmori, F., Yamada, M., and Abe, H. (1979). J. Biol. Chem. 254: in press.
Tada, M., Yamamoto, T., and Tonomura, Y. (1978b). Physiol. Rev. 58:1-79.
Will, H., Blanck, J., Smettan, G., and Wollenberger, A. (1976). Biochim. Biophys. Acta 449:295-303.

FEED-BACK REGULATION OF SYNAPTIC TRANSMISSION BY ATP AND
ADENOSINE DERIVATIVES IN MAMMALIAN BRAIN

Yoichiro Kuroda
Kazuo Kobayashi

Department of Neurochemistry
Tokyo Metropolitan Institute for Neurosciences
Fuchu-shi, Tokyo 183, Japan

It is of interest to note that ATP is released from various secreting cells together with other physiologically active materials (neurotransmitters, hormones) by electrical or chemical stimulation and detected as a mixture of adenosine derivatives (ATP, ADP, 5'-AMP and adenosine) in the extracellular medium.
Recycling of ATP.
In mammalian brain, experiments using synaptosome beds indicate that ATP is released from presynaptic site and degraded to adenosine in the synaptic cleft; adenosine is then taken up by nerve endings and re-phosphorylated to ATP for reutilization (1).
Role of the released ATP and adenosine derivatives.
Experimental evidence on olfactory cortex slices or synaptosome preparations from guinea pig brain suggests the following feed-back regulatory role of the adenosine derivatives on neurotransmission in mammalian brain.
1) Inhibition mediated by Ca^{2+} movement.
Adenosine derivatives inhibited the postsynaptic potentials observed in olfactory cortex slices (2), presumably by the inhibition of Ca^{2+} influx into nerve endings which resulted in the reduction of the transmitter release (3).
2) Facilitation mediated by cyclic AMP.
Adenosine derivatives increased the level of cyclic AMP in synaptosomes (4). Cyclic AMP analogues caused facilitation in olfactory cortex slices. The adenosine-induced increase of cyclic AMP in presynaptic terminals can mediate the facilitation (5).

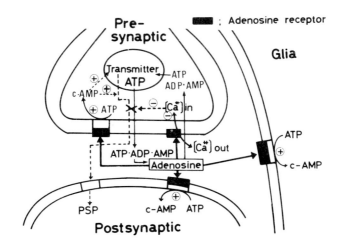

Figure. A possible "Samsara" model of regulatory actions by adenosine derivatives during neurotransmission.

3) <u>Stimulation of catecholamine synthesis mediated by cyclic-AMP</u>.

An adenosine analogue activated tyrosine hydroxylase in a synaptosomal preparation from guinea pig striatum (6). Possibly, adenosine in the synaptic cleft is the first messenger, and cyclic AMP in the presynaptic terminals is the second messenger for the activation of catecholamine synthesis after nerve stimulation.

ATP is released with the transmitter and degraded to adenosine derivatives in the synaptic cleft. Those act on adenosine receptors on presynaptic membrane and result in inhibition, facilitation and stimulation of transmitter (catecholamine) synthesis via Ca^{2+} or cyclic AMP. Possible localization of other adenosine receptors are shown at postsynaptic and glial membrane, both are coupled with cyclic AMP accumulation.

REFERENCES

(1) Kuroda, Y., and McIlwain, H. (1974). J. Neurochem. 22:691.
(2) Kuroda, Y., Saito, M., and Kobayashi, K. (1976). Brain Res 109:196.
(3) Kuroda, Y., Saito, M., and Kobayashi, K. (1976). Proc. Japan Acad. 52:86.
(4) Kuroda, Y., Kobayashi, K. (1978). Proc. Japan Acad. Ser. B 54:243.
(5) Kuroda, Y. (1978). J. Physiol. Paris, in press.
(6) Kuroda, Y., and Kobayashi, K. (1978). Proc. Japan Acad. in press.

STRUCTURAL STUDIES OF SARCOPLASMIC
RETICULUM IN VITRO AND IN SITU

Sidney Fleischer
Cheng-Teh Wang
Akitsugu Saito
Maria Pilarska
J. Oliver McIntyre

Department of Molecular Biology
Vanderbilt University
Nashville, Tennessee

Sarcoplasmic reticulum (SR), is a specialized membrane system of muscle which mediates muscle contraction and relaxation by regulating the Ca^{++} concentration of the sarcoplasm. SR has been isolated from muscle in the form of membranous vesicles which are capable of energized Ca^{++} uptake. SR is currently under intensive study in many laboratories both as a prototype for membrane transport function and because of its vital role in muscle physiology (1,2,3).

A molecular biology approach was initiated by us with the following aims (4): 1) to isolate highly purified SR and to characterize it; 2) to disassemble SR into its components and to characterize the components in terms of the functional characteristics of SR; 3) to reconstitute membrane vesicles similar to normal SR, and capable of transport function; and 4) to vary the composition of the membrane and relate composition with structure and structure with function. We will review some of our progress in this program.

The isolation of highly purified normal SR (N-SR) and in sizeable quantity (hundreds of mg protein) was achieved by centrifugation using zonal rotors (5). In our preparation of SR, three proteins truly predominated as judged by polyacrylamide gel electrophoresis using dissociating conditions (PAGE). The calcium pump protein, 118,000 daltons (6), the calcium binding protein and a polypeptide of approximately 55,000 daltons, designated M_{55}, comprise about 75%, 10% and

10% respectively, of the protein of the purified SR fraction (Fig. 1, gel 2) (5).

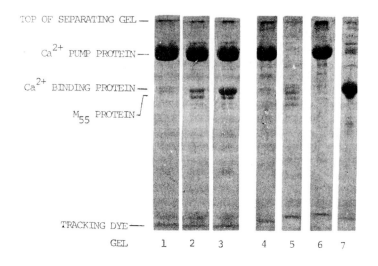

Fig. 1. Polyacrylamide gel electrophoresis (PAGE) of SR. The samples in the gels are of: 1) light SR; 2) N-SR; 3) heavy SR; 4 and 5, membranous fraction and soluble extract of light SR, respectively; 6 and 7 membranous fraction and soluble extract of heavy SR, respectively. The extract was prepared by extraction with 1mM EDTA at pH 8.5 (9,25).

The morphology of N-SR can be viewed by electron microscopy using three different methods of sample preparation (Fig. 2). In thin sections (Fig. 2A), the trilaminar appearance of the membrane can be visualized. Negative staining, reveals small particles (∼40 A°) at the outer surface of the membrane (Fig. 2B) (7). Freeze-fracture electron microscopy shows an asymmetric distribution of particles in the hydrophobic fracture face with most of the particles observable in the outer leaflet (concave face) of the membrane (Fig. 2C)(8).

Highly purified SR can further be subfractionated into light and heavy SR vesicles (Fig. 3). Heavy SR contains electron dense matter within its compartment whereas light SR is devoid of electron opaque contents. The electron opaque contents are composed mainly of calcium binding protein (Fig. 1, gel 3). Light SR is essentially all membrane, of which the calcium pump protein comprises 90% or more of the membrane protein (Fig. 1, gel 1). Indeed, SR is a highly specialized membrane system (9)!

Structural Studies of Sarcoplasmic Reticulum

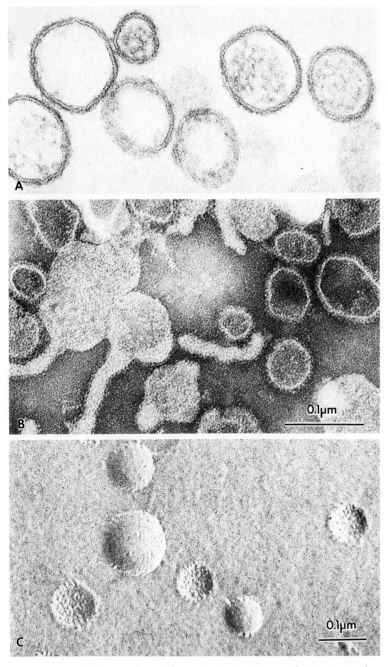

Fig. 2. N-SR as visualized by electron microscopy in thin sections (A), using negative staining (B), and by freeze-fracture (C).

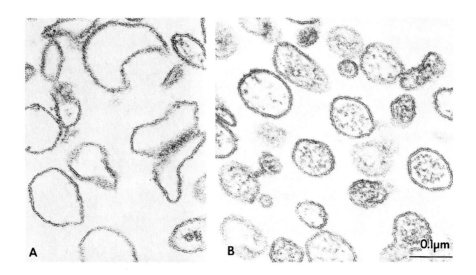

Fig. 3. Electron micrographs of (A) light and (B) heavy SR, prepared from N-SR by isopycnic centrifugation (9).

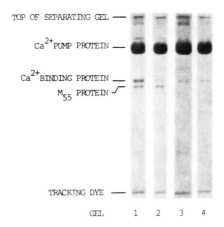

Fig. 4. PAGE patterns of (A) N-SR; (B) R-SR; (C) purified calcium pump protein; and (D) functional membrane vesicles prepared from purified calcium pump protein (12).

Purification of the membrane components can be accomplished more readily from purified SR (5). The calcium pump protein has two specific Ca^{++} binding sites and one ATP binding site and is the membrane component which forms the phosphoenzyme intermediate (10,11). The calcium binding protein has a remarkable high capacity to bind Ca^{++} although the binding is non-specific i.e., the binding is high affinity ($K_D \sim$ 1mM), but the ability to bind Ca^{++} is greatly diminished in the presence of isotonic KCl (5).

Normal SR can be dissociated using detergents (deoxycholate) and then reconstituted to form functional membrane vesicles by removing the perturbing conditions (12,13). Conditions are stringent for reassembly of the membrane to form functional reconstituted SR (R-SR). Reconstitution must be carried out at 15-20° in order to achieve good energized accumulation of Ca^{++}. The protein of R-SR like N-SR membranes consists mainly of calcium pump protein. With the procedure used, most of the M_{55} but little of the calcium binding protein is retained in R-SR. Reconstitution can also be achieved with purified calcium pump protein to yield functional membrane vesicles (Fig. 4). R-SR has a lipid content which is comparable to that of the N-SR membrane. The reconstitution procedure which we have described yields membranes which differ from that reported by others (14,15) in that: 1) the R-SR membrane like N-SR is mainly protein (60% of the mass) and not liposomes mildly doped with calcium pump protein; 2) R-SR does not require trapping agents such as oxalate within its compartment in order to achieve energized calcium uptake; and 3) R-SR is capable of active calcium accumulation in the absence of a trapping agent.

Recently, we have developed a procedure to visualize the asymmetry of the SR membrane in thin sections using tannic acid to enhance contrast (16). The outer layer of the trilaminar membrane is 70 A° wide as compared with 20 A° wide for the middle and inner layers (Fig. 5A - contrast appearance with that in Fig. 2A). The asymmetry of the normal SR membrane can thus be visualized in thin sections as well as by negative staining and by freeze-fracture electron microscopy (Fig. 2B and 2C). The asymmetry of the SR membrane can now also be visualized for the first time in tissue sections (Fig. 6).

R-SR differs from N-SR in one key aspect, the asymmetry of the membrane has not been retained in membranes formed in the test tube. This can be seen with each of the three methods of sample preparation used for electron microscopy. In thin sections the trilaminar appearance of R-SR is symmetric (70,20,70 A° wide) (Fig. 5C). The surface particles, observed by negative staining can be seen on the inner as well as outer

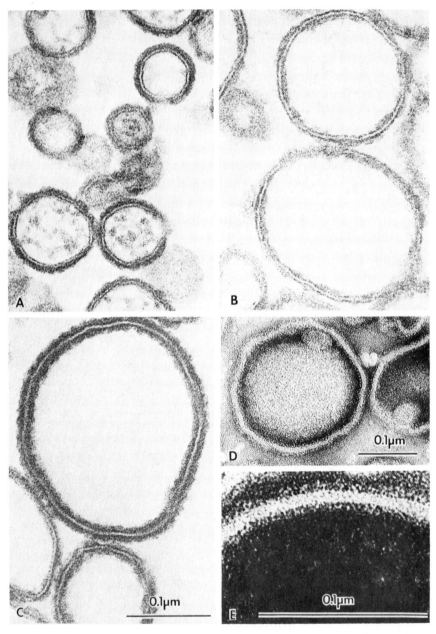

Fig. 5. SR as visualized in thin sections (A-C) and by negative staining (D-E). (A-C) Tannic acid was used to achieve enhanced contrast of the SR membranes. A is N-SR; B and C are R-SR of lower and higher protein content, respectively. The 0.1μ bar in C gives the enlargement for A-C. Fig. D and E are negative staining of R-SR (16).

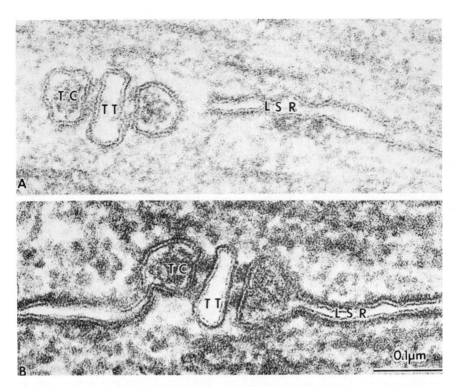

Fig. 6. Rabbit skeletal muscle in thin sections with emphasis on the triad region. A) fixed without and B) with tannic acid enhancement. The asymmetry of the SR membrane can readily be visualized in B. TT - T-tubule; TC - terminal cisternae of SR; LSR-lateral cisternae of SR (16).

faces of R-SR (Fig. 5D and 5E). With freeze-fracture, there is a symmetrical distribution of particles on the inner and outer fracture faces (17) (Fig. 9C).

We have also developed a method to measure phospholipid accessible on the surface of membranes (18). The method consists of inserting D-β-hydroxybutyrate apodehydrogenase (BDH) into the outer surface of membrane vesicles. BDH apodehydrogenase is a lipid-requiring enzyme which readily inserts into the surface of phospholipid or membrane vesicles. When BDH is inserted, it is reactivated by endogenous phospholipid of the membrane. The saturating amount of BDH which inserts in excess BDH is proportional to accessible lipid on the surface of the membrane (see below). We find using this method that there is four-fold more accessible phospholipid on the inner surface as compared with the outer surface of the SR membrane (Fig. 7).

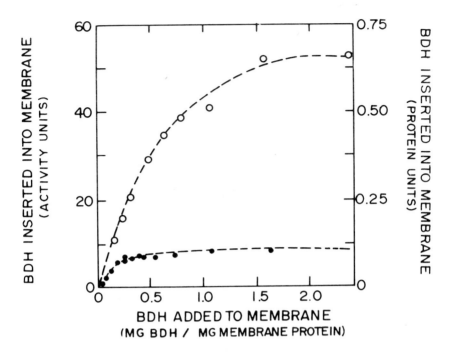

Fig. 7. The insertion of D-β-hydroxybutyrate apodehydrogenase (BDH) into normal (●- - -●) and leaky (o- - -o) SR vesicles. Vesicles were made leaky with 1mM EDTA at pH 8.5 (18).

The studies described above suggest molecular detail for the orientation of the calcium pump protein in the SR membrane (Fig. 8). The main component of the SR membrane is the calcium pump protein. Therefore, the asymmetry which is observable in thin sections and by negative staining must be referable to the asymmetric orientation of the calcium pump protein in the membrane i.e., a portion of the calcium pump protein extends beyond the lipid from the outer surface of the membrane. The calcium pump protein must be transmembrane since it pumps Ca^{++} across the membrane from the outer compartment to the inside. In R-SR, the anisotropy of the pumps has been randomized giving rise to the symmetrical appearance both in thin sections as well as by negative staining (Fig. 5).

The reconstitution procedure has been modified to prepare R-SR membranes with varying lipid content, with both greater and lesser phospholipid than that of N-SR (19). The variation in lipid content which we have found most practical for correlative studies ranges from half to twice the lipid

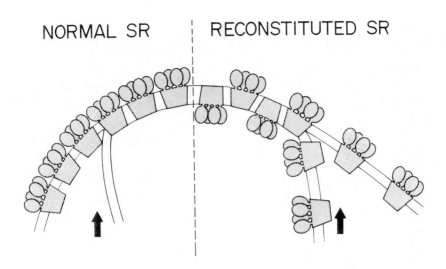

Fig. 8. Diagrammatic representation of N-SR and R-SR vesicles and the effect of the freeze-fracture process on these membranes (16).

content of N-SR. Such a series of preparations has permitted correlation of composition with structure in freeze-fracture electron microscopy. There is a direct correlation of the number of particles at the hydrophobic surface, observed by freeze-fracture electron microscopy, with the protein content of the R-SR membrane and the extent of smooth surface with the lipid content of the membrane (Fig. 9) (20).

The availability of R-SR preparations of varying lipid content helped to make possible the development of the procedure to measure accessible phospholipid on the surface of membranes (referred to above) by serving as a calibration series. The insertion of BDH (Fig. 10) into R-SR of varying lipid content was found to be proportional to phospholipid which is in excess of $\sim 7 \mu g$ P/mg protein. This amount of lipid, which is inaccessible for insertion of BDH is interpreted as that which surrounds the calcium pump protein (~ 27 moles phospholipid/mole calcium pump protein) and may have some relevance to "boundary lipid" (21) or "lipid annulus" (22). The proportion of BDH inserted into excess phospholipid of R-SR is the same as that which can be inserted into phospholipid vesicles in aqueous medium (not shown) (18).

The ability to prepare functional membrane vesicles of defined composition, and with varying phospholipid content in the range of high protein content of the type described here, makes possible detailed study of membrane structure including motion. Such studies are in progress together with Leo

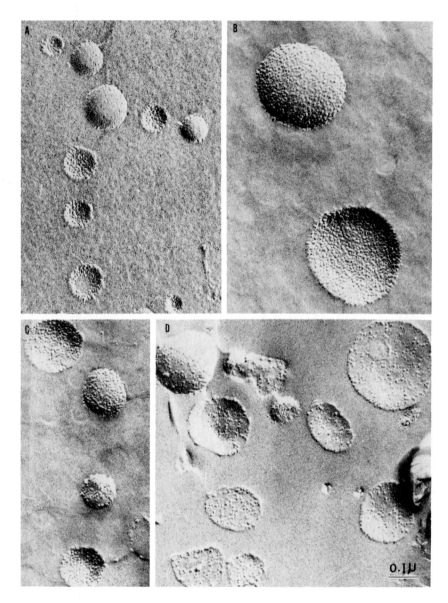

Fig. 9. Freeze-fracture electron microscopy of N-SR (A) and R-SR of varying phospholipid content (B) 0.35; (C) 0.77; and (D) 1.16 μmoles phospholipid per mg protein; the value for N-SR is 0.77 (20).

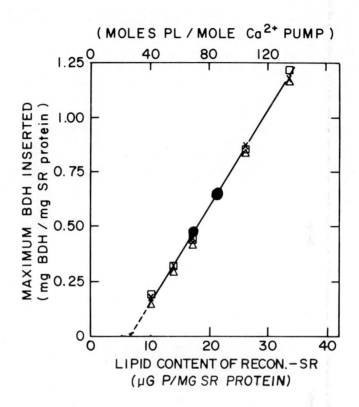

Fig. 10. Insertion of BDH into R-SR membranes of varying lipid content. BDH inserts into accessible phospholipid at the outer surface of the membrane.

Herbette, Antonio Scarpa and Kent Blasie of the Univ. of Pennsylvania (23) and Joachim Seelig and his colleagues at the Univ. of Basel (24).

These studies were supported in part by grants from the National Institutes of Health, AM 14632, and the Muscular Dystrophy Associations of America, Inc. M. Pilarska was a Fogarty Fellow (SF05 TW02328) from the Nencki Institute of Experimental Biology, Warsaw, Poland.

REFERENCES

1. Tada, M., Yamamoto, T., and Tonomura, Y. (1978). Phys. Rev. 58:1.
2. MacLennan, D.H., and Holland, P.C. (1975). Ann. Rev. Biophys. and Bioeng. 4:377.
3. Martonosi, A. (1971) in Biomembranes (Manson, L.A., ed.) Vol. 1, pp. 191-256, Plenum Press, New York.
4. Meissner, G. and Fleischer, S. in Symposium on "Calcium Binding Proteins", (1974) (W. Drabikowski and E. Carafoli, eds), p. 281, Polish Scientific Publishers and Elsevier Publishing Co. Warsaw, Poland.
5. Meissner, G., Conner, G., and Fleischer, S. (1973). Biochim. Biophys. Acta 298:246.
6. Rizzolo, L., LeMaire, M., Reynolds, J. and Tanford, C., (1976). Biochemistry 15:3433.
7. Ikemoto, N., Sreter, F., and Nakamura, A. (1968). J. Ultrastruct. Res. 23:216.
8. Deamer, D. and Baskin, R. (1969). J. Cell Biol. 42:296.
9. Meissner, G. (1975). Biochim. Biophys. Acta 389:51.
10. Meissner, G. (1973). Biochim. Biophys. Acta 298:906.
11. Meissner, G., and Fleischer, S. (1971). Biochim. Biophys. Acta 241:356.
12. Meissner, G., and Fleischer, S. (1973). Biochem. Biophys. Res. Commun. 52:913.
13. Meissner, G., and Fleischer, S. (1974). J. Biol. Chem. 249:302.
14. Racker, E. (1972). J. Biol. Chem. 245:4508.
15. Warren, G., Toon, P., Birdsall, N., Lee, A., and Metcalfe, J. (1974). Proc. Nat. Acad. Sci., USA 71:622.
16. Saito, A., Wang, C.-T., and Fleischer, S. (1978). J. Cell Biol. 79:601.
17. Packer, L., Mehard, C., Meissner, G., Zahler, W., and Fleischer, S. (1974). Biochim. Biophys. Acta 363:159.
18. McIntyre, J., Wang, C.-T., and Fleischer, S. (In press). J. Biol. Chem.
19. Wang, C.-T. (1976). Fed. Proc. 35:1664.
20. Wang, C.-T., Saito, A., and Fleischer, S. (Submitted) J. Biol. Chem.
21. Jost, P., Griffith, O., Capaldi, R., and Vanderkooi, G. (1973). Proc. Natl. Acad. Sci., USA 70:480.
22. Metcalfe, J., and Warren, G. (1977) in International Cell Biology 1976-77. (Brinkley, B.R., and Porter, K.R., eds.) pp. 15-23, Rockerfeller Press, New York.
23. Herbette, L., Wang, C.-T., Scarpa, A., Fleischer, S., Schoenborn, S. and Blasie, K. (1978), Abstract Biophysical Society Meeting 21:205a.

24. Fleischer, S., Wang, C.-T., Seelig, J., and Brown, M. (1979), Abstract Biophysical Society Meeting (In Press).
25. Duggan, P., and Martonosi, A. (1970) J. Gen. Physiol. 56:147.

PART II

ATP-GENERATING SYSTEMS

FLASH-INDUCED INCREASE OF ATPASE ACTIVITY
IN RHODOSPIRILLUM RUBRUM CHROMATOPHORES[1]

Margareta Baltscheffsky
Arne Lundin

Department of Biochemistry
Arrhenius Laboratory
University of Stockholm
Stockholm, Sweden

I. INTRODUCTION

Light-induced activation of ATPase activity was first observed in spinach chloroplasts (Petrack and Lipmann, 1961, Petrack et al., 1965). Some characteristics of the light activated chloroplast ATPase is a stabilization of the activity by phosphate (Carmeli and Lifshitz, 1972) and an acceleration of the return to the latent state by ADP and by un-couplers (Carmeli, 1969). The activated state is maintained in the dark by the energization caused by the hydrolysis of ATP (Hoch and Martin, 1963). Ryrie and Jagendorf (1971) demonstrated that a light-induced conformation change in the ATPase enzyme, CF_1, is strictly related to the energetic state, and in all likelihood is a prerequisite for photophosphorylation.

In photosynthetic bacteria, light-induced activation of ATPase activity has been shown to occur in chromatophores from Rhodopseudomonas capsulata (Melandri et al., 1972). The dark activity in these chromatophores is stimulated by the presence of phosphate, and the light-induced activation is prevented by the presence of ADP.

---

[1]Supported by a grant from the Swedish Natural Science Research Council.

In chromatophores from Rhodospirillum rubrum no comparable light-induced activity has hitherto been found, although Edwards and Jackson (1976) have reported that light may reactivate the ATPase activity when it is inhibited by uncoupler. Horiuti et al., (1968) have reported that light stimulates the ATPase activity in the presence of PMS (phenazine methosulfate) but inhibits it in the absence of PMS. The basic, dark ATPase in R. rubrum is higher than has been reported in other photosynthetic bacteria. It is stimulated by low concentrations of uncoupler and inhibited by the energy transfer inhibitors oligomycin (Baltscheffsky et al., 1967) and DCCD (dicyclohexylcarbodiimide) (Baltscheffsky, unpublished). When membrane bound, the activity is elicited by Mg^{++}, and to a lesser degree by Ca^{++}. Solubilization of the F_1 entity yields an enzyme with five subunits (Johansson and Baltscheffsky, 1975) which is only activated by Ca^{++}. Mg^{++} is a competitive inhibitor to the solubilized Ca^{++}-ATPase activity (Johansson et al., 1973). In this respect it behaves similarly to the chloroplast enzyme.

In this paper we will describe an increase of the membrane bound ATPase activity, which is induced by short light flashes at low ATP concentrations (10^{-7} - 10^{-6} M). This increased activity appears to be due to a decrease of the apparent K_m for ATP of the enzyme.

II. RESULTS

A. General Properties

During the study of flash-induced ATP synthesis, using a luciferase assay, which gives a comparatively stable luminescence (Lundin et al., 1976), making it possible to monitor the ATP concentration for several minutes after an addition of ATP to the medium, we observed that the ATP formed after single or multiple light flashes was more rapidly hydrolyzed than ATP, added in the dark, as a calibration. The stimulation was 4-8 fold and the activation lasted for 30-40 seconds after the flashes.

In Fig. 1 is seen a trace from a typical experiment. ATP is added and gives rise to a luminescence increase (downwards in the figure). The luminescence slowly decreases due to the dark hydrolysis of the added ATP. After a 1 ms flash, which causes about two turnovers of the cyclic electron transport system, the rate of ATP hydrolysis increases approxi-

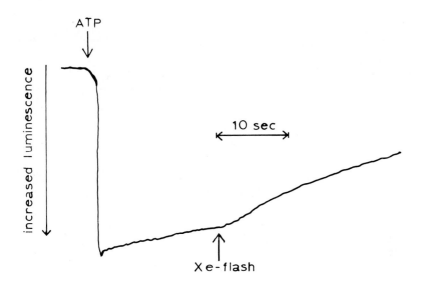

FIGURE 1. Stimulation of ATPase activity by a 1 ms flash. ATP to a final concentration of 0.56 µM was added where indicated. Reaction mixture contained in a final volume of 2 ml: 0.1 M clycyl-glycine pH 7.85, 10 mM $Mg(Ac)_2$, 1 mM Pi, 0,1 mM Na-succinate, 0.14 mM luciferin, about 10 µg of luciferase and chromatophores equivalent to 38 µM bacteriochlorophyll.

mately 4-fold, and proceeds at a stimulated rate until the end of the trace.

The degree of stimulation is dependent on the number of turnovers applied to the chromatophores. Fig. 2 shows that the stimulation becomes maximal only after 6-7 flashes. In this case very short 5 µs flashes, each causing only one turnover, were used, so the number of flashes corresponds to the number of turnovers.

The flash-induced stimulation of the ATPase activity is dependent on the presence of phosphate in the reaction medium. Fig. 3 shows that maximal activity is obtained at about 0.5 mM Pi. Higher phosphate concentration decreases both the flash activated and the dark activity, but at the level where the light induced activity is maximally stimulated, the dark activity is only marginally inhibited. Both the dark and the light-induced activities become approximately 50% inhibited at high phosphate concentrations.

The apparent K_m for ATP of the membrane bound ATPase activity is around 0.2 mM, when measured in the absence of Pi

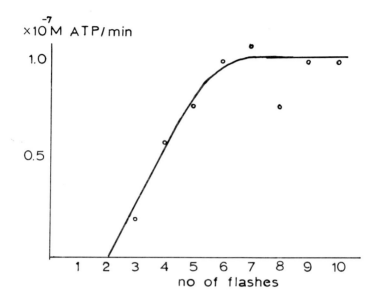

FIGURE 2. Rate of ATPase activity as a function of the number of single turnover (5 µs) light flashes. Conditions as in Fig. 1 except that chromatophores were equivalent to 10 µM Bchl, and the initial ATP concentration 0.2 µM.

with conventional colorimetric techniques (Johansson et al., 1973). Since the luciferase system employed here permits the exact and easy measurement of ATPase activity also in the presence of Pi, an investigation of the apparent K_m in the presence of 0.5 mM Pi was done. The result is seen in Fig. 4. The K_m of the dark ATPase is in the presence of Pi 5 µM, instead of 0.2 mM in the absence of added Pi. An about 4-fold decrease of the apparent K_m for ATP, from 5 µM to 1.3 µM is obtained after a single 1 ms flash. This decrease is seen as the increase of activity when the substrate concentration is kept at, or below, the K_m.

B. Effects of Uncoupler and Ionophores

The uncoupler FCCP (fluorocarbonyl cyanide phenylhydrazone) has several different effects on the flash stimulated

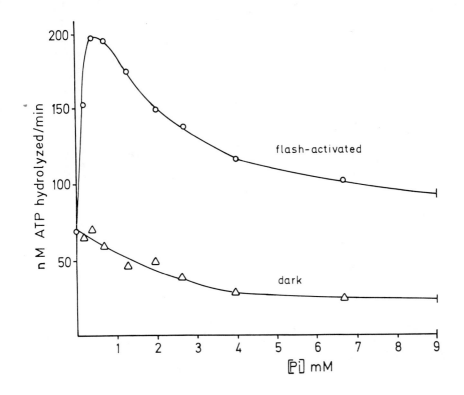

FIGURE 3. Influence by Pi concentration on the light stimulated and dark ATPase activities. Conditions as in Fig. 1 except that the Pi concentration was as indicated in figure.

ATPase activity. The most prominent is that the duration of the activated state is progressively shortened in the presence of increasing concentrations of FCCP as is shown in Fig. 5. The activation process itself does not seem to be affected, but at higher concentrations of FCCP, there is a very abrupt cut off after approximately 1 s of increased ATPase activity, At higher concentrations of FCCP, there is also, in the presence of both ADP and Pi, a further stimulation of the initial activity, even if the duration is shortened. This is evident only in the presence of both ADP and Pi. In the absence of ADP the effect is the opposite: an inhibition occurs at higher FCCP concentrations. The curve in Fig. 6 of the flash

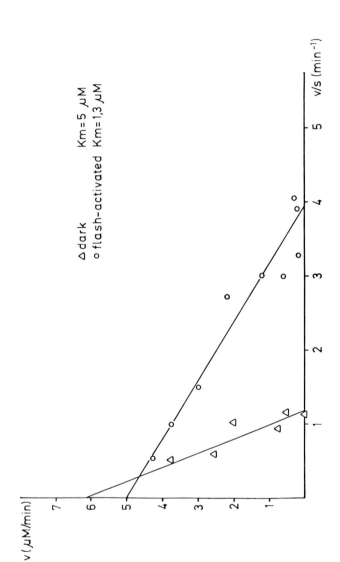

FIGURE 4. Eadie-Hofstee plot for determining the apparent K_m of dark and flash activated ATPase activities. Conditions as in Fig. 1 except that the Pi concentration was 0.5 mM and the bacteriochlorophyll was 56 μM. Each measuring point was taken as the initial activity of ATPase after a 1 ms flash.

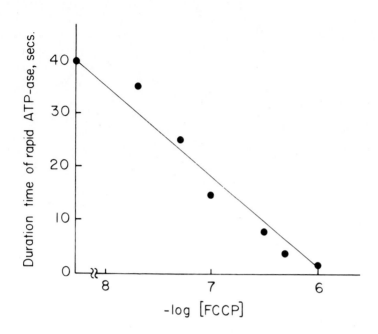

FIGURE 5. Effect of increasing concentrations of FCCP on the duration of the flash stimulated ATPase activity. Conditions as in Fig. 1.

stimulated ATPase activity in the presence of ADP is very similar to the stimulation obtained on the dark ATPase by FCCP, measured at mM ATP concentrations in the absence of added ADP and Pi.

Valinomycin plus K^+ prevents the flash-induced stimulation. Nigericin on the other hand has an effect similar to that of FCCP, indicating that the maintenance of a proton gradient is essential in sustaining the increased activity. These recent data will be presented and discussed in detail in a forthcoming publication (Baltscheffsky and Lundin, in preparation).

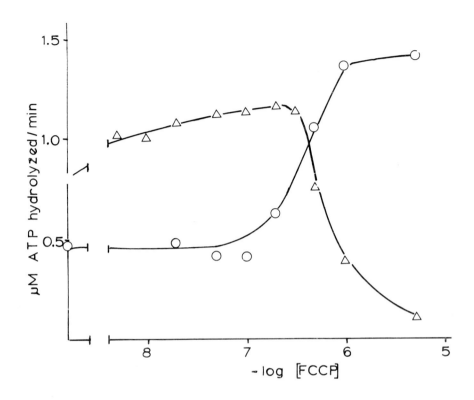

FIGURE 6. Effect of FCCP on light stimulated ATPase activities in presence (-O-) and absence (-△-) of ADP. Conditions as in Fig. 1.

III. DISCUSSION

In many respects the light stimulated ATPase in R. rubrum shows similarities with the light triggered ATPase activity in chloroplasts. Notable is the effect by P_i, which in R. rubrum is a prerequisite for eliciting the light stimulation, in chloroplasts the presence of P_i stabilizes the activated state, and the effect of uncouplers which in both cases accelerate the return to the latent state (Carmeli, 1969, and

Fig. 5), although on quite a different time scale. Also the
concept that the dark hydrolysis of ATP producing an energization of the membrane maintains the activity in the dark
is common for both systems. ADP accelerates the return to
the latent state of the chloroplast ATPase (Carmeli and
Lifshitz, 1972) and prevents the light-induced reversal of
the uncoupler inhibition of the R. rubrum system (Edwards
and Jackson, 1976). Both ATPases are in the solubilized form
converted from predominantly Mg^{++} dependent membrane bound
activity to Ca^{++} dependent enzymes.

It is not difficult to visualize that the change in K_m
(Fig. 4) is the result of a conformational change in the
ATPase enzyme, especially since such changes have been shown
to occur upon illumination both in chloroplasts and chromatophores from Rps. capsulata. The drastic decrease of the
apparent K_m of the dark activity in the presence of Pi as
compared to the apparent K_m in its absence seems to indicate
that the binding of Pi to the ATPase enzyme causes a substantial modification. It is possible that this lower apparent K_m
is closer to the physiological one than the earlier measured
value, as the intracellular Pi concentration may well be in
the mM range in R. rubrum cells. The intracellular Pi concentration may thus serve as a regulator of the ATPase activity both in the light and the dark.

The above described light-induced activation of the
ATPase, which in reality constitutes a lowering of the K_m for
ATP, i.e., an increased affinity for the substrate, is by
many criteria an energy-linked phenomenon. Since valinomycin
+ K^+ totally abolishes the activation, its onset appears to
be dependent on the membrane potential. The maintenance of
the activated state on the other hand cannot be sustained by
the potential after a single flash (or a series of flashes)
since, as judged by the decay of the flash-induced carotenoid
shift, the potential has decayed already after 3-4 seconds.
However, it has been shown that the hydrolysis of ATP is
associated with the inward translocation of protons (Scholes
et al., 1969). Thus the ATPase may via the formed H^+ gradient be self-sustaining for a limited period of time, even
at low ATP concentration. The short-circuiting of protons
which would occur in the presence of FCCP then does not prevent the activation of the ATPase, but shortens the time that
a sufficient membrane potential is maintained after a flash
and thereby the duration of the activated state (as is seen
in Fig. 5).

REFERENCES

Baccarini-Melandri, A., Fabbri, E., Fistater, E., and Melandri, B.A. (1975). Biochim. Biophys. Acta 376:72.
Baltscheffsky, M., Baltscheffsky, H., and von Stedingk, L.-V. (1967). Brookhaven Symp. Biol. 19:246.
Carmeli, C. (1969). Biochim. Biophys. Acta 189:256.
Carmeli, C., and Lifshitz, Y. (1972). Biochim. Biophys. Acta 267:86.
Edwards, P., and Jackson, J.B. (1976). Eur. J. Biochem. 62:7.
Hoch, G., and Martin, I. (1963). Biochem. Biophys. Res. Commun. 12:223.
Horiuti, U., Nishikawa, K., and Horio, T. (1968). J. Biochem. (Japan) 64:577.
Johansson, B.C., and Baltscheffsky, M. (1975). FEBS Lett. 53:221.
Johansson, B.C., Baltscheffsky, M., Baltscheffsky, H., Baccarini-Melandri, A., and Melandri, B.A. (1973) Eur. J. Biochem. 40:109.
Lundin, A., Rickardsson, A., and Thore, A. (1976). Anal. Biochem. 75:611.
Melandri, B.A., Baccarini-Melandri, A., and Fabbri, E. (1972). Biochem. Biophys. Acta 275:383.
Petrack, B., and Lipmann, F. (1961). In "Light and Life" (W.D. McElroy and H.B. Glass, eds.) p. 621. The Johns Hopkins Press, Baltimore.
Petrack, B., Craston, A., Sheppy, F., and Farron, F. (1965). J. Biol. Chem. 240:906.
Ryrie, I.J., and Jagendorf, A.T. (1971). J. Biol. Chem. 246:582.
Scholes, P., Mitchell, P., and Moyle, J. (1969). Eur. J. Biochem. 8:450.

BIOENERGETICS OF THE EARLY EVENTS
OF BACTERIAL PHOTOPHOSPHORYLATION[1]

Bruno A. Melandri
Assunta Baccarini Melandri

Institute of Botany
University of Bologna
Bologna, Italy

I. ON THE THERMODYNAMICS AND KINETICS
OF PHOSPHORYLATION

A widespread agreement exists at present that a chemiosmotic mechanism of coupling is operating in membrane associated ATP synthesis. The main experimental evidences obtained recently, which integrate with the fundamental demonstration by Jagendorf and Uribe (1) of ATP synthesis induced by acid base transition, are concerned with reconstitution experiments, pioneered by Racker and Stoeckenius (2), in which artificial heterologous systems containing bacterial rhodopsin and an ATPase preparation are utilized. The extensive purification of a stable ATPase from a thermophilic aerobic bacterium (3), which allows the studies of all extrinsic and intrinsic subunits of the coupling complex, and its successful utilization for the reconstitution of light driven ATP synthesis in association with bacterial rhodopsin (4), support very strongly the involvement of a single specific enzyme complex in the direct coupling of protonic fluxes to ATP synthesis.

The original proposal of the chemiosmotic hypothesis (5) postulates that the mechanism of coupling is mediated by the protons interacting with the various proton pumps present in the membrane (driven by redox reactions or by ATP hydrolysis) and in diffusion equilibrium with the protons present in the bulk aqueous phases of the inner and outer compartments. This

---

[1]Supported by CNR grant 77.01412.04

view allows the thermodynamic treatment of the bioenergetics of phosphorylation in terms of average proton electrochemical potential in the bulk phases, but disregards any possible coupling mediated by rapid diffusion processes of protons at the interphases, not in equilibrium with the bulk phases. This situation, which is of course much more difficult to be dealt with in quantitative terms, both experimentally and theoretically, could lead to conditions in which the degree of coupling between redox reactions and ATP synthesis is higher than that expected on the basis of the average transmembrane protonic potential (6) or to apparent kinetic inconsistency between ATP synthesis and the formation of the protonmotive force.

This latter aspect of the problem has been considered in a series of papers by Ort et al. (7,8,9). These authors have studied the onset of phosphorylation and of electron transport in spinach chloroplasts in the presence of valinomycin and permeant buffers and concluded that, under certain experimental conditions, photophosphorylation can take place before a sufficient proton potential difference is established across the thylakoid membrane. They have suggested that the compartments in which the light reactions induce the high protonic potential, driving ATP synthesis, is smaller than the whole inner compartment of chloroplasts, a conclusion which could have only a kinetic and not a structural meaning. This view has not been substantiated by direct measurements of the protonmotive force but is mainly based on theoretical evaluations.

In bacterial chromatophores flash-induced phosphorylation has been evaluated coupling ATP formation to the luciferine-luciferase reaction and monitoring the luminescence immediately after a single flash (10,11). This approach offers a very high sensitivity which allows the detection of the ATP synthesized in one flash, even so short to elicit a single turnover of the photosynthetic apparatus, thus avoiding the repetitive technique, used in chloroplasts (7), which could produce cumulative energization in bacterial chromatophores. The early events of bacterial photophosphorylation can thus be studied and correlated with the formation of the protonic electrochemical potential difference, determined by spectroscopic techniques (12) under identical conditions.

Analogous experiments in continuous light were previously performed in our laboratory: by this approach we observed that when electron flow was partially inhibited by antimycin A or reduced by decreasing the intensity of the actinic light, the rate of ATP synthesis was drastically reduced in spite of the formation of a high protonmotive force across the membrane (12, 13). We interpreted these data as evidence of a short range interaction between the electron transport chains and the ATP synthetase complexes, interaction involving the mechanism of energy transduction itself (e. g. localized proton diffusion as discussed above) or a kinetic control on the coupling enzyme. The stationary conditions under which these measurements were performed did not allow to discriminate between these two possibilities; the studies of the transient events of ATP synthesis and of the formation of the protonic gradient, described in the following section, are an attempt to overcome these difficulties of interpretation.

II. TRANSIENT EVENTS IN THE FORMATION OF THE PROTONMOTIVE FORCE AND OF ATP SYNTHESIS

The amount of ATP formed in chromatophores of Rhodopseudomonas capsulata following flashes of variable duration (from 2 to 500 ms) is compared in Fig. 1 with the formation of both components of the protonmotive force ($\Delta\psi$ and $-Z \Delta pH$). Synthesis of ATP can be observed already with the shortest flash used (2 ms), consistently with the very rapid rise of the membrane potential, which exceeds a value of 100 mV within few milliseconds of illumination. The appearance of a protonic concentration difference ($-Z \Delta pH$) is, on the other hand, much slower than that of the membrane potential and its extent is still very small following a 500 ms flash, as compared to the value reached in the steady state (12).

When the same type of experiments are performed in the presence of ionophores, the results shown in Fig. 2 are obtained. Addition of nigericin (50 mM KCl present) does not affect markedly the yield of ATP per flash, neither delays the onset of phosphorylation. In agreement with these observations, the formation of the protonmotive force and its extent are not influenced by this antibiotic, since at this early stage Δp is formed mainly by $\Delta\psi$. Conversely addition of valinomycin, which dissipates $\Delta\psi$, induces a marked lag in photophosphorylation and slows the building up of the protonmotive force, which is under these conditions formed mainly by ΔpH (not shown). A similar delaying effect by valinomycin has been reported previously in spinach chloroplasts (7).

The effect of the partial inhibition of the electron transport by antimycin A has also been examined (Fig. 3). This inhibitor, added alone, reduces the ATP yield per flash, but does not induce any lag in phosphorylation (curve a and b). When associated with valinomycin (curve d and e), however, it causes a marked increase in the delay induced by the ionophore (curve c), with an effect depending on the degree of inhibition of electron flow. Again these observations are in agreement with the measured rates of formation of Δp: in the absence of valinomycin, antimycin A, at the concentration used in the experiments of Fig. 3, has a negligible effect on the rise and

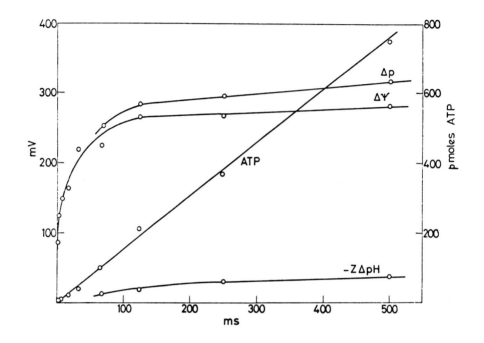

FIGURE 1. ATP synthesis and Δp, $\Delta \psi$ and ΔpH induced in chromatophores of Rps. capsulata by single flashes of variable duration.
Assay mixture: Na-glycylglycine, pH 8.0, 100 mM; Mg-acetate, 10 mM; bovine serum albumin, 1 mg/ml:KCl, 50 mM; P_i, 2 mM; Na-succinate 0.2 mM; ADP, 0.02 mM; luciferine, 0.06 mM; luciferase, 3 μg/ml and chromatophores corresponding to 13 μM bacteriochlorophyll. ATP synthesis was monitored as described in Ref. 10. $\Delta \psi$ was evaluated from the extent of the carotenoid shift (12) and ΔpH measured by 9 aminoacridine fluorescence (12).

the extent of $\Delta \psi$, the predominant component of Δp. When $\Delta \psi$ is collapsed by valinomycin, on the other hand, inhibition of the electron flow results in a marked delay in the formation of ΔpH (Fig. 4). As already observed in previous work (12,13), however, a more prolonged illumination (15-30 s) of chromatophores partially inhibited by antimycin A, elicits a steady state value of ΔpH practically convergent with that measured in the uninhibited control vesicles.

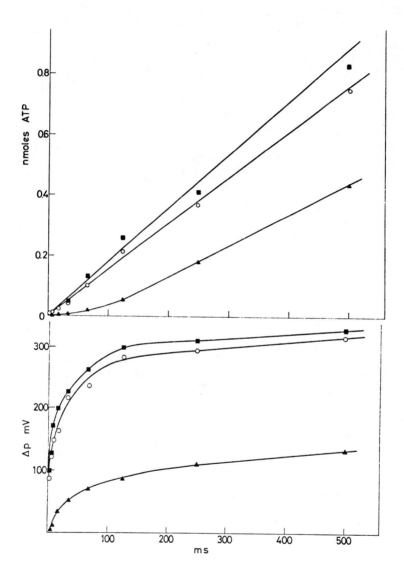

FIGURE 2. The effect of ionophores on the early events of bacterial photophosphorylation (a) and of the formation of Δp (b). —O—, control; —■— plus 1.0 µg/ml negericin; —▲— plus 1.0 µg/ml valinomycin.

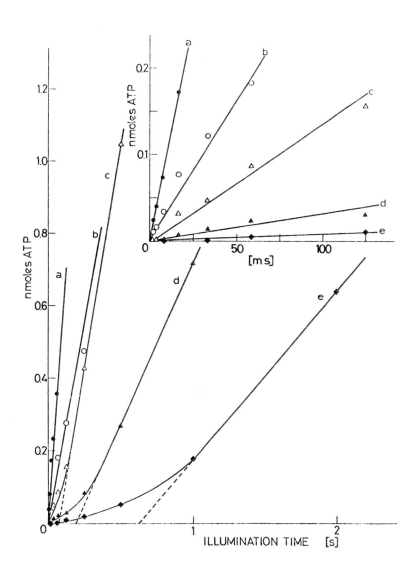

FIGURE 3. The effect of antimycin A, alone or in association with valinomycin, on the early events of bacterial photophosphorylation. Curve a, control; curve b, plus 1.0 µg/ml antimycin A; curve c, plus 1.0 µg/ml valinomycin; curve d, as c plus 0.5 µg/ml antimycin A; curve e, as c plus 1.0 µg antimycin A.

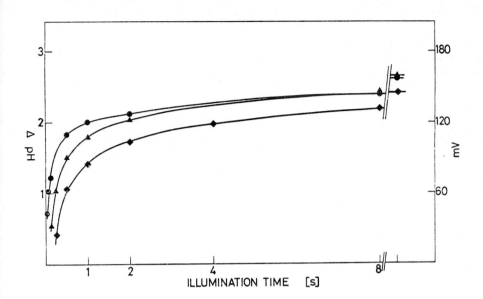

FIGURE 4. Time course of formation of ΔpH in the presence of valinomycin and 50 mM KCl. ——●——, control; ——▲——, plus 0.5 µg/ml antimycin A; ——◆——, plus 1.0 µg/ml antimycin A.

III. CONCLUSIONS

In bacterial chromatophores from Rps. capsulata ATP synthesis starts at full rate within 1-2 ms after the onset of illumination; this observation is consistent with pioneer results by Minton and Keister (14), obtained by isotopic techniques, and with data by Duysens et al. (11) and Baltscheffsky and Lundin (15) who reported that, during a train of single turnover falshes, already in the first flash 50-70% of the ATP synthetised in the subsequent flashes is formed. It is clear therefore that the energy threshold necessary for ATP synthesis (which can be evaluated to about 11 Kcal.mole under our experimental conditions) is overcome very rapidly in illuminated chromatophores, possibly, in a large fraction of the chromatophore population, after a single turnover. Membrane potential is the obvious condidate to be considered as the prevalent energetic component of the protonic gradient in the earliest events of photophosphorylation, as confirmed experimentally by the inhibition by valinomycin of the flash induced ATP synthesis. Theoretical considerations on the electric capacity of an average chromatophore indicate that the translocation of very few electronic charges, corresponding to 1-2 turnovers of all photosynthetic units are sufficient for building up a $\Delta\psi$ exceeding 100 mV (16,17), close to the energy threshold observed in most of the experiments reported here. The fast onset of photophosphorylation is therefore consistent with the chemiosmotic mechanism of energy transduction and with the electrical properties of the membrane.

The study of the time course of formation of the protonmotive force in bacterial chromatophores indicates that a static head value of Δp of the order of 300-350 mV is reached after an illumination time (100-150 ms) much longer than that required to establish the maximal yield of ATP per flash (1-2 ms). Although the relatively slow response of luciferase ($t_{1/2}$ for maximal emission of luminescence $\simeq$200 ms (10)) does not allow to establish if ATP is formed during the flash or in the postillumination period, these results indicate clearly that the extent of Δp is not the factor limiting energetically the synthesis of ATP for a large range of membrane energies, since for $\Delta p \geq 100-120$ mV the amount of ATP formed is strictly linear with the duration of the flash.

Partial inhibition of electron flow by antimycin A, in the absence of valinomycin, does not delay significantly the onset of phosphorylation in the millisecond range (Fig. 3) but drastically affects the steady state rate of ATP synthesis (Fig. 3 and ref. 12). These results are only partially explained by the low electric capacity of the membrane, which can be promptly charged at relatively high Δp also by a partially inhibited electron flow; this concept in fact fails in explaining the effect of antimycin on the steady state rate of ATP synthesis (12) since a negligible inhibition by the same concentration of this antibiotic is observed on the membrane potential (as evaluated from the extent of the carotenoid signal). When valinomycin is added to chromatophores in the presence of 50 mM K^+ the onset of the maximal yield of ATP per flash is markedly delayed ($\simeq$70 ms). Under these conditions the static head value of Δp is drastically reduced by the dissipation of the electrostatic component due to K^+ fluxes; the time course of formation of the protonmotive force is also markedly delayed. These results are again fully consistent with the properties of chromatophores, whose endogenous buffer capacity, empirically evaluated, amounts to 400-700 H^+ per chromatophore per pH unit (16); the building up of a Δp of 100-120 mV, formed prevalently by ΔpH would require therefore some 10-15 turnovers of all photosynthetic units present in an average vesicle. This figure is in agreement with the time course of formation of Δp in the presence of valinomycin. Under this condition, when the buffer capacity of chromatophores becomes a decisive factor in the time constant of energization of chromatophores, a prompt and marked effect of the inhibitors of electron flow becomes apparent. Partial inhibition by antimycin A in the presence of valinomycin results in a large delay in the formation of the minimally required ΔpH (Fig. 4) and in a large lag before the maximal yield of ATP per flash is established (Fig. 3, curves d and e). The same amounts of antimycin A cause however an inhibition in the steady state rate of ATP synthesis (ref. 12) in spite of the observation that eventually, after some 20 s of illumination in the presence of valinomycin, ΔpH is build up to a maximal value, scarcely affected by the partial inhibition of electron flow (Fig. 4).

The main conclusion which can be drawn by the study of the transients of photophosphorylation and of Δp formation is concerned with the size of the pool of the high energy state for ATP synthesis. There is no doubt, on the basis of the response of the carotenoid shift and of the quenching of 9-aminoacridine to antimycin, that the membrane potential and the proton concentration gradient across the chromatophore membrane are largely delocalized. This pool of energy is formed cooperatively by all photosynthetic units (13,18), working across the same membrane and facing the same inner compartment, the number of which can be affected by electron flow inhibitors. The same cooperative behaviour is observed in the present experiments for the formation of the high energy state required for the maximal yield of ATP per flash; this indicates that also the high energy precursor of ATP is formed cooperatively by the working photosynthetic units and therefore is largely delocalized as well (12).

The effect of electron transport inhibitors on the steady state rate of phosphorylation, on the other hand, indicates that the kinetic parameters of ATP synthesis are not directly related to the value of Δp, but rather seem somewhat correlated to the number of working photosynthetic units in the average chromatophore (12,13). These observations, as a whole, suggest that a strong kinetic control of ATP-synthetase is superimposed to a chemiosmotic mechanism of energy transduction. This kinetic control seems to operate through a short range interaction between a working photosynthetic unit and the ATPase complexes associated to it, and does not appear to be mediated by the over all delocalized protonmotive force. It can be speculated that active-inactive transition of the ATP synthetase are related with the activation of an ATPase activity, which is now well documented in chromatophores (19,20).

REFERENCES

1. Jagendorf, A. T., and Uribe, E., Proc. Natl. Acad. Sci. U. S. 55:170 (1966).
2. Racker, E., and Stoeckenius, W., J. Biol. Chem. 249:662 (1974).
3. Sone, N., Yoshida, M., Hirata, H., and Kagawa, Y., J. Biol. Chem. 250:7917 (1975)
4. Yoshida, M., Sone, N., Hirata, M., Kagawa, Y., and Ohno, K., Biochem. Biophys. Res. Commun. 67:1295 (1975).
5. Mitchell, P., "Chemiosmotic Coupling and Energy Transduction", Glynn Res. Ltd., Bodmin, 1968.
6. Padan, E., and Rottenberg, H., Eur. J. Biochem. 40:431 (1973).
7. Ort, D. R., Dilley, R. A., and Good, N. E., Biochim. Biophys. Acta 449:95 (1976).
8. Ort, D. R., Dilley, R. A., and Good, N. E., Biochim. Biophys. Acta 449:108 (1976).
9. Ort, D. R., FEBS Lett. 69:81 (1976).
10. Lundin, A., Thore, A., and Baltscheffsky, M., FEBS Lett. 79:73 (1977)
11. Duysens, L. M. N., van Grondelle, R., and del Valle Tescon, S., in "Proc. IV Int. Congress on Photosynthesis" (D. O. Hall et al., eds.), p. 173. Biochemical Society, London, 1978.
12. Baccarini Melandri, A., Casadio, R., and Melandri, B. A., Eur. J. Biochem. 78:389 (1977).
13. Casadio, R., Baccarini Melandri, A., and Melandri, B. A., FEBS Lett. 87:323 (1978).
14. Keister, D. L., and Minton, N. J., Proc. Natl. Acad. Sci. U. S. 63:489 (1969).
15. Baltscheffsky, M., and Lundin, A., Abst. IV Intern. Congr. Photosynt., Reading, p. 17. 1977.
16. Melandri, B. A., Casadio, R., and Baccarini Melandri, A., in "Proc. IV Int. Congress on Photosynthesis" (D. O. Hall et al., eds.), p. 601. Biochemical Society, London, 1978.
17. Packham, N. K., Berriman, J. A., and Jackson, J. B., FEBS Lett. 89:205 (1978).
18. Saphon, S., Jackson, J. B., Lerbs, V., and Vitt, H. T., Biochim. Biophys. Acta 408:58 (1975)
19. Melandri, B. A., and Baccarini Melandri, A., J. Bioenergetics 8:109 (1976).
20. Webster, G. D., and Jackson, J. B., Biochim. Biophys. Acta 503:135 (1978).

ESTIMATION OF THE SURFACE POTENTIAL IN
PHOTOSYNTHETIC MEMBRANES*

Shigeru Itoh
Katsumi Matsuura
Kazumori Masamoto
Mitsuo Nishimura

Department of Biology
Faculty of Science
Kyushu University
Fukuoka, Japan

I. INTRODUCTION

Since the proposal of the chemiosmotic hypothesis (Mitchell, 1966), electrochemical potential gradient of proton across the membrane has been extensively studied. It plays a crucial role in energy coupling between electron transfer and ATP synthesis. The generation and the use of the electrochemical free energy are performed in the membrane which has low permeabilities for proton and other ions. The electrochemical potential is expressed as

$$\mu_i = \mu_i' + RT \ln a_i + Z_i F \varphi,$$

where μ_i' is the potential in the arbitrarily defined reference state, a_i and Z_i are the activity and the valence of the species, i, and φ is the electrical potential. Other symbols have their usual meanings. In the aqueous phases on each side of the membrane, electrochemical potential of the ions of any

---

* This paper was presented in the Symposium as two separate papers by Itoh and by Matsuura, Masamoto, Itoh and Nishimura. Supported by the Ministry of Education, the Toray Science Foundation and the Ito Science Foundation.

species is uniform from the surface of the membrane to the bulk phase far from the surface. However, each term of the electrochemical potential, i. e., chemical potential and electrical potential, is generally not uniform throughout the aqueous phase.

Biological membranes have generally net negative charges immobilized on their surface at physiological pH. According to the Gouy-Chapman diffuse double layer theory (Overbeak,1950, McLaughlin, 1977) a point close to the negatively charged surface is expected to be at a negative electrical potential with reference to the bulk aqueous phase. This electical potential difference is compensated by the difference in the chemical potential for each ion. Activities of cations close to the surface are expected to be higher than those in the bulk phase, while the activities of anions are expected to be lower. The interchange between the electrical and chemical terms of the electrochemical potential at the membrane surface suggests the introduction of significant effects on the membrane reactions since the components on the membrane experience electrical and chemical environments different from those in the bulk phase. Recent studies by Rumberg and Muhle (1976) and by Barber et al. (1977) in chloroplast membrane pointed out the significance of these phenomena.

Change in the chemical term of the electrochemical potential at the surface will change the apparent reactivity of the membrane components with the ionic reagent added in the aqueous phase (Itoh, 1978a, 1978b) since the reactivity will depend on the surface activity but not on the bulk activity of the reagent. This effect was analyzed in the first part of this paper by studying the effect of salts on the reactivity of the primary electron donor of system I, P700, which has a midpoint potential of about +430 mV (Kok, 1961), to ferrocyanide in sonicated spinach chloroplasts. From the surface activity of ferrocyanide, which was estimated by the change in reactivity, surface potential value of the thylakoid membrane was calculated under various conditions.

On the other hand, change in the electrical potential at the surface of one side of the membrane is expected to cause change in the intramembrane electrical field. In the second part of the paper, this effect was studied by the measurement of the shift of carotenoid absorption spectrum which is known to be an intrinsic indicator of the intramembrane electrical field (Schmidt et al., 1972), in the chromatophore membrane of photosynthetic bacterium, Rhodopseudomonas sphaeroides. Surface potential value was estimated from the salt-induced change of the intramembrane electrical field, which was calibrated by the diffusion potential of proton.

The results of these two methods applied to different membrane systems indicate the validity of the use of the Gouy-Chapman theory in the photosynthetic membranes and give reliable and convenient means of probing the surface potential in biological membranes.

II. THEORY

The Gouy-Chapman Theory

Immobilized charges on the membrane surface give rise to a difference in electrical potential at the surface with reference to the bulk aqueous phase. The relation between the net surface charge density, q, and the electrical potential of the surface, ψ_o, is given as follows according to the Gouy-Chapman diffuse double layer theory (Overbeak, 1950, McLaughlin, 1977).

$$q = (\frac{2RT\epsilon}{\pi})^{1/2} C_b^{1/2} \sinh (\frac{ZF\psi_o}{2RT}) \tag{1}$$

where Z and C_b represent valence and the bulk concentration of symmetrical salt. ϵ is the dielectric constant of water.

For low values of ψ_o ($\psi_o < 50/Z$ mV), Eqn. 1 reduces to

$$\psi_o \simeq (\frac{2\pi RT}{F^2 \epsilon})^{1/2} \frac{q}{Z} C_b^{-1/2}. \tag{2}$$

Activity of the i-th ion at the membrane surface, a_{is}, with respect to that at the bulk phase, a_{ib}, can be obtained by the Boltzmann distribution of the ion.

$$a_{is} = a_{ib} \exp (- \frac{Z_i F \psi_o}{RT}) \tag{3}$$

Thus, the cations (anions) of higher valence are expected to be more concentrated at the negatively (positively) charged membrane surface and to affect the ψ_o value even at a low bulk concentration. The concentrations of the anions of higher valence are expected to be very low at the negative surface.

If the surface and the bulk activities of the i-th ion can be measured, ψ_o can be calculated according to Eqn. 3.

Schematic representation of the relation between the surface potential, surface ion activities and surface pH are shown in Fig. 1.

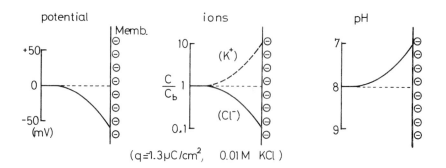

Fig. 1. Schematic representation of the relation between the surface potential, surface activities of ions and surface pH.

III. ESTIMATION OF THE SURFACE POTENTIAL BY THE KINETIC METHOD

Surface Potential and the Apparent Rate Constant

Rate of the reaction between an electron transfer component on the membrane and a redox reagent in the aqueous phase is expected to be proportional to the surface activity of the reagent with the rate constant, $k°$. On the other hand, an apparent rate constant, k, is usually calculated with respect to the concentration of reagent in the bulk phase. Thus, from Eqn. 3, the relation between k and $k°$ can be obtained.

$$k/\delta_b = k° \exp(-ZF\psi_o/RT) \qquad (4)$$

or

$$\ln(k/\delta_b) = \ln k° - ZF\psi_o/RT \qquad (5)$$

Z is the valence of the reagent. δ_b is the activity coefficient of the reagent in the bulk phase. Activity coefficient of the membrane component embedded in the matrix can be assumed to be constant.

In the case of ferrocyanide, numerical substitutions give,

$$\log(k/\delta_b) = \log k° - \frac{(-4)}{60}\psi_o \quad (\text{at } 25°, \psi_o \text{ in mV}) \qquad (6)$$

or

$$\psi_o = -15(\log k° - \log k/\delta_b) \qquad (7)$$

With every 15-mV positive shift of the surface potential, about 10-fold increase of the apparent rate constant is expected as far as the rate-limiting step is the electron transfer at the surface. At relatively low surface potentials in the presence of monovalent salt, a linear numerical approximation equation is obtained by replacing ψ_o in Eqn. 6 with that in Eqn. 2,

$$\log (k/\delta_b) = \log k^\circ - 0.074 \times (-4) \underline{q} \, C_b^{-1/2} \quad (8)$$

where $\underline{q}$ is the net surface charge density expressed in $\mu C/cm^2$, C_b is the bulk concentration of the salt in M.

Eqn. 8 indicates that the plot of log k/δ_b versus $C_b^{-1/2}$ should give a straight line. The value of $\underline{q}$, and hence ψ_o, calculated from the slope of the line will give the electrostatic feature of the surface in the vicinity of the reaction site. On the other hand, k° value, calculated from the intercept, will depend on the nature of both the regeant and the membrane component.

Bulk phase activity coefficient, δ_b, at a given ionic strength, I, can be calculated according to an extension of the Debye-Hückel expression (ref. 8),

$$\log \delta_b = - \frac{0.51 \, z^2 \, \sqrt{I}}{1 + 0.33 \, \underline{A} \, \sqrt{I}} \quad (9)$$

where $\underline{A}$ is the effective ionic radius of the reagent.

Details of the above treatise have been published elsewhere (Itoh, 1978b).

Salt Effect on the Reduction Rate of P700 by Ferrocyanide

Effects of various salts on the apparent rate constant of reaction between oxidized P700 and ferrocyanide were studied. P700 in the sonicated chloroplast membrane prepared from spinach leaves (Itoh, 1978a) was oxidized by excitation light in the presence of 0.2 mM ferrocyanide, and the reduction rate after the cessation of the light was measured. The rate depended on the ferrocyanide concentration and on the concentration of indifferent electrolyte added to the medium (Fig.2). The apparent rate constant was calculated from the reduction rate by subtracting the endogenous rate (the rate without ferrocyanide) which was very low and was neither affected by the additions of salts nor by an electron transfer inhibitor, 3-(3',4'-dichlorophenyl)-1,1-dimethylurea (DCMU).

Fifty- to hundred-times increase in the apparent rate constant was usually observed when salts were added. Divalent

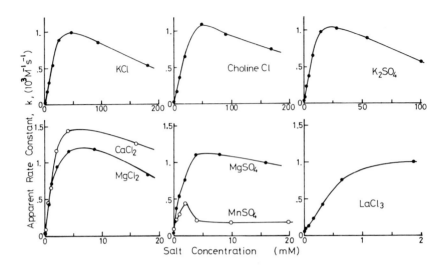

Fig. 2. Effects of various salts on the reduction rate of $P700^+$ by ferrocyanide. Reaction mixture contained 5 mM sodium tricine buffer, pH 7.8, 19 μM DCMU, 8 μM methyl viologen and the sonicated chloroplasts equivalent to 68 μg chlorophyll/ml. The redox state of P700 was monitored by measuring absorption change at 705 nm with a reference at 730 nm. Apparent rate constant was calculated from the rate of $P700^+$ reduction in the presence of 0.2 mM $K_4Fe(CN)_6$ corrected for the endogenous rate.

cation salts were more effective than monovalent ones and increased k in lower concentration ranges. Ca^{2+} was a little more effective than Mg^{2+}. $MnCl_2$ was not effective in the higher concentration range. Trivalent cation salt, $LaCl_3$, increased the rate constant at much lower concentrations than the other salts tested.

It can be concluded that the increase in the apparent rate constant by salts was cation specific (K_2SO_4 was not more effective than KCl). The higher the valence of the cation, the more effective in increasing the k value. These characteristics of the salt effect are explained by considering the surface activities of the cations and ferrocyanide according to the Gouy-Chapman theory. Electrostatic interaction between the negative membrane surface (isoelectric point of the membrane is about pH 4.2 according to Mercer et al., 1955) and ferrocyanide seems to control the reactivity of P700 situated in the membrane. The salt-induced increase of the apparent rate constant can be explained by the increase of the surface activity of ferrocyanide due to the screening of the negative charges on the surface by salt.

Nonelectrostatic interactions between ions and the membrane surface may also affect the reaction rate in some cases as seen with $MnCl_2$. This remains to be studied further.

Calculation of the Surface Potential from the Apparent Rate Constant Change in the Presence of Various Salts

Values of log k/δ_b were plotted against inverse square root of ionic strength from data with KCl and choline chloride (Fig. 3). The plot gave almost identical straight lines within the range tested as predicted by Eqn. 8. From the slope of the line with KCl a surface charge density of -0.84 $\mu C/cm^2$ was calculated. The $k°$ value of 3.1×10^5 $M^{-1}s^{-1}$ was obtained from the intercept at the infinite ionic strength. On the right-hand ordinate of the figure, values for the surface potential corresponding to the values of log k/δ_b calculated using Eqn. 7 were also shown. Thus, membrane surface in the vicinity of P700 molecule had a surface potential of -36 mV at

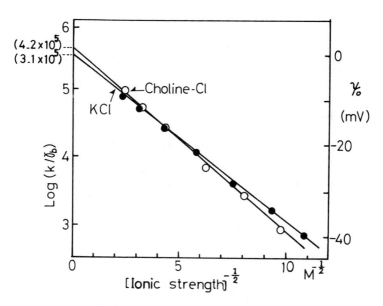

Fig. 3. Plot of log k/δ_b versus inverse square root of ionic strength in the reaction between $P700^+$ and ferrocyanide. The apparent rate constants in the presence of KCl and choline chloride were obtained from data in Fig. 2. In the calculation of the ionic strength, and hence, of the δ_b values, contributions from the buffer and other additives were also included.

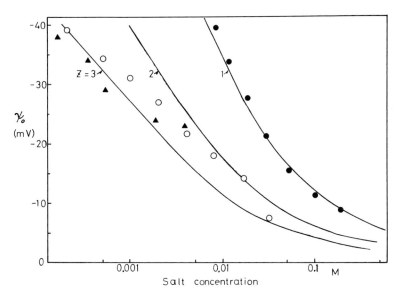

Fig. 4. Values of surface potential calculated from the apparent rate constants in the presence of various salts. Surface potential was calculated from the corresponding data in Fig. 2 using the same $k^°$ value obtained with KCl in Fig. 3. Closed circles, KCl. Open circles, $MgSO_4$. Closed triangles, $LaCl_3$. Solid lines indicate theoretical dependences with q = -0.84 $\mu C/cm^2$ with symmetrical salts.

0.01 M monovalent salt in the sonicated chloroplasts.

Fig. 4 shows the dependence of the surface potential, calculated from the k values in Fig. 2 using Eqn. 7, on the concentrations of KCl, $MgSO_4$ and $LaCl_3$. In the calculations, the value of $k^°$ obtained with KCl was used. The theoretical curves, calculated using the nonlinear equation (Eqn. 1) in the cases of Z-Z symmetrical salts, were also shown. No corrections for the contributions from co-existing buffer, ferrocyanide and other additives were made in the calculation of the curves with Z = 2 and 3. The data with KCl fitted well the theoretical line with z = 1. The data with $MgSO_4$ or $LaCl_3$ also showed a fair agreement with the theoretical lines. Nonideal behavior of the higher valent ions, which may become significant in calculations of both the surface potential and the bulk activity coefficient of ferrocyanide, may be responsible for the quantitative discrepancies.

Analysis of the salt-induced change in the reactivity of P700 to another reductant, ascorbate (Z = -1), gave a similar value of the surface potential (-38 mV at 0.01 M monovalent salt concentration). On the other hand, if the medium pH was

changed, different values of the surface potential were obtained (Itoh, unpublished data), suggesting the change in the surface charge density due to protonation. Similar analyses of the reactivities of cytochrome $\underline{f}$, cytochrome $\underline{b}_{559}$ and the primary acceptor of system II in class II chloroplast membranes to ferricyanide gave the ψ_o values of -66 mV, -69 mV and -60 mV, respectively, at I = 0.01 M. These values probably indicate the electrostatic characteristics of the different parts of the outer surface of the chloroplast membrane.
A little lower value of surface potential in the vicinity of P700 in sonicated chloroplasts obtained in the present study may represent the electrostatic feature of the inner surface.

Analysis of the reactivity of various membrane components differently localized in the energy transducing membranes to the ionic reagents according to the method presented in this study will give the electrostatic characteristics of the various parts of the membrane surface.

IV. ESTIMATION OF SURFACE POTENTIAL BY INTRINSIC PROBE FOR INTRAMEMBRANE ELECTRICAL FIELD IN PHOTOSYNTHETIC MEMBRANE OF BACTERIA

The surface potential was estimated by measuring the change in the intramembrane electrical field in chromatophores from a photosynthetic bacterium, Rhodopseudomonas sphaeroides (Matsuura, Masamoto, Itoh and Nishimura, 1978, submitted to Biochim. Biophys. Acta). In this bacterium, the spectral shift of carotenoid has been used as a good indicator of the membrane potential (Jackson and Crofts, 1969; Matsuura and Nishimura, 1977). The carotenoid shift is considered to respond to intramembrane electrical field (Schmidt et al., 1972). The shift can be used to estimate the surface potential change.

The Relationship between the Surface Potential and the Intramembrane Electrical Field

The relationship between membrane potential (potential difference between bulk aqueous phases), surface potential and intramembrane electrical field is schematically shown in Fig. 5. If the membrane potential is kept constant, a change in the surface potential on one side should cause a change of the same magnitude in the potential difference between two membrane surfaces. If the field in the membrane is uniform, the potential difference between surfaces is proportional to the intramembrane electrical field.

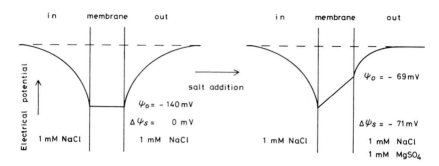

Fig. 5. Schematic diagram of the potential profile across charged membrane before and after a salt addition to the outer phase. The charge density is -2.8 $\mu C/cm^2$, a value estimated in the chromatophore membrane at pH 7.8 as shown in the text. $\Delta\psi_s$ is the potential difference between two surfaces of the membrane.

Estimation of the Surface Potential by the Carotenoid Absorption Change

Chromatophores were prepared in a low-salt buffer from photosynthetically grown R.sphaeroides. Concentrated salt solution was added to the chromatophore suspension in order to change the surface potential. Only the outside surface potential will change under the present experimental conditions, due to the low permeability of the membrane to ions. Carbonylcyanide m-chlorophenylhydrazone (CCCP) was added before the salt addition to maintain the uniform electrochemical potential of H^+. pH changes in the outer solution by the salt additions were less than 0.1 and the changes in the intravesicular pH were also estimated to be small. Therefore, we can assume that the membrane potential was kept almost constant before and after the salt additions. Under this situation, the carotenoid shift induced by the salt addition can be ascribed to the surface potential change on the outside of vesicles.

Fig. 6 shows the carotenoid change by NaCl and $MgSO_4$ additions to chromatophore suspension at various pH. At neutral pH, the salt additon induced a decrease in the difference absorbance (488-minus-506 nm) corresponding to the blue shift of the carotenoid spectrum. The blue shift indicates the outside positive potential change as shown by the diffusion potential of K^+ or H^+ (Jackson and Crofts, 1969; Matsuura and Nishimura, 1977). This outside-positive potential change can be explained by the salt-induced decrease of negative surface potential on the outer surface. At about pH 5.2 the salt

Surface Potential in Photosynthetic Membranes

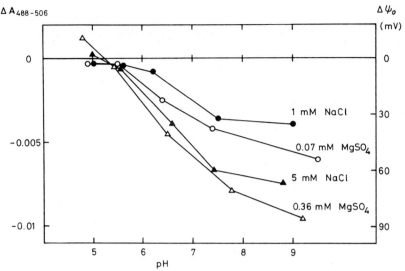

Fig. 6. pH dependence of the carotenoid absorbance change induced by additions of MgSO₄ and NaCl. Chromatophores (10 μM bacteriochrolophyll) were suspended in distilled water and pH was adjusted with H_2SO_4 or KOH before the salt additions. 1 μM CCCP was present. The absorbance changes were measured with the dual wavelength mode (488-minus-506 nm). Change in the potential corresponding to the carotenoid change was calibrated by the diffusion potential of H^+ and appears on the right-side ordinate.

addition induced no change in the carotenoid spectrum, suggesting the electrical neutrality of the membrane surface. This pH is almost the same as the isoelectric pH of the <u>Chromatium</u> chromatophore reported by Case and Parson (1973). At pH's higher than 5.2, the net surface charge and the surface potential should become negative. The similarity between the pH dependences of the potential changes induced by NaCl and MgSO₄ additions supports the idea that the carotenoid change is caused by the change in surface potential and not by the diffusion potential across the membrane.

The absorbance change was dependent on the valence of cation as well as the concentration of salt added at a given pH. Fig. 7 shows the carotenoid change induced by the addition of various salts of mono- and divalent ions at pH 7.8. Except for the valence of cations, little ionic-species-dependent difference was observed. The salts of divalent cations were effective at concentrations lower than monovalent ones by a factor of about 50. When 0.1 mM MgSO₄ (MgCl₂, CaCl₂) or 5 mM NaCl (KCl, Na₂SO₄) were added, the potential at the outside surface

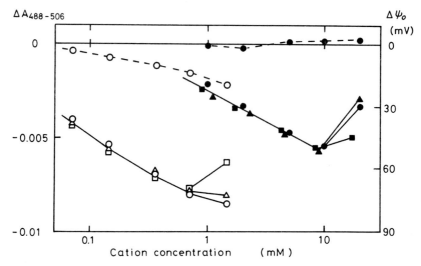

Fig. 7. Dependence of carotenoid absorbance change on concentration of salts added. Chromatophores were suspended in 0.6 mM Tricine-NaOH buffer, pH 7.8. o—o, $MgSO_4$; △—△, $MgCl_2$; □—□, $CaCl_2$; o--o, $MgSO_4$ in the presence of 10 mM Na_2SO_4; ●—●, NaCl; ▲—▲, Na_2SO_4; ■—■, KCl; ●--●, NaCl in the presence of 5 mM $MgCl_2$.

rose by about 45 mV. The salt-induced change became small in the presence of another salt. These observations are expected from the Gouy-Chapman theory.

Using a set of concentrations of mono- and divalent salts which give the same surface potential, we can obtain two equations for the surface charge density and the surface potential according to Eqn. 1,

$$\psi_o = \pm \frac{RT}{F} \cosh^{-1}(\frac{C'}{2C''} - 1)$$

$$q = \pm \sqrt{\frac{(C'^2 - 4 C'' C') RT\epsilon}{2 C'' \pi}}$$

where C' and C" are concentrations of mono- and divalent salts, respectively, which give the same surface potential. In Fig.7 5 mM NaCl gave the same potential change as 0.1 mM $MgSO_4$. From these concentrations we obtained the surface potential of -98 mV and the charge density of -2.8 $\mu C/cm^2$. These values explain well the dependence of the potential change on the concentration of added divalent cations (Fig. 7) with and without Na_2SO_4. If the surface charge density is not affected by the salt concentration, the surface potential in any ionic solution can be calculated. For example, in 0.1 M monovalent

Surface Potential in Photosynthetic Membranes

salt solution, the surface potential was calculated to be -36 mV.

A part of the electrical field change induced by illumination in chromatophores may be caused by the surface potential change. If the ion concentration or the surface charge density is changed by illumination, the surface potential will change. The surface charge density on the inner surface of chromatophore membrane will change significantly by the energization which accompanies the pH change.

V. CONCLUSION

As shown in the present study in the photosynthetic membranes of chloroplasts and bacteria, immobilized charges on the membrane surface change the electrical potential at the surface and result in the distributions of ions at the surface different from those in the bulk phase. The extents of surface potential estimated were tens of millivolts in both membranes. If one can measure either the electrical or chemical term of the electrochemical potential at the surface, the other term can be calculated as shown in this paper.

The interchange between these two terms at the membrane surface does not produce any free energy change. However, the processes which are affected primarily by one of these terms will respond to the interchange. The change in the chemical potential as seen in the surface-activity change of ferrocyanide will also determine the extent of protonation of the surface groups, which depends on the surface pH. The surface pH will regulate the energy-transducing machinery in the membrane. Fluxes of ions and the rates of other processes coupled to ion translocation such as ATP synthesis or the rate or elctron transfer will depend not only on available amount of free energy but also the concentration of ions at the surface. Then they will be affected by the surface potential. On the other hand, the change in the intramembrane electrical field induced by the surface potential change will induce a shift of redox equilibrium between the membrane components (Hinkel and Mitchell, 1970, Matsuura and Nishimura, 1978).

It should be noted that both the electrical and chemical terms of the electrochemical potential must be measured at the same point in the aqueous phases on either side of the membrane to get the correct electrochemical potential difference across the membrane. When the surface potentials of the two membrane surfaces are different, the value of electical potential difference between the membrane surfaces such as that measured by the intramembrane probes, cannot be used in conjunction with the value of chemical potential difference such as pH difference between the bulk phases.

REFERENCES

1 Barber, J., Mills, J. D. and Love, J. (1977). FEBS Lett. 74, 174-181.
2 Case, G. D. and Parson, W. W. (1973). Biochim. Biophys. Acta 292, 677-684.
3 Hinkel, P. and Mitchell, P. (1970). Bioenergetics 1, 45-60
4 Itoh, S. (1978a). Plant Cell Physiol. 19, 149-166.
5 Itoh, S. (1978b). Biochim. Biophys. Acta, in press.
6 Jackson, J. B. and Crofts, A. R. (1969). FEBS Lett. 4, 185-189.
7 Kok, B. (1961). Biochim. Biophys. Acta 48, 527-533.
8 Lange's Handbook of Chemistry (1973). (Dean, J. A., ed.), Section 5. McGraw-Hill Inc., New York.
9 Matsuura, K. and Nishimura, M. (1977). Biochim. Biophys. Acta 459, 483-491.
10 Matsuura, K. and Nishimura, M. (1978). J. Biochem. 84, 539-546.
11 McLaughlin, S. (1977). In"Current Topics in Membrane and Transport,"vol. 9 (Bonner, F. and Kleinzeller, A. eds.), pp. 71-144. Academic Press, New York.
12 Mercer, F. V., Hodge, A. J., Hope, A. B. and McLean, J. D. (1955). Austral. J. Biol. Sci. 8, 1-18.
13 Mitchell, P. (1966)."Chemiosmotic Coupling in Oxidative and Photosynthetic Phosphorylation."Glynn Research, Bodmin, Cornwall.
14 Overbeak, J. T. G. (1950).In"Colloid Science"(Kruyt, H. R., ed.),vol. I, pp. 115-193, Elsevier, Amsterdam.
15 Rumberg, B. and Muhle, H. (1976).Bioelectrochem. Bioenergetics 3, 393-403.
16 Schmidt, S., Reich, R. and Witt, H. T. (1972).In"Proceedings of the IInd International Congress on Photosynthesis Research"(Forti, G., Avron, M. and Melandri A. eds.),pp. 1087-1095. The Hague.

CONVERSION OF Ca^{2+}-ATPase ACTIVITY INTO Mg^{2+}- AND Mn^{2+}-ATPase ACTIVITIES WITH COUPLING FACTOR PURIFIED FROM *RHODOSPIRILLUM RUBRUM* CHROMATOPHORES[1]

Gilbu Soe, Nozomu Nishi, Tomisaburo Kakuno
Jinpei Yamashita and Takekazu Horio

Division of Enzymology
Institute for Protein Research
Osaka University
Suita-shi, Osaka

I. INTRODUCTION

It is known that chromatophores from light-grown cells of *Rhodospirillum rubrum* show ATPase and ATP-Pi exchange activities in the dark as partial reactions of the ATP formation coupled with the cyclic electron flow in the light (Baltscheffsky et al., 1958; Horio et al., 1962; Horio et al., 1964; Horio et al., 1965; Horiuti et al., 1968), the key enzyme for the activities described above being the coupling factor bound to the chromatophore membrane (Avron, 1963; Vambutas et al., 1965; Johansson, 1972).

Johansson et al. (1973) reported that the coupling factor, if purified from acetone-dried powder of chromatophores according to the method of Baccarini-Melandri et al. (1971), shows ATPase activity in the presence of Ca^{2+}, but not in the presence of Mg^{2+}. Thereafter, the purified coupling factor is called Ca^{2+}-ATPase.

Earlier, Hosoi et al. (1975) studied effects of several pH indicators on various activities with chromatophores, and found that some of them including 2,4-dinitrophenol and ethyl orange remarkably stimulate the Mg^{2+}-ATPase activity. Recently, Soe et al. (1978) studied effects of pH indicators and

---

[1]Supported by grant (No. 311909) from the Ministry of Education, Science and Culture of Japan.

detergents on the activity of Ca^{2+}-ATPase. Their findings are as follows.

1) Appropriate pH indicators can convert Ca^{2+}-ATPase into Mg^{2+}- and Mn^{2+}-ATPase. The efficiencies for the conversion are ethyl orange > tropaeolin 000 ≥ metanil yellow > tropaeolin 00 > ethyl red ≥ bromthymol blue, their pK_a's ranging from 1.3 to 12.0.

2) Appropriate detergents can induce a similar conversion. Dodecylsulfonate ($C_{12}H_{25}SO_3Na$) is most effective, whereas dodecylpyridinium chloride is moderately effective. The concentrations of these effective detergents for the maximum conversion are significantly lower than their critical micell concentrations. With $C_nH_{2n+1}SO_3Na$, the highest extent of the conversion into Mg^{2+}-ATPase is obtained with the C_{11} and C_{12}, whereas that into Mn^{2+}-ATPase with the C_{14}.

3) 2,4-Dinitrophenol alone stimulates approximately 2-fold the Ca^{2+}-ATPase activity, but it is not effective in the conversion. However, in the presence of dodecylpyridinium chloride, it is remarkably effective. Methyl red and ethyl red show similar effects.

They speculated the conversion mechanism of Ca^{2+}-ATPase into Mg^{2+}- and Mn^{2+}-ATPase as follows.

In the presence of Ca^{2+}, the ATP-hydrolyzing reaction can proceed with repeated turnover because of the reaction (Ca^{2+} + Pi + H^+ → $CaHPO_4$). On the other hand, the ATP-hydrolyzing reaction in the presence of Mg^{2+} or Mn^{2+} can not proceed with repeated turnover, because the proton produced by a single turnover of ATP hydrolysis is tightly held by the catalytic site buried in a hydrophobic region of the ATPase. When appropriate pH indicators or detergents are present, the tertiary structure of the hydrophobic region is modified so as to form a channel through which the proton can leak out of the catalytic site; thus, the ATP-hydrolyzing reaction can proceed with repeated turnover. If the channel is partially opened so that other appropriate pH indicators such as 2,4-dinitrophenol are accessible to the site, the leakage of the proton is stimulated by the function of the dyes as proton carriers; thus, the Mg^{2+}- and Mn^{2+}-ATPase activities are stimulated.

The present report deals with the isolation from chromatophores of a factor capable of converting Ca^{2+}-ATPase into Mg^{2+}- and Mn^{2+}-ATPase (conversion factor).

II. MATERIALS AND METHODS

See Soe et al. (1978).

III. RESULTS

Fig. 1 shows effects of acetone concentrations on various activities of chromatophores. Chromatophore suspensions in 0.1 M Tris buffer (pH 8.0) were supplemented with indicated concentrations of acetone (final $A_{880nm} = 50$), sonicated for 90 sec and then centrifugally separated into the precipitates and supernatants. The resulting precipitates were centrifugally washed with the buffer. The washings thus obtained were mixed to the supernatants. At 40% acetone with the washed precipitates (treated chromatophores), the ATP formation (O) and

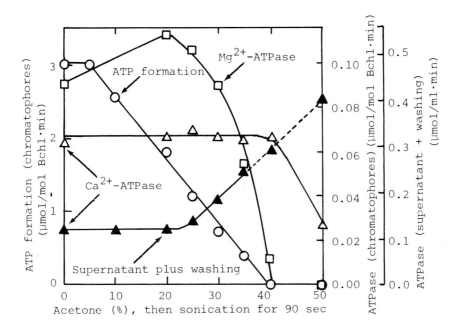

Fig. 1. Effect of acetone treatment of chromatophores on their various activities.

the Mg^{2+}-ATPase activity (□) were completely depressed, whereas the Ca^{2+}-ATPase activity (△) was hardly influenced. With the supernatants plus the washings, the Ca^{2+}-ATPase activity (▲) increased with increasing concentrations of acetone from 25% to 50%, indicating that Ca^{2+}-ATPase was solubilized from chromatophores. It was otherwise determined that the enzyme once solubilized was precipitated when the acetone concentrations were higher than 35%; thus, the precipitated enzyme was solubilized by the washings (▲---▲). This suggests that a factor capable of binding the enzyme to the chromatophore membrane was removed in the acetone solutions. In fact, conversion factor was solubilized in 40-50% acetone with high efficiencies.

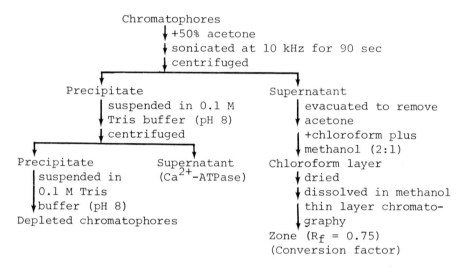

Fig. 2. Purification procedure of conversion factor, Ca^{2+}-ATPase and depleted chromatophores from intact chromatophores.

Fig. 2 shows the scheme for purification of conversion factor and Ca^{2+}-ATPase from chromatophores. The Ca^{2+}-ATPase thus obtained had the same specific activity as the enzyme obtained at the last step of the purification procedure by Baccarini-Melandri et al. (1971) (approximately 10 µmol Pi liberated from ATP/mg protein·min), composed of five different subunits of 54,000, 50,000, 32,000, 13,000 and 7,500 daltons in accordance with Johansson et al. (1975). Thin layer chromatography was carried out on 2-mm thick silica gel plates with chloroform/methanol/water (130/50/8 in v). The zone of R_f = 0.75 was scraped off, vacuum-dried and dissolved in methanol. The absorbance spectrum of the resulting solution (conversion factor) has two main peaks at 236 and 410 nm, a shoulder at 275 nm, and three minor peaks at 500, 540 and 580 nm (Fig. 3). In a preliminary study, the conversion factor thus obtained does not contain phospholipid, iron and protein.

Effects of concentrations of conversion factor on the activities of Ca^{2+}-ATPase were examined (Fig. 4). The Mg^{2+}- and Mn^{2+}-ATPase activities increased with its increasing concentrations, almost in parallel with the decrease of the Ca^{2+}-

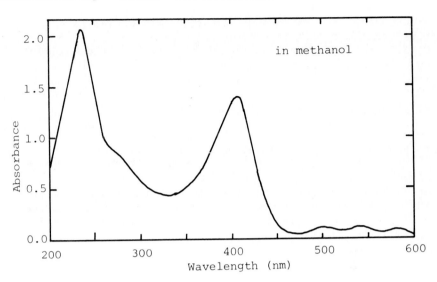

Fig. 3. Absorbance spectrum of conversion factor.

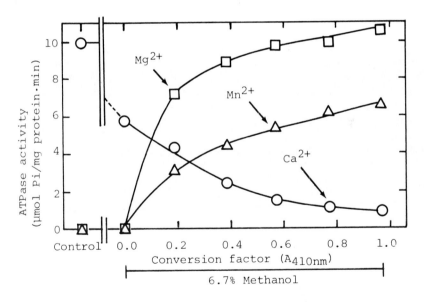

Fig. 4. Effect of concentrations of conversion factor on conversion of Ca^{2+}-ATPase into Mg^{2+}- and Mn^{2+}-ATPase.

ATPase activity.

Acetone, ethanol and methanol can convert Ca^{2+}-ATPase into Mg^{2+}- and Mn^{2+}-ATPase if their concentrations are higher than

7%. However, the extents of the Mg^{2+}- and Mn^{2+}-ATPase activities induced by 10% the respective organic solvent in addition of 1.3% methanol are significantly lower than those induced by conversion factor (A_{410nm} = 0.2) plus 1.3% methanol.

IV. DISCUSSION

Konings et al.(1973), Oren et al. (1977), and Schneider et al. (1978) reported the solubilization of Mg^{2+}-ATPase from chromatophores by sonication in the presence of dithiothreitol and ethylenediaminetetraacetate or with the aid of Triton X-100. It is now under investigations whether the Mg^{2+}-ATPase described above is a complex of Ca^{2+}-ATPase and conversion factor.

REFERENCES

Avron, M. (1963). Biochim. Biophys. Acta 77:699.
Baccarini-Melandri, A., and Melandri, B.A. (1971). In "Methods in Enzymology" (S.P.Colowick and N.O.Kaplan, ed.), vol. 23, p. 556. Academic Press, New York.
Baltscheffsky, H., and Baltscheffsky, M. (1958). Acta Chem. Scand. 12:1333.
Horio, T., and Kamen, M.D. (1962). Biochemistry 1:44.
Horio, T., Nishikawa, K., Katsumata, M., and Yamashita, J. (1965). Biochim. Biophys. Acta 94:371.
Horio, T., Nishikawa, K., and Yamashita, J. (1964). J. Biochem. 55:327.
Horiuti, T., Nishikawa, K., and Horio, T. (1968). J. Biochem. 64:577.
Hosoi, K., Soe, G., Kakuno, T., and Horio, T. (1975). J. Biochem. 78:1331.
Johansson, B.C. (1972). FEBS Lett. 20:339.
Johansson, B.C., and Baltscheffsky, M. (1975). FEBS Lett. 53:221.
Johansson, B.C., Baltscheffsky, M., Baltscheffsky, H., Baccarini-Melandri, A., and Melandri, B.A. (1973) Eur. J. Biochem. 10:109.
Konings,A.W.T., and Guillory, R.J. (1973). J. Biol. Chem. 248:1045.
Oren, R., and Gromet-Elhanan, Z. (1977). FEBS Lett. 79:147.
Schneider, E., Schwülera, U., Müller, H.W., and Dose, K. (1978). FEBS Lett. 87:257.
Soe, G., Nishi, N., Kakuno, T., and Yamashita, J. (1978). J. Biochem. 84:805.
Vambutas, V.K., and Racker, E. (1965). J. Biol. Chem. 242:2660.

DIVALENT METAL IONS AS MODIFIERS
OF THE NONLINEAR INITIAL RATES OF ATPase ACTIVITY
IN PHOTOSYNTHETIC COUPLING FACTORS

C. Carmeli[1]
Y. Lifshitz
M. Gutman

Department of Biochemistry
The George S. Wise Center of Life Sciences
Tel Aviv University
Tel Aviv, Israel

A nonlinear initial rate of ATP hydrolysis is obtained on the addition of divalent metal ion-ATP complex to a heat activated isolated coupling factor 1 from chloroplasts. The acceleration of the initial rate follows a first order kinetics. The observed first order kinetic constant (Kobs) changes with the concentration of the substrate reaching half maximal value at the Km for ATP hydrolysis. Preincubation of the enzyme with the divalent metal ions decreases the Kobs from 1 sec^{-1} to 0.04 sec^{-1}. Saturation of the divalent metal ion effect was obtained at the micromolar range when $CaCl_2$ was used. Similar results were obtained with coupling factor from the photosynthetic bacteria chromatium vinosum.

It is suggested that some early stages in ATP hydrolysis induce a conformational change in the enzyme which is expressed as kinetic changes. Binding of divalent metal ions in the absence of ATP slows down this change.

I. INTRODUCTION

The terminal steps of photophosphorylation are catalyzed by coupling factor 1 (CF_1) in chloroplasts. In the presence of

[1]Present address: Membrane Bioenergetics Group, Lawrence Berkeley Laboratory, University of California, Berkeley, California, USA.

sulfhydryl reagents the enzyme can be light induced to catalyze ATP hydrolysis (1). The acquired reversibility is probably due to a conformational change (2,3,4) which is caused by the light induced generation of an electrochemical potential of protons across the chloroplast membrane. It was suggested (5,6) that these conformational changes in the coupling factor are the way through which the energy released during oxidation reduction drives ATP synthesis.

The isolated soluble CF_1 is activated by heat treatment (7) or by a mild tryptic digestion (8) to catalyze ATPase activity. It was suggested that the soluble enzyme also undergoes changes in conformation. The slow time-dependent changes in Ki of Co^{3+}-phenathroline-ATP complex (9) is an indicator for conformational changes caused by the binding of the reagent to CF_1. Slow changes in the intrinsic fluorescence of CF_1 and the fluorescence of ethano-ADP upon its binding to the enzyme (10) might reflect the same process. We have reported (11) that ATPase activity in soluble CF_1 undergoes a transient state on addition of divalent metal ions-ATP as substrate. Similar results were obtained in ATPase from yeast mitochondria (12). The data presented here indicate that the kinetic changes could be caused by conformational changes which take place during the early events in ATP hydrolysis. These changes are modified by divalent metal ions.

II. MATERIALS AND METHODS

Coupling factor 1 was prepared from lettuce chloroplasts according to the method used for spinach chloroplasts (13) and stored as ammonium sulfate suspension at 4°. Following heat activation (7) the protein catalyzed ATP hydrolysis at a rate of 35 μmoles x mg^{-1} x min^{-1}, indicating high purity of the enzyme.

The stored enzyme was freed from ammonium sulfate by passage on a Sephadex G-50 column (1 x 25 cm) equilibrated and eluted with 80mM tricine-NaOH (pH 8) and 1mM EDTA. The enzyme was heat activated (7) then passed on a second Sephadex column equilibrated with 0.1mM EDTA and 5mM tricine-NaOH (pH 8).

ATPase activity was measured spectrophotometrically using cresol red for monitoring the acidification of the low buffered solution. Stoichiometry of 0.96 moles proton release per mole ATP hydrolyzed at pH 8 (14) was used for quantitation of the catalytic rate. The difference in the absorbance changes at 580/630 nm were monitored by Amico DW-2 spectrophotometer equipped with a stopped flow apparatus at time resolution of 5 msec. The reaction was started by mixing equal volumes of $[Ca-ATP]^{2-}$ at the indicated concentration and 50 µM cresol red (pH 8) with 80 µg per ml of CF_1 in 1mM tricine-NaOH (pH 8) and 0.1mM EDTA at 37°. Protein concentration was determined according to the method of Hartree (15).

III. RESULTS AND DISCUSSION

A. Initial Rates of ATPase Activity In Chromatium vinosum Coupling Factor

ATPase activity in coupling factor from Chromatium vinosum was assayed in order to determine whether the nonlinear initial rate we have earlier observed (11) is unique to the chloroplasts coupling factor. It seems that it is a more general phenomenon since an acceleration of the initial rate was also observed in trypsin activated coupling factor isolated from these photosynthetic bacteria (Fig. 1). In order to determine the kinetic order and the kinetic constant of this change we

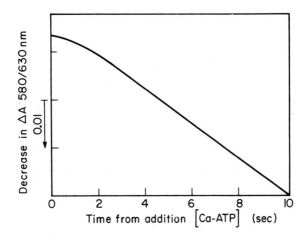

Fig. 1. A nonlinear initial rate of ATPase activity in coupling factor from Chromatium vinosum. Coupling factor was isolated and trypsin activated according to a published procedure (16). The reaction was started by addition of the substrate to the enzyme solution in a stopped flow apparatus and monitored as absorbancy changes of the pH indicator cresol red caused by acidification of the medium during ATP_2 hydrolysis. The reaction mixture contained: 1.2 mM $[Ca-ATP]^{2-}$, 0.5 mM free Ca^{2+}, 50 µM EDTA, 25 µM cresol red, 1 mM tricine-NaOH (pH 8) at 37°.

assumed that the enzyme can be at two interchangeable forms a more active and less active one. The rate of activity at a given time (V_t) represents the concentration of the more active form of the enzyme. A first order reaction was indicated from the linear line obtained by plotting the log of $Vmax-V_t$ (Fig. 2) when Vmax is the maximal rate which is assumed to represent the fully activated enzyme. A first order kinetic constant of 0.69 sec^{-1} was calculated from these data.

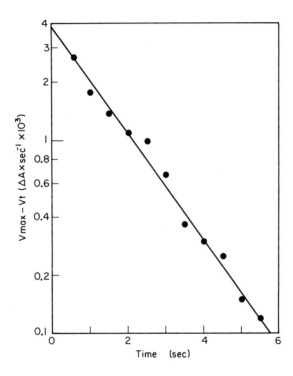

Fig. 2. Determination of the kinetic order of the change in initial rate of ATPase activity in coupling factor from Chromatium vinosum. The maximal rate (Vmax) and the rate at a given time (V_t) were calculated from spectroscopic data similar to those which are shown in Fig. 1.

B. Control of the Change in the Initial Rates Of ATPase Activity in CF_1 by Divalent Metal Ions

Similar analysis showed that the acceleration of the rate of ATPase activity in heat treated CF_1 followed a first order rate having kinetic constant of 1.1 sec^{-1}. However, when the enzyme was preincubated with $CaCl_2$ the length of acceleration period increased (Fig. 3), while the observed rate constant decreased. The effect was concentration dependent; however the exact range of concentration was difficult to determine. In the experiment enzyme solution containing 100 μM EDTA was titrated with $CaCl_2$. It seems from the data (Fig. 3) that some of the EDTA was chelated with metal ions impurities in the solution. Full saturation of the effect was reached within 15 μM of $CaCl_2$. However, following preincubation of the enzyme for 1.5 h saturation was reached approximately within 3μM.

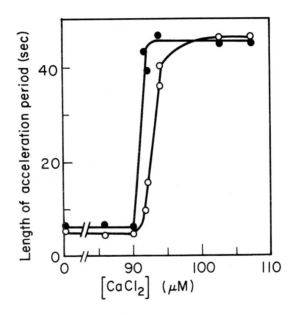

Fig. 3. The effect of $CaCl_2$ on the length of the acceleration period of ATPase activity in CF_1. Heat activated CF_1 was preincubated with various concentrations of $CaCl_2$ before 2.5 mM $[Ca-ATP]^{2-}$ was added. The reaction was measured 5 min (-O-) or 1.5 h (-●-) after the addition of $CaCl_2$.

The time course of the increase in the length of acceleration period in the presence of $CaCl_2$ is given in

Fig. 4. It can be seen that the change proceeded slowly at 1 μM $CaCl_2$, but it was considerably faster at higher concentrations of $CaCl_2$ as can be expected from the data presented in Fig. 3.

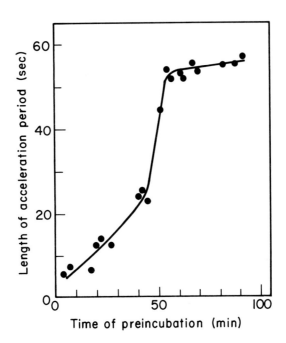

Fig. 4. Time course of the increase in the length of the acceleration period of ATPase activity in CF_1. Heat activated CF_1 was preincubated with approximately 1 μM free $CaCl_2$. ATPase activity was assayed at the indicated time intervals of preincubation as described under Fig. 3.

The effect of $CaCl_2$ on the length of the acceleration period is reversible. Thus, addition of EDTA to enzyme solution which was preincubated with $CaCl_2$ caused a decrease in the length of the acceleration period (Fig. 5) probably because EDTA greatly decreased the concentration of free Ca^{2+} ions in the solution. It should be pointed out that other divalent metal ions such as Mg^{2+} and Mn^{2+} which were tested had similar effects to that of Ca^{2+} ions.

Divalent Metal Ions as Modifiers

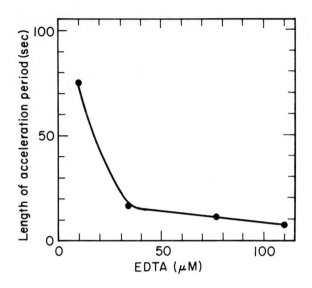

Fig. 5. Decrease in the length of the acceleration period by addition of EDTA to enzyme which was preincubated with $CaCl_2$. CF_1 was preincubated for 30 min with 50 µM $CaCl_2$. The indicated concentrations of EDTA were added before ATPase activity was assayed as indicated under Fig. 3.

A decrease in the concentration of Ca^{2+} ions after preincubation was also achieved by dilution of the enzyme solution. This treatment also resulted in a decrease in the length of the acceleration period.

Binding of divalent metal ions to the enzyme in the absence of ATP decreased K_{obs} to 0.047 sec^{-1} (Table I).

TABLE I. Effect of Free Ca^{2+} Ion in Preincubation and in the Reaction on the K_{obs} of the Acceleration of ATPase Activity

Free Ca^{2+} Present in		$[Ca-ATP]^{2-}$	
		1.5 mM	2 mM
	Concentration	$K_{obs} (\text{sec}^{-1})$	$K_{obs} (\text{sec}^{-1})$
Reaction	None	0.45	0.8
Reaction	10 mM	0.37	0.68
Reaction	15 mM	0.23	0.5
Preincubation	50 µM	0.047	0.047

As shown in Fig. 3 the concentration of $CaCl_2$ which gave half maximal effect was in the micromolar range similar to the dissociation constant for binding of divalent metal ions to the enzyme. This value is significantly lower than the Ki for competitive inhibition of free Ca^{2+} ions which is 7 mM (17). Although the difference in dissociation constants could be interpreted as indicating two different sites of binding, it is also possible that binding of Ca^{2+} ions to the active site induces the decrease in Kobs of the change in the initial rate. The dissociation constant could be changed as a result of the conformational change which takes place during the early events of ATP hydrolysis. Indeed, in the presence of ATP, the concentrations of free Ca^{2+} ions required to decrease Kobs of the change in initial rates were at the range of Ki for competitive inhibition (Table I).

C. The Effect of Substrate on the Observed Kinetic Constant of the Change in the Initial Rate of ATPase Activity in CF_1

One way to explain the changes in the initial rates involves an assumed existence of two interchangeable forms of the enzyme, a more active (E') and a less active (E) one (Equation 1.1). Most of the heat activated CF_1 is in the less active form.

$$E \rightleftarrows E' \qquad (1.1)$$

$$E' + S \rightleftarrows E'S \rightarrow E' + P \qquad (1.2)$$

$$E + M^{2+} \rightleftarrows E \cdot M \qquad (1.3)$$

When substrate (S), which was in these experiments $[Ca-ATP]^{2-}$, binds to the more active form of the enzyme, it shifts the equilibrium toward higher concentrations of E' (1.2). The observed rate of the change in the rate of activity represents the rate of the conversion of E to E'. The binding of the substrate results also in ATP hydrolysis (equation 1.2). Binding of divalent metal ion (M^{2+}) to the enzyme prior to the addition of ATP shifts the equilibrium toward a population of less active enzyme to which divalent metal ions are bound (E·M). The change in rate of activity under this condition will depend on the rate of change of E·M to $E + M^{2+}$ (equation 1.3).

The site of binding of divalent metal ions could be the site of binding of Mn^{2+} to the enzyme as determined by us by the method of EPR spectroscopy (17). It seems that the site of binding of the substrate could be the active site of the enzyme. This is indicated by the finding that the observed rate constant for the change in the initial rate depends on the concentration of substrate (Fig. 6).

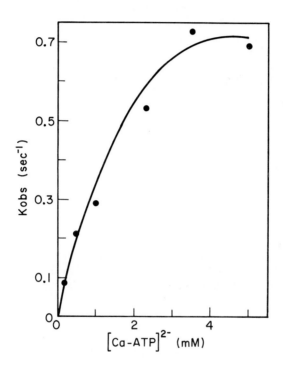

Fig. 6. The dependence of the observed rate constant of the change in initial rate of ATPase activity on the concentration of substrate. The rates of ATPase activity were measured at the indicated concentrations of $[CaATP]^{2-}$ in the presence of 0.1 mM free Ca^{2+} ions. Values of Kobs were calculated as in Fig. 2.

The fact that the concentration of substrate which gave half maximal Kobs value is similar to the Km for the hydrolytic activity favors the suggestion that the change occurs through binding of the substrate to the active site of the enzyme.

In the absence of divalent metal ions ATP did not shift the enzyme to its more active form. This could indicate that it is not the binding of $[M \cdot ATP]^{2-}$ itself but rather some stage or intermediate during the hydrolysis of ATP which causes a change to the more active form of the enzyme. Indeed $[M \cdot ADP]^{1-}$ which could be considered as a non hydrolyzable analogue of ATP also did not change the state of activity of the enzyme. It would be expected that $[M \cdot ADP]^{1-}$ will cause the change if it was caused only by the binding of the substrate to the enzyme. If this were the case, the formulation of the process of activation would have to be changed as the following:

$$E + S \rightleftarrows ES \rightarrow E' + P \qquad (2.1)$$

$$E' + S \rightleftarrows E'S \rightarrow E' + P \qquad (2.2)$$

$$E + M^{2+} \rightleftarrows E \cdot M \qquad (2.3)$$

Equations 2.1 and 2.3 instead of 1.1 and 1.3 in the previous formulation represent the control processes while 2.2 represents the steady state catalytic activity. The slow hydrolysis of ATP by the less active enzyme will be accompanied by the conversion of E to E' (2.1). The possible existence of a conformational change induced by ATP hydrolysis is analogous to the light induced conformational changes observed in the membrane bound CF_1. If kinetically adequate this finding could support the suggested (5,6) role of conformational changes in energy transduction. Further experiments designed to verify the various kinetic and dissociation constants are presently under way.

REFERENCES

1. Petrack, B., Carston, A., Sheppy, F. and Farron, F. (1965). J. Biol. Chem. 248, 2049-2055.
2. Ryrie, I. J. and Jagendorf, A. T (1971). J. Biol. Chem. 246: 582.
3. McCarty, R. E. and Fagan, J. (1973). Biochemistry 12,1503.
4. Harris, D. A. and Slater, E. C. (1975). Biochem. Biophys. Acta 387, 335-348.
5. Boyer, P. O. (1977). Biophys. Biochem. Acta 13, 289-301.
6. Slater, E. C. (1974). Biophys. Biochem. Acta 13, 379-384.
7. Farron, F. and Racker, E. (1970). Biochemistry 9, 3829-3836.
8. Vambutas, V. K. and Racker, E (1965). J. Biol. Chem. 240, 2660-2667.
9. Werber, M. M., Danchin, A., Hochman, Y., Carmeli, C. and Lanir, A. (1977) in: Metal-ligand interactions in organic chemistry and biochemistry, Part 1, pp. 283-290 (Pullman, B. and Goldblum, N. eds., Reidel Publishing Co., Dvidsecht-Holland).

10. Girault, G. and Galmiche, J. M. (1977). J. Eur. Biochem. 77, 501-510.
11. Carmeli, C., Lifshitz, Y. and Gutman, M. (1978). FEBS Letters 89, 211-214.
12. Recktenwald, D. and Hess, B. (1977). FEBS Letters 80, 187-189.
13. Lien, S. and Racker, E. (1971). Methods in Enzymology 23, 547-556.
14. Nishimura, M., Ito, I. and Chance, B. (1962). Biochem. Biophys. Acta 59, 179-182.
15. Hartree, E. F. (1972). Anal. Biochem. 48, 422-427.
16. Gepshtein, A., Carmeli, C. and Nelson, N. (1978). FEBS Letters 85, 219-223.
17. Hochman, Y., Lanir, A. and Carmeli, C. (1976). FEBS Letters 61, 255-259.

CHANGES IN SUBUNIT CONSTRUCTION OF CHLOROPLAST COUPLING FACTOR 1 WITH DETACHMENT FROM THE MEMBRANE AND ADDITION OF DIVALENT CATION

Yasuo Sugiyama
Yasuo Mukohata[1]

Department of Biology
Faculty of Science
Osaka University
Toyonaka, Osaka

I. INTRODUCTION

Chloroplast coupling factor 1 (CF_1[2]) *in situ* synthesizes ATP by using the electrochemical potential gradient across the thylakoid membrane. Isolated CF_1 is inactive, but after activation by heat or trypsin (1) hydrolyzes ATP. CF_1 is an allosteric protein composed of eight subunits which are grouped (2) and named according to their molecular weights (α_2, β_2, γ, δ, ε_2; the subscript 2 represents two of each subunit are involved in one CF_1)(3,4). It is now believed that the α and β subunits involve the regulatory (3) and the active (5) sites, respectively, the δ subunit is a link (6) for binding CF_1 to the membrane and the ε subunit is an ATPase inhibitor (7).
However, it is not known how these subunits function in the process of ATP synthesis (or hydrolysis), in collaboration with others, or whether the active site of membrane-bound CF_1 for ATP synthesis is identical to that for ATP hydrolysis of CF_1 activated as ATPase.

---

[1] Supported in part by a grant (#311909) from the Ministry of Education, Science and Culture of Japan.
[2] Abbreviations: CF_1, chloroplast coupling factor 1; PLP, pyridoxal phosphate; SDS, sodium dodecylsulfate; DTT, dithiothreitol.

Chemical modification sometimes leads to useful information about the active site of an enzyme. Pyridoxal phosphate (PLP) has been used with several enzymes interacting with phosphate-containing ligands (8,9) in order to detect lysyl residue(s) in the active site or its vicinity.

Here, we report the results of PLP modification of both membrane-bound and isolated CF_1 and suggest that:1) the essential lysyl residue is involved in the activities related to CF_1;2) the arrangement of CF_1 subunits differs depending on whether CF_1 is located on the membrane or dissolved in solution;3) rearrangement of CF_1 subunits, $ie.$, exposure of the active site of isolated-CF_1 ATPase, occurs upon addition of divalent cation, such as Mg^{2+} or Ca^{2+}.

II. MATERIALS AND METHODS

Spinach chloroplasts (thylakoid stacks) were prepared in a choline medium by a method described previously (10). CF_1 was isolated according to the method of Lien and Racker (11) with a slight modification (sucrose density gradient centrifugation was omitted). [^{3}H]PLP was synthesized according to Stock et al. (12).

Chlorophyll was determined by the method of Arnon (13). Protein was determined according to Lowry et al. (14) with bovine serum albumin as standard. The value of 325,000 (15) was used for the molecular weight of CF_1. The concentration of PLP was determined by the absorbance at 388 nm with the value of $\varepsilon=6,600$ $M^{-1}cm^{-1}$ in 0.1N NaOH (16).

PLP modification of chloroplasts: Chloroplasts (equivalent to 400μg chlorophyll/ml) were incubated in a medium containing 0.1M sucrose, 10mM tricine-NaOH (pH 8.3) and 5mM $MgCl_2$ with a given concentration of PLP at 15°C in the dark for 10 min. Next, $NaBH_4$ was added to the incubation mixture to reduce (17) possible Schiff base complexes and fix PLP covalently on chloroplasts (CF_1). The modified chloroplasts were washed with the above sucrose medium and assayed for photosynthetic activities. If necessary, crude [^{3}H]PLP-bonded CF_1 was isolated by the chloroform extraction method (6).

PLP modification of isolated CF_1: CF_1 (200-300μg/ml) was incubated in 20mM tricine-NaOH (pH 8.0) and 1mM EDTA with a given concentration of PLP at 15°C in the dark in the presence or absence of 10mM $MgCl_2$. After incubation for more than 60 min [modification almost reached equilibrium after 30min (18)], the possible Schiff base of the PLP-CF_1 complex was reduced by $NaBH_4$. The modified CF_1 was separated by passage through a column of Sephadex G-25 (1x13 cm).

Mole ratio of bonded PLP to CF_1: The total numbers of reduced Schiff bases (the PLP bonded to CF_1) were determined either from the absorbance at 325 nm with the value of $\varepsilon = 1 \times 10^4$ $M^{-1}cm^{-1}$ (19) for N^{ε}-phosphopyridoxyl lysine or from the radioactivity of [^{3}H]PLP bonded to CF_1. In the latter case, the [^{3}H]PLP-CF_1 after $NaBH_4$ reduction was collected by TCA precipitation then solubilized in 0.1M Tris-acetate (pH 9.0) with 1mM EDTA and 2% SDS, and mixed into modified Bray's cocktail (+10% Triton X-100) (20).

Distribution of PLP among CF_1 subunits: The [^{3}H]PLP-CF_1 was electrophoresed on polyacrylamide disc gel (10%) in the presence of SDS (0.1%) for 3 hours at 5 mA/tube. Gels were stained with Coomassie blue, destained in a mixture of methanol, acetic acid and water (2:3:20) then sliced. Each slice (1 mm thick) was digested with 30% H_2O_2 (0.5 ml) then added to the above Bray's cocktail. The radioactivity was measured with a liquid scintillation counter.

Non-cyclic photophosphorylation was assayed in a mixture (2 ml) of 0.1M sucrose, 10mM tricine-NaOH (pH 8.3), 5mM $MgCl_2$, 600μM potassium ferricyanide, 500μM ADP, 1mM [^{32}P]Pi and chloroplasts equivalent to 40μg chlorophyll at 15°C. γ-[^{32}P]ATP formed was determined by the method of Asada et al. (21) with a slight modification. Light-DTT-activated Mg^{2+}-ATPase (22) was prepared by illuminating chloroplasts in a medium composed of 50mM Tris-Cl (pH 8.0), 50mM NaCl, 5mM $MgCl_2$, 5mM DTT and 20 μM phenazine methosulfate for 10 min at 15°C. Immediately after activation, a portion (0.2 ml) was transferred into a reaction medium (1 ml) containing 50mM Tris-Cl (pH 8.0), 5mM $MgCl_2$ and 5mM ATP then incubated for 10 min at 37°C. Heat-activated Ca^{2+}-ATPase (23) of isolated CF_1 was prepared in a medium composed of 20mM tricine-NaOH (pH 8.0), 1mM EDTA, 25mM ATP and 2mM DTT by heating the mixture at 62°C for 2.5 min. The Ca^{2+}-ATPase was assayed at 37°C in a medium (0.5 ml) of 20 mM tricine-NaOH (pH 8.5), 10mM $CaCl_2$ and 10mM ATP with 3μg of activated CF_1. In both ATPase assays, the liberated inorganic phosphate was determined by the method of Taussky ans Shorr (24).

III. RESULTS

Chloroplasts were incubated with PLP in the dark for 10 min and collected by centrifugation after reduction of the postulated Schiff base by $NaBH_4$, then resuspended in a medium for assay of phosphorylation or ATPase. Figure 1A shows that the PLP-modified chloroplasts lost their activities of photophosphorylation and light-DTT-activated Mg^{2+}-ATPase, which seems to be more sensitive to PLP than phosphorylation. Similar results have been reported for the phenylglyoxal modification of chloroplasts (25). In contrast, the basal electron transport (26) and the light-induced proton uptake were not damaged by PLP modification. These results show that PLP modified the membrane-bound CF_1 itself and inhibited both ATP synthesis and hydrolysis. When 100mM of lysine was added prior to the $NaBH_4$ reduction, the activities of these modified chloroplasts were not different from those of the unmodified chloroplasts, indicating that the PLP modification is reversible.

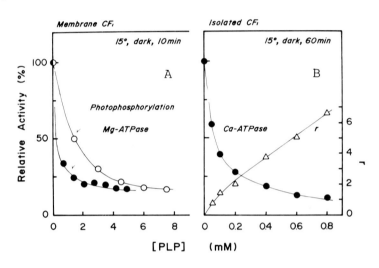

Fig. 1. Effects of PLP modification on the activities related to photophosphorylation. A) Photophosphorylation and light-DTT-activated Mg^{2+}-ATPase of chloroplasts. B) Heat-activated Ca^{2+}-ATPase of isolated CF_1, and also the mole ratio (r) of the bonded PLP to CF_1. The rates of phosphorylation, Mg^{2+}-ATPase and Ca^{2+}-ATPase of the control were 90μmoles ATP formed/mg chlorophyll-hr, 481μmoles Pi released/mg chlorophyll-hr and 12.3μmoles Pi released/mg CF_1-min, respectively.

PLP also forms a Schiff base complex with isolated CF_1. The Schiff base formation reaches an equilibrium after incubation for more than 30 min and can be reversed by dialysis or dilution (18). Thus the Schiff base must be reduced by $NaBH_4$ to allow PLP to covalently bond to CF_1. After reduction of the Schiff base, the modified CF_1 was separated from the unreacted PLP, then thermally activated (23) to the Ca^{2+}-ATPase and the amounts of bonded PLP to CF_1 were determined. Figure 1B shows that the CF_1 modified with PLP in the presence of 10 mM $MgCl_2$ lost its heat-activated ATPase activity in inverse relationship with the amounts of PLP bonded to CF_1. When CF_1 was first thermally activated as Ca^{2+}-ATPase then modified by

PLP, the results were identical with those in Fig. 1B. Therefore, the specific functional site(s) modified by PLP would have been exposed even before the heat activation. The degree of ATPase inactivation had almost reached saturation at 600μM of PLP but the amount of bonded PLP still increased even at 800μM. This indicates that modification of non-specific sites continued at higher PLP concentrations.

$MgCl_2$ was required for the PLP modification. According to our tests thus far, both Ca^{2+} and Mn^{2+} can take the place of Mg^{2+}. In the absence of such divalent cations, both the degree of Ca^{2+}-ATPase inactivation and the amount of bonded PLP were very small, e.g., CF_1 modified with 120μM of PLP in the absence of divalent cation retained 90% of the Ca^{2+}-ATPase activity and 0.7 moles of PLP were bonded to a mole of CF_1.

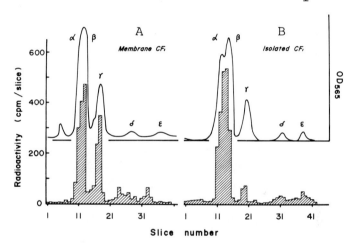

Fig. 2. Distribution pattern of [^{3}H]PLP among CF_1 subunits. CF_1 was electrophoresed on polyacrylamide (10%) disc gel in the presence of SDS (0.1%). A) CF_1 isolated after [^{3}H]PLP modification in chloroplasts. B) CF_1 modified by PLP in solution. Three moles of PLP were found in one mole of CF_1 in A and 1.5 moles in B.

The distribution of [^{3}H]PLP among subunits of CF_1 was examined by separating the latter by SDS disc gel electrophoresis. Fig. 2 shows the distribution patterns of [^{3}H]PLP in membrane-bound (A) and isolated (B) CF_1. When crude CF_1 was isolated from [^{3}H]PLP-modified chloroplasts which retained about 50% of their phosphorylation activity, three moles of PLP were found per CF_1. These three moles of PLP were evenly distributed among α, β and γ subunits but few were found in the δ and ε subunits (26)(to separate the α and β subunits, 12% polyacrylamide gel was used for electrophoresis). In isolated

CF_1, which retained 35% of the Ca^{2+}-ATPase activity after [^{3}H] PLP modification, 1.5 moles of PLP were bonded per CF_1. These 1.5 moles of PLP were almost evenly found in the α and β subunits. Only a fractional (0.15) mole of PLP was found in the γ subunit even when a total of 7 moles of PLP were found per mole CF_1.

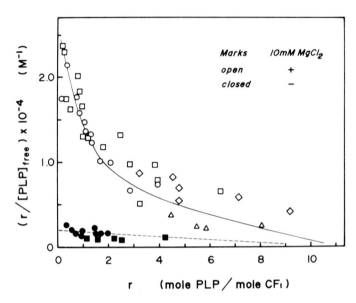

Fig. 3. Scatchard plots of PLP binding to CF_1 in the presence (open marks) or absence (closed marks) of 10mM $MgCl_2$. For the calculated curves, see RESULTS.

Figure 3 shows Scatchard plots of PLP binding to isolated CF_1 in the presence or absence of 10mM $MgCl_2$. Here, we assume that the amounts of the reduced Schiff base (PLP covalently bonded to CF_1) are equal to those of Schiff base (PLP reversiblly binding to CF_1) in equilibrium with free PLP and free CF_1. This is based on the fact that the rate of Schiff base reduction by $NaBH_4$ is very fast relative to the rate needed to reach an equilibrium after dilution with $NaBH_4$ solution (18). The plots show that more than ten binding sites would be present on CF_1 independent of the presence of $MgCl_2$. In the presence of $MgCl_2$, the plots suggest that there are multiple groups of binding sites on CF_1 differing in their affinity to PLP. Assuming two independent groups of PLP binding sites on CF_1 as the simplest case, the binding constant and the maximum number for those binding sites can be calculated with the following equation (27).

$$r = \frac{r_H K_H [PLP]}{1 + K_H[PLP]} + \frac{r_L K_L [PLP]}{1 + K_L[PLP]}$$

where r is the ratio of moles of bound PLP per mole CF_1, [PLP] is the concentration of free PLP, K is a binding constant (in M^{-1}, hereafter) of the ligand, and subscripts H and L denote the high and low affinity sites, respectively. We assume $r_H=1$ at $K_H=2\times10^4$ and $r_L=10$ at $K_L=650$, and the solid line in Fig. 3 is obtained. In the absence of $MgCl_2$, the plots show that PLP binding sites on CF_1 would not be grouped. The binding constant and the maximum number of binding sites may be obtained from the dashed line as being 250 and 10, respectively.

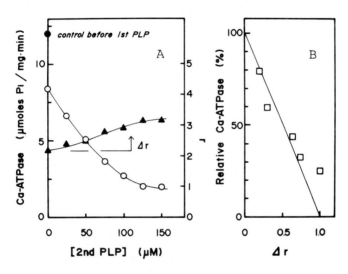

Fig. 4. Two-step PLP modification of CF_1. A) Changes in the Ca^{2+}-ATPase activity and the mole ratio of PLP/CF_1 by the second PLP modification. CF_1 was incubated with 250μM of PLP in the presence of 2mM ATP and 10mM $MgCl_2$ for 60 min then reduced by $NaBH_4$. After removal of unreacted PLP and ATP, CF_1 was incubated again with an indicated concentration of PLP without ATP for 60 min then reduced, and the ATPase activity and the ratio were determined. B) Plots of the Ca^{2+}-ATPase activity (%) against the increment of the ratio (Δr) due to the second modification.

ATP and ADP protected the isolated CF_1 from the Ca^{2+}-ATPase inactivation due to PLP modification (18). Half maximum protection was obtained with 0.2 mM ATP (or ADP). It has been strongly suggested (18) that ATP competes with PLP at the one high affinity site with $K_H=2\times10^4$ and that this site is catalytic. Therefore, CF_1 was first incubated with 250μM of PLP in the presence of 2mM ATP to modify the low affinity (non-catalytic) sites, then collected by ammonium sulfate (final 50% saturation) precipitation. This PLP-modified CF_1, with 2.2 moles of PLP per mole of CF_1 and retaining 70% of its Ca^{2+}-ATPase activity, was incubated again with PLP in the absence of ATP. This differential PLP modification (Fig. 4A) shows that the ATPase activity was lost after the second modification at a fairly low concentration of PLP and the PLP/CF_1 ratio increased by not more than one. Figure 4B shows the plots of the remaining Ca^{2+}-ATPase activity after the second modification against the increment of bonded PLP due to the second modification (Δr), indicating that one mole of PLP is bound per mole of CF_1 at 100% inactivation of Ca^{2+}-ATPase.

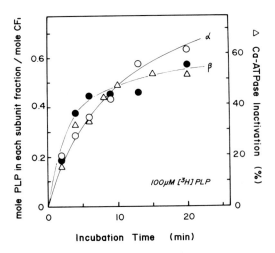

Fig. 5. Time courses of the increase of PLP in the α (O) and β (●) subunit fractions and the inactivation of Ca^{2+}-ATPase (△). For details, see RESULTS.

Figure 5 shows the time courses of PLP modification of the α and β subunits of CF_1 and of the Ca^{2+}-ATPase inactivation. After CF_1 was incubated with [^{3}H]PLP (100μM) for the indicated time, $NaBH_4$ was added, then the modified CF_1 was assayed for the ATPase inactivation and also electrophoresed on polyacrylamide (12.5%) disc gel in the presence of 0.1% SDS for 18 hrs. No remarkable differences were found in the rate of modification of these two subunits. When ATPase was inactivated by 50%, one half mole of PLP was found in both α and β subunit fractions of a mole of CF_1. Which one of these subunits carries the active site of Ca^{2+}-ATPase could not be determined, even using different conditions.

IV. DISCUSSION

PLP modification inhibited the activities related to CF_1 in isolated chloroplasts, synthesis and hydrolysis of ATP (Fig. 1), but not the activity of electron transport. This indicates that CF_1 is one of the components modified by PLP. PLP was actually found in CF_1 (Fig. 2) isolated from the PLP-modified chloroplasts which showed energy transfer inhibition (26). Since PLP is rather hydrophillic, our results suggest that the site(s) essential to the function of CF_1 is located near its surface. This site(s) would be amino group(s) (17), most probably ε-amino group(s) of lysine residue(s) (8,9).

The distribution of PLP among the subunits of CF_1 markedly differed depending on whether CF_1 was modified on the membrane or in a solution (Fig. 2). The absence of PLP in the γ subunit of CF_1, which was modified in a solubilized form, would be related to the observation that the anti-γ-antibody inhibited photophosphorylation but not Ca^{2+}-ATPase of isolated CF_1 (3). Burial of the γ subunit among other subunits would make the anti-γ-antibody and PLP unable to reach their sites of interaction.

The amounts of bonded PLP required for complete inactivation of photophosphorylation seemed to be larger than those of Ca^{2+}-ATPase of isolated CF_1. To the membrane-bound CF_1, at least two PLP molecules [probably on the two β subunits (28)] seem to be responsible for inactivation, while one is likely to be enough to inactivate solubilized Ca^{2+}-ATPase (Fig. 4). These results also suggest that the detaching process of CF_1 from the membrane has some effect on the subunit conformation and the inter-subunit interaction of this multi-subunit enzyme.

Isolated CF_1 was used to study the kinetics of PLP binding to CF_1 more quantitatively. In the presence of $MgCl_2$, isolated CF_1 possesses at least one site with high affinity for PLP and a number of sites with lower affinities (here we tentatively fixed their number at ten with identical lower affinity for PLP). The site with the highest affinity for PLP has 50% protection from PLP modification with ATP (or ADP) at 0.2mM. This value is much higher than that (a few µM) needed to protect chloroplasts (CF_1) from N-ethylmaleimide inhibition and the dissociation constant of ATP binding to the regulatory [inhibition (29)] site, and is rather close to the Km value for ATP in the Ca^{2+}-ATPase reaction (28). Furthermore, the binding of protective ATP was competitive with that of PLP(18). These results strongly suggest that the site with the highest affinity for PLP is involved in the catalytic site of the ATPase. This is well supported by the results of two-step modification (Fig. 4). At the first modification in the presence of ATP, PLP bonded mostly non-catalytic sites and at the second modification in the absence of ATP, it bonded only one site with a large loss of the Ca^{2+}-ATPase activity. Thus we concluded that the Ca^{2+}-ATPase of isolated CF_1 is completely inactivated by PLP modification of a single site, which is consistent with the result obtained by other methods of analysis (18).

In contrast, in the absence of Mg^{2+}, this highest affinity site for PLP, $ie.$, the catalytic site, is not exposed or does not reveal its high affinity for PLP (Fig. 3). Since the experimental K_L value for non-catalytic sites lowered from 650 in the presence of Mg^{2+} to 250 in its absence, the K_H value for the catalytic site should be expected to be at least 7×10^3 in the absence of Mg^{2+}, if we simply assume the difference in the binding affinity between Mg-PLP and PLP. Besides, formation of Schiff base itself, in general, does not require Mg^{2+} (37). Thus it would be more conceivable that Mg^{2+} acts as an effector to change the conformation of CF_1 and expose the active site. Mg^{2+} could be replaced by Ca^{2+} or Mn^{2+}, although the effective concentration somewhat differed; $Mn^{2+} < Mg^{2+} < Ca^{2+}$. Results suggesting such non-specific divalent-cation dependence of enzyme conformation have been reported (30,31). Furthermore, with isolated CF_1 which binds three moles of nucleotides (ATP or ADP) per one mole of CF_1, the binding mode of nucleotide has been shown to change depending on the presence or absence of Mg^{2+} (and Ca^{2+})(32). In the presence of Mg^{2+}, the binding of the third nucleotide is positively cooperative to the second one, while in its absence, it is slightly negatively cooperative. Although we have not determined which one of three nucleotide-binding sites in isolated CF_1 is (latently) catalytic, the change in the binding coopearativity would be related to the observed Mg^{2+}-dependent exposure-burial of the active site.

As shown in Fig. 5, the location of the active site could not be determined. Even under other (improved) experimental conditions, no distinction could be made. However, the present results indicate that one site is catalytic and the other is not. There is one catalytic site on one subunit (probably β), and many non-catalytic sites on the other subunit which are modified at random but seemingly in parallel to the catalytic site modification.

These results obtained by PLP modification of CF_1 require rather sophisticated consideration of the features of this multi-subunit enzyme. The simplest model for the present results is illustrated in Fig. 6. The subunit conformation and inter-subunit interaction depend on the situation of CF_1. By detachment from the membrane in the absence of Mg^{2+} [$ie.$, in a dilute EDTA solution (11)], the γ subunit of CF_1 would be buried among other subunits. The PLP binding sites on the α and β subunits would also be hidden. Addition of divalent cation exposes these binding sites. However, the two catalytic sites on the (β) subunit pair would be so close to each other that the binding of PLP to one site would interfere with the binding of PLP or ATP to the other.

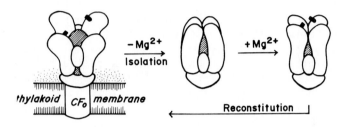

Fig. 6. A simple model illustrating the results of PLP modification. The catalytic sites (black dots) with high affinity for PLP are located on one of the two large subunit pairs (probably 2β) and many non-specific PLP binding sites (not shown) on the other large subunit pair (probably 2α). The γ subunit to which small ε subunit(s) is attached is shadowed. The δ subunit forms a link to CF_0. For the explanation of the conformation change and the inter-subunit interaction, see DISCUSSION.

In mambrane-bound CF_1, the γ subunit is exposed and maintains a distance between two catalytic sites. Energization of the membrane by light induces a conformation change of CF_1, which involves further exposure of the γ subunit (33,34). The reaction steps of phosphorylation, such as binding of ADP and Pi, condensation of these ligands and release of ATP are conducted by this activated CF_1 in cooperation with two catalytic (β) subunits and also other (α and γ) subunits. The binding of effector Mg^{2+} and its release, probably in exchange for the translocated protons, and the binding and release of nucleotides (effector, substrate and product), which are regulated by Mg^{2+} (32), would participate in the conformation changes and the cooperative interaction of subunits in the course of phosphorylation. In isolated CF_1, since ATP hydrolysis is a downhill reaction, subunit interaction may not be as necessary but exposure of either one of the catalytic sites would be required.

More complicated models, such as that of half of the site reactivity (35,36) are possible and may be closer to the real features of the enzyme.

REFERENCES

1. Vambutas, V. K. and Racker, E., J. Biol. Chem. 240:2660 (1965)
2. Racker, E., Hauska, G. A., Lien, S., Berzborn, R. J. and Nelson, N., In "Proceedings of the 2nd International Congress of Photosynthesis Research" (Forti, G. et al eds.) Vol. 2. pp1097. Dr. W. Junk N. V. Publishers, The Hague. (1971).
3. Nelson, N., Deters, D.W., Nelson, H. and Racker, E., J. Biol. Chem. 248:2049 (1973).
4. Nelson, N., Kamienietzky, A., Deters D. W. and Nelson, H. In "Electron Transfer Chains and Oxidative Phosphorylation" (Quagliariello, E. et al. eds.) pp 149. North-Holland Publishing Co., Amsterdam (1975).
5. Deters, D. W., Racker, E., Nelson, N. and Nelson, H., J. Biol. Chem. 250:1041 (1975).
6. Younis, H. M., Winget, G. D. and Racker, E., *ibid*. 252: 1814 (1977).
7. Nelson, N., Nelson, H. and Racker, E., *ibid*. 247:7657 (1972).
8. Colombo, G. and Murcus, F., Biochemistry 13:3085 (1974).
9. Milhausen, M. and Levy, H. R., Eur. J. Biochem. 50:453 (1975).
10. Mukohata, Y., Yagi, T., Matsuno, A., Higashida, M. and Sugiyama, Y., Plant Cell Physiol. 15:163 (1974).
11. Lien, S. and Racker, E., Method Enzymol. 23:547 (1971).
12. Stock, A., Ortanderl, F. and Pfleiderer, G., Biochem. Z. 344:353 (1966).
13. Arnon, D. I., Plant Physiol. 24:1 (1948).
14. Lowry, O. H., Roseborough, N. J., Farr, A. L. and Randall, R. L., J. Biol. Chem. 193:265 (1951).
15. Farron, F., Biochemistry 9:3823 (1970).
16. Peterson, E. A. and Sober, H. A., J. Am. Chem. Soc.76: 169 (1954).
17. Fischer, E. H., Kent, A. B., Snyder, E. R. and Krebs, E. G. *ibid*. 80:2906 (1958).
18. Sugiyama, Y. and Mukohata, Y., FEBS Lett., in press (1979).
19. Forry, A. W., Olsgaard, R. B., Nolan, C. and Fischer, E. H., Biochemie 53:269 (1971).
20. Bray, G. A., Anal. Biochem. 1:279 (1960).
21. Asada, K., Takahashi, M. and Urano, M., *ibid*.48:311 (1972).
22. Datta, D. B., Ryrie, I. J. and Jagendorf, A. T., J. Biol. Chem. 249:4404 (1974).
23. Farron, F. and Racker, E., Biochemistry 9:3829 (1970).
24. Taussky, H. and Shorr, E., J. Biol. Chem. 202:675 (1953).
25. Vallejos, R. H., Viale, A. and Andreo, C. S., FEBS Lett. 84:304 (1977).
26. Sugiyama, Y. and Mukohata, Y., *ibid*. 85:211 (1978).

27. Edsall, J. T. and Wyman, J., "Biophysical Chemistry" Vol. 1. pp591, Academic Press Inc., New York (1958).
28. Nelson, N., Biochim. Biophys. Acta, 456:314 (1976).
29. Mukohata, Y., Yagi, T., Sugiyama, Y., Matsuno, A. and Higashida, M., J. Bioenerg., 7:91 (1975).
30. Edwards, P. A. and Jackson, J. B., Eur. J. Biochem., 62:7 (1976).
31. Vandermeullen, D. L. and Govindjee, *ibid.*, 78:585 (1977).
32. Higashida, M. and Mukohata, Y., J. Biochem., 80:1177 (1976).
33. Ryrie, I. J. and Jagendorf, A. T., J. Biol. Chem., 246:3771 (1971).
34. McCarty, R. E., Pittman, P. R. and Tsuchiya, Y., *ibid.* 247:3048 (1972).
35. Lażdunski, M., Prog. Bioorg. Chem., 3:81 (1974).
36. Levitzki, A. and Koshland Jr., D. E., Curr. Topics Cell. Regul., 10:1 (1976).
37. Schnackerz, K. D. and Noltmann, E. A., Biochemistry, 10:4837 (1971).

EFFECTS OF DIVALENT CATIONS ON MEMBRANE ATPase FROM A STRICTLY ANAEROBIC SULFATE-REDUCING BACTERIUM DESULFOVIBRIO VULGARIS

Michiko Takagi
Kunihiko Kobayashi
Makoto Ishimoto

Faculty of Pharmaceutical Sciences
Hokkaido University
Sapporo, Japan

Strictly anaerobic sulfate-reducing bacteria use inorganic sulfur compounds as electron acceptors for the oxidation of organic substrates and oxidative phospholylation is supposed to couple to the electron transports. They are thought to carry out a primitive or ancestral type of respiration (1). Membrane ATPase has been solubilized from a sulfate-reducing bacterium Desulfovibrio vulgaris (2). The present paper describes the uneven effects of various divalent cations on D. vulgaris ATPase in membrane-bound and solubilized states.

Membrane ATPase from D. vulgaris was solubilized by non-ionic detergent Emulgen 810 (2), and was further purified by gel filtration on Ultrogel AcA 22 and by DEAE-cellulose column chromatography. ATPase activity was assayed by liberation of inorganic phosphate from ATP (5 mM) at pH 8.0. The Mg^{2+}-stimulated ATPase activity of the most purified enzyme was 7 μmoles/min/mg protein and was 18-fold higher than that of membrane fraction.

The activity in both membrane and solubilized fractions required divalent cation; in the reaction with 5 mM ATP, maximum activity with Mg^{2+} was attained at 2.5 mM of the cation (Fig. 1). Mn^{2+} and Cd^{2+} were 2-3 times more effective than Mg^{2+} at the same concentration. Ca^{2+}, Sr^{2+}, or Cu^{2+} scarcely stimulated the activity at 2.5 mM, and these ions antagonized Mg^{2+} in ATPase activity. Maximum activity with Ca^{2+} was attained at 25 mM. Solubilization and purification did not essentially change these properties. Activities depending on metals increased more remarkably than that depending on Mg^{2+} on solubilization and purification (Fig. 1). Optimum pH's for Mg^{2+}-and Ca^{2+}-stimulated activities were both 8-9, in contrast to the earlier observation on D. gigas membrane ATPase (3) which had the optimum activity at pH 6.5 in the presence of 2.5 mM Ca^{2+} in spite of the optimum at pH 8 in the presence of Mg^{2+}.

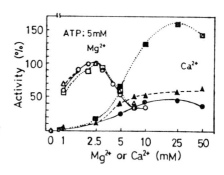

Fig. 1. Effects of Mg^{2+} and Ca^{2+} at various concentrations on ATPase activity. Mg^{2+} (○,△,□), Ca^{2+} (●,▲,■); in membrane (○,●), solubilized (△,▲), and purified (□,■) preparations.

Mg^{2+}-ATPase activity in membrane fraction was inhibited by N,N'-dicyclohexylcarbodiimide and tri-n-butyltin chloride(TBT) at 10 μM and the activity became 10-100 fold less sensitive to both inhibitors on solubilization. Ca^{2+}-ATPase was inhibited by TBT to lesser extent in membrane and solubilized fractions. Extent of inhibition of Mn^{2+}, Cd^{2+}, or Co^{2+}-stimulated ATPase by TBT was similar to that of Mg^{2+}-activated ATPase in both fractions. Sodium azide inhibited Mg^{2+}-ATPase activity in membrane and solubilized preparations to the same extent (60 % inhibition at 0.1 mM) but Ca^{2+}-ATPase was less sensitive to azide (30-40 % inhibition at 1 mM). In contrast to D. gigas ATPase (3), 2,4-dinitrophenol(DNP) stimulated neither Mg^{2+}-ATPase nor Ca^{2+}-ATPase of D. vulgaris membrane fraction, but inhibited both activities. ITP and GTP were hydrolyzed three times faster than ATP in the presence of Mg^{2+}. On the contrary, ATP was most efficiently hydrolyzed in the presence of Ca^{2+}. Km for ATP was 1 mM in the presence of Mg^{2+}.

The specific features of cation effects, i.e., high stimulation by Mn^{2+} and Cd^{2+}, different velocities of GTP and ITP hydrolysis in the presence of Mg^{2+} and Ca^{2+}, or different susceptibilities to inhibitors of Mg^{2+}-dependent and Ca^{2+}-dependent activities were also found with purified enzyme. It would support the idea that divalent cation interacts with the enzyme not only at the substrate binding site but also at the regulatory site(s), resulting in the cation-specific conformational change of the enzyme.

REFERENCES

(1) Kobayashi, K., Seki, Y., Katsura, E., and Ishimoto, M. (1978). In "Origin of Life" (H. Noda, ed.), p. 421. Japan Scientific Societies Press, Tokyo.
(2) Kobayashi, K., Okuoka, Y., and Ishimoto, M. (1976). Seikagaku, 48: 608. (in Japanese).
(3) Guarraia, L.J., and Peck, H.D., Jr. (1971). J. Bacteriol. 106: 890.

RESOLUTION AND RECONSTITUTION
OF PROTON TRANSLOCATING ATPASE

Nobuhito Sone
Masasuke Yoshida
Hajime Hirata
Yasuo Kagawa

Department of Biochemistry
Jichi Medical School
Minamikawachi-machi, Tochigi-ken

I. INTRODUCTION

Biomembranes that synthesize ATP contain a reversible H^+-translocating ATPase, named $F_o \cdot F_1$ as a coupling device (1). We have shown that upon reconstitution into proteoliposomes the purified enzyme (TFo·F_1) from a thermophilic bacterium PS3 can catalyze the formation of an electrochemical proton gradient by ATP hydrolysis (2,3) and carried out ATP synthesis coupled to H^+ conduction (4,5). The coupling device (from the thermophilic bacterium) is composed of an enzyme (TF_1) with ATPase activity and a membrane moiety (TFo) with H^+ conduction which is sensitive to energy transfer inhibitors including DCCD, N,N'-dicyclohexylcarbodiimide (6). Stability of TF_1 and TFo makes it possible to isolate five individual subunits of TF_1 which are active in reconstituting ATPase activity (7,8) and highly purified TFo which is active in reconstituting vesicles with H^+ conduction and TF_1-binding (9,10).

Figure 1 shows the subunit composition of $TF_o \cdot F_1$ and the role of them which we shall described herein.

II. COMPOSITION OF $TF_o \cdot F_1$

Subunit composition of $TF_o \cdot F_1$ is summarized in Table 1. The molar ratio of subunits was estimated from distribution pattern of radioactivity among subunits of $TF_o \cdot F_1$ obtained

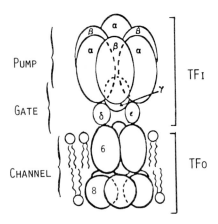

Fig. 1. Proposed model of subunit structure of H^+-translocating ATPase $(TF_O \cdot F_1)$. "6" and "8" indicate TF_1-binding protein and DCCD-binding protein, respectively.

from the bacteria cultured in [U-^{14}C]amino acid mixture (11). Molecular weights of these subunits are also listed.

Subunit composition of F_1-ATPases is still controversial. Our results on the stoichiometry of $\alpha_3\beta_3\gamma\delta\epsilon$ for TF_1 is supported by the following facts; that the molecular weight of TF_1 (380,000 daltons) is in good agreement with this stoichiometry, that one SH-group is found in one α subunit and three SH-groups in one TF_1 molecule (Table II), and that two dimensional crystal of TF_1 is arrayed in hexagonal pattern confirmed by the image reconstruction technique (12).

TABLE I. Estimation of Molecular Weight of Subunit Proteins of $TF_O \cdot F_1$

	Subunit of $TF_O \cdot F_1$		No. of copies	M.W. by SDSPAGE	M.W. by other method[a]
No.	TF_1	TF_O			
1	α	–	3	56,000	52,000
2	β	–	3	53,000	51,000
3	γ	–	1	32,000	33,000
4	–	+	0–1	19,000	
5	δ	–	1	15,000	20,000
6	–	+	2–3	13,500	15,500
7	ϵ	–	1	11,000	16,000
8	–	+	4–6	5,400	6,500

[a] Gel filtration in the presence of 6 M guanidine hydrochloride for TF_1 subunits and amino acid analysis for TF_O subunits.

TABLE II. Titration of Sulfhydryl Groups in TF_1, α Subunit, and β Subunit.

Sample	TNB released	Protein	-SH/protein
	nmol	nmol	
TF_1	6.88	2.23	3.03
TF_1 (reduced)	6.87	2.14	3.21
α	4.88	6.84	0.70
α (reduced)	4.14	4.46	0.93
β	0.31	9.92	0.03
β (reduced)	0.16	3.22	0.05

Reduction was carried out in the presence of 50 mM dithiothreitol and 1 % sodium dodecylsulfate and assayed with DTNB (5,5'-dithiobis(2-nitrobenzoic acid)) after removal of dithiothreitol by gel filtration (13).

TFo, obtained by treating $TFo \cdot F_1$ with urea is composed of 2-3 kinds of subunits (bands 4,6 and 8). It is noteworthy that multiple copies of band 6 and 8 proteins were found in one molecule. On the contrary the content of band 4 protein, less than 1, varied in different preparations without any change in TFo activities.

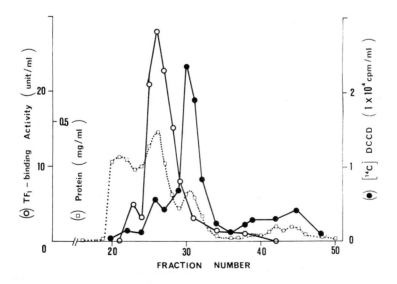

Fig. 2. Resolution of TFo into TF_1-binding and DCCD-binding proteins with a column of Sephacryl S-200 in the presence of sodium dodecylsulfate (10).

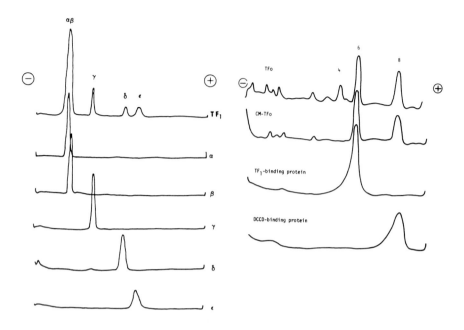

Fig. 3. Patterns of SDSPAGE of TF_1 and it subunits (left) and TFo and TF_1-binding (band 6) protein and DCCD-binding (band 8) protein (right).

III. RESOLUTION OF $TF_0 \cdot F_1$

TF_1 was dissociated in 6 M guanidine hydrochloride and each subunits were isolated by ion-exchanging chromatography in the presence of 8 M urea. Only γ was adsorbed on CM-cellulose and ε, δ, α and β were eluted (in this order) from DEAE-cellulose column (8). After removal of urea each subunit was concentrated by ammonium sulfate precipitation.

TF_O was obtained by treating $TF_O \cdot F_1$ with urea by dissociating TF_1 moiety (9). A preparation of TFo, in which band 4 protein was removed by passing TF_O through CM-cellulose in the presence of Triton X-100 and urea, was as active as original TF_O (10). Band 6 and band 8 proteins were separated by gel filtration using Sephacryl S-200 in the presence of 0.2 % sodium dodecylsulfate (Fig.). Vesicles reconstituted from the fractions containing band 6 protein were able to bind TF_1, while [^{14}C]DCCD was incorporated in the fractions which contained band 8 protein. Band 8 protein was also prepared by extracting TF_O with $CHCl_3$-methanol, and it was thus assumed to be a proteolipid. Patterns of polyacrylamide gel electrophoresis in the presence of sodium dodecylsulfate (SDSPAGE) of TF_1, its five subunits, TFo and its two subunits are shown in Fig. 3.

IV. RECONSTITUTION OF TF$_1$

From isolated individual subunits, functional TF$_1$ can be reconstituted (8). ATP hydrolyzing activity is an essential property of the pump. This activity was reconstituted by simple mixing of the subunits at 20-45°C at pH 6.3-7.0. During the incubation, Mg^{2+} stimulated the reassemly, while ATP and dithiotheitol had no effect. Another fundamental activity of TF$_1$ is gate-like activity, ability to block H$^+$ conduction through TFo. Table III summarizes the reconstitution of TAPase activity and inhibition of H$^+$ conductivity of TFo-vesicles by various combination of TF$_1$ subunits. Combinations of $\alpha+\beta+\gamma$ and $\beta+\gamma$ (partially) had ATPase activity and combinations of $\gamma+\delta+\varepsilon$ blocked H$^+$ conduction in TFo-vesicles. The sequential binding of these subunits to TFo is illustrated as follows (7):

$$\text{TFo} \begin{array}{c} \xrightarrow{\delta} \text{TFo}\cdot\delta \xrightarrow{\varepsilon} \\ \xrightarrow{\varepsilon} \text{TFo}\cdot\varepsilon \xrightarrow{\delta} \end{array} \text{TFo}:{}^{\delta}_{\varepsilon} \xrightarrow{\gamma} \text{TFo}:{}^{\delta}_{\varepsilon}:\gamma \xrightarrow{\alpha}_{\beta} \text{TFo}:{}^{\delta}_{\varepsilon}:\gamma:{}^{\alpha_3}_{\beta_3}$$

On the contrary all five subunits of TF$_1$ were necessary besides TFo for the reconstitution of DCCD-sensitive ATPase activity and energy transforming reactions catalyzed by H$^+$-translocating ATPase (see Section VIII).

TABLE III. Reconstitution of ATPase and Gate-like Activity from Various Combinations of TF$_1$ Subunits

α	β	γ	δ	ε	ATPase activity unit^a/mg	Inhibition of H$^+$ conductivity unit^b/mg
+	+	+	+	+	4.44	3.0
	+	+	+	+	1.72	-
+		+	+	+	0.08	-
+	+		+	+	0.20	-
+	+	+		+	4.01	0.4
+	+	+	+		4.28	0.0
+	+	+			4.27	0.4
		+	+	+	0.00	7.8
	+	+			1.25	-
+	+				0.02	-
+		+			0.17	-
	+	+			1.25	-
			+	+	-	0.9

^{a}Units are in μmol Pi released per mg.
^{b}Arbitrary unit due to an initial velocity of 9-aminoacridine fluorescence change as shown in Fig. 4.

V. RECONSTITUTION OF TFo-VESICLES

TFo is highly hydrophobic and obtained as aggregates. For measurement of F_O activity the preparation was reconstituted by the dialysis method (2) into proteoliposomes (TFo-vesicles). TFo-vesicles thus prepared can conduct H^+ passively and can bind TF_1. Fig. 4 shows the most sensitive method to evaluate H^+ conductivity. The addition of valinomycin to TF_O-vesicles loaded with KCl resulted in generation of a membrane potential (inside negative), which caused an influx of H^+. Change of pH inside the vesicles was followed by 9-aminoacridine, an indicator of pH difference across the membranes. The initial velocity of quenching was roughly proportional to the amount of TFo added within a certain range. This passive H^+ conduction was blocked by either TF_1 or DCCD. Protons were conducted in both directions in response to the membrane potential imposed. The pH profile of the rate revealed that H^+, not OH^-, was the true substrate conducted (9).

TFo preparations containing only band 6 and band 8 proteins were also active as described above. However, the vesicles reconstituted from isolated band 6 protein and band 8 protein by the gel filtration in the presence of sodium dodecylsulfate could bind TF_1, but did not conduct H^+. Thus it seems likely that the subunit protein(s) of TFo was inactivated in sodium dodecylsulfate, or that the reassembly of TFo molecule did not occur under the experimental condition.

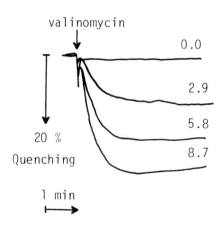

Fig. 4. Measurement of H^+ conductivity using 9-aminoacridine. Vesicles were reconstituted in the ratio shown for each tracing (0 to 8.7 μg TFo/ mg of phospholipids) and loaded with KCl (9).

VI. DCCD-BINDING PROTEIN AND TF_1-BINDING PROTEIN

Two kinds of subunit proteins plays different roles in TFo. Band 6 protein isolated by the gel filtration in the presence of sodium dodecylsulfate is able to bind TF_1, while band 8 protein, to which DCCD binds covalently with concomitant loss of H^+ conduction, is responsible for H^+ conduction across the membrane. Table IV shows amino acid compositions of band 6 protein and of band 8 protein. Band 8 protein contained an extremely high percentage of hydrophobic amino acids (0.29 in polarity) as DCCD-binding protein from other sources (15-17). On the contrary band 6 protein was not so hydrophobic (0.39 in polarity).

It is of interest if isolated band 8 protein mediates H^+ conduction when reconstituted into vesicles. Nelson et al. (18) and Criddle et al. (19) have reported some evidence for H^+

TABLE . IV. Amino Acid Compositions of Band 6 Protein and Band 8 Protein.

Amino acid	Band 6 protein Composition	Residue	Band 8 protein Composition	Residue
	mol %		mol %	
Asp	5.65	8	1.92	1
Thr	5.34	8	4.51	3
Ser	4.73	7	4.36	3
Glu	12.28	17	8.08	5
Pro	1.43	2	1.85	1
Gly	7.98	11	15.51	10
Ala	11.64	16	13.45	9
Cys	0.00	0	0.00	0
Val	7.57	11	11.12	7
Met	3.83	6	3.12	2
Ile	7.31	10	10.23	7
Leu	12.36	17	12.63	8
Tyr	2.86	4	1.61	1
Phe	5.16	7	4.76	3
His	2.07	3	0.39	0
Lys	5.07	7	0.23	0
Trp	n.d.	–	n.d.[a]	0
Arg	5.10	7	6.13	4
Total	99.98	141	99.90	64
Polarity[b]		0.39		0.29

[a] Not determined, but another sample hydrolyzed with 6 % thioglycolic acid did not contained.

[b] By the formulation of Capaldi and Vanderkooi (14).

conduction by proteolipid fraction extracted from lettuce chloroplasts or yeast Fo·F_1, respectively. However, band 8 protein isolated by a gel filtration in the presence of sodium dodecylsulfate or extracted with $CHCl_3$-methanol did not conduct H^+ when reconstituted into vesicles. Thus at least there are two possibilities: the proteolipid may be denatured during extraction with organic solvents or alternatively another component of TF_O, band 6 protein, is necessary to span the membrane and to conduct H^+. It seems probable that H^+ conduction may occur when this proteolipid is abundantly incorporated into liposomes. However, band 6 protein is necessary to block H^+ conduction with TF_1.

VII. CHEMICAL MODIFICATION OF TF_O AND MECHANISM OF H^+ CONDUCTION

To obtain more information on the structure of TF_O and the mechanism of H^+ conduction, effects of various treatments on two fundamental activities of TF_O were examined. Table V

TABLE V. Effects of Protease Digestion or Chemical Modification on H^+ Conduction and TF_1-binding Activity.

Exp.	Addition (conc.)	Percent of control	
		H^+-conductivity	TF_1-binding
1	Trypsin (81 µg/ml)	78	5
	Nagarse (81 µg/ml)	109	20
2	Acetic anhydride (79 mM)	104	7
	Succinic anhydride (75 mM)	417	8
	Diazobenzene sulfonate (7.4 mM)	117	61
3	DCCD (0.5 mM)	6	98
	EDC (0.21 M)	76	85
	EDC plus ethylenediamine (0.5 M)	12	66
4	Tetranitromethane (4 mM)	38	91
	Tetranitromethane (8 mM)	7	96
	I_2 (0.23 mM in 0.7 mM NaI)	28	92
	Lactoperoxidase with NaI (10 mM)	106	88
5	Glyoxal (69 mM)	36	93
	Glyoxal (345 mM)	26	67
	Phenylglyoxal (79 mM)	12	19

EDC ; 1-ethyl-3(3-dimethylaminopropyl)carbodiimide.
Incubation: Exp. 1, 2°C, 60 min; Exp. 2, 25°C, 30 min; Exp.3, 25°C, 90 min; Exp. 4, 25°C, 10 min; Exp. 5, 25°C, 30 min.

summarizes the results. Protease digestion (exp. 1) and modification of amino groups of TFo (exp.2) resulted in the loss of TF_1-binding activity, while H^+ conductivity was not altered or even stimulated. On the contrary H^+ conductivity was specifically blocked by a low concentration of hydrophobic carbodiimide DCCD and by hydrophilic carbodiimide EDC at a high concentration with ethylenediamine (exp.3) and by hydrophobic tyrosine-modifiers such as tetranitromethane and iodine (exp.4). It seems likely from these results and thier amino acid compositions that the band 6 protein is located at the surface of the membrane so as to bind the hydrophilic TF_1, while the band 8 protein is integrated deeply in the hydrophobic region of the membrane and responsible for H^+ conduction.

Arginine modifiers such as phenylglyoxal inhibited H^+ conduction as well as TF_1-binding (exp.5). However, since TF_1-binding activity decreased faster than H^+ conduction (not shown), phenylglyoxal may attack two different essential arginines, one probably in band 6 protein and another in band 8 protein. The amino acid composition (Table IV) indicates

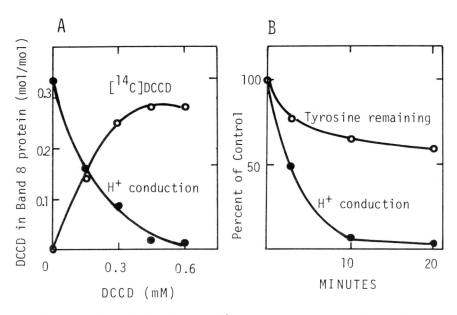

Fig. 5. A. Inhibition of H^+ conduction on binding of DCCD to band 8 protein. TFo (0.56 mg protein) was treated with [^{14}C]DCCD at 2°C for 3 hours. B. Time course of nitration of a tyrosine residue in band 8 protein and inhibition of H^+ conduction by tetranitromethane. TFo (0.39 mg protein) was treated with 5 mM tetranitromethane at 25°C. Tyrosine remaining in band 8 protein was followed with an amino acid analyzer.

that the number of dissociatable residues in band 8 protein are rather small; 5 glutamic acid, 1 aspartic acid (some of them may be amides), 4 arginine and 1 tyrosine. It is noteworthy that modifications of these residues always resulted in the blocking of H^+ conduction.

Another interesting result is shown in Fig. 5. Titration of [^{14}C]DCCD to TFo resulted in the   loss of its H^+ conductivity which was directly related to the amount of DCCD bound to band 8 protein (A). This stoichiometry indicates that one third or one fourth of band 8 protein is modified with DCCD resulted in the total loss of the activity. Similarly when one-third of a tyrosine residue of band 8 protein is nitrated, almost all H^+ conduction ceased (B). These results may be interpreted as follows; only one-third or one-fourth of the H^+ channel is active. However, since H^+ conductivity in TFo is stable, it seems unlikely that 2/3 - 3/4 of band 8 protein is always inactivated (20). Thus the H^+ channel is probably made of multiple copies of band 8 protein and modification of one band 8 protein in a channel seems sufficient for blocking of H^+ conduction.

The inhibition of H^+ conduction by DCCD obeyed first order kinetics and reaction order with respect to DCCD concentration was about 1 (Fig. 6), indicating that one DCCD attacks one DCCD-binding site without co-operativity.

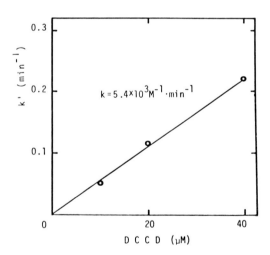

Fig. 6. Plot of pseudofirst order rate constant (k') obtained by plotting semilogarithmically of the inhibition of H^+ conduction by DCCD versus DCCD concentration. k, second order rate constant for enzyme-inhibitor interaction .

TABLE VI. Reconstitution of the Energy Transforming Reactions from Individual Subunits of TF_1 and TFo-vesicles

Exp.	Addition		^{32}Pi-ATP exchange	$\Delta F/F$ [a]
			nmol/mg TFo	%
1	TFo-vesicles	TF_1	420	40
		$\alpha\beta\gamma\delta\epsilon$	212	24
		$\beta\gamma\delta\epsilon$	59	–
		$\alpha\beta\gamma\delta$	9	0
		$\alpha\beta\gamma\epsilon$	13	3
		$\alpha\beta\gamma$	4	0
		None	4	0
2	TFo-vesicles	TF_1	133	79
	CM-TFo-vesicles	TF_1	114	66
	CM-TFo-vesicles	None	11	0

[a] ATP-dependent fluorescence change of 8-anilinonaphthalene-1-sulfonate.

VIII. RECONSTITUTION OF ENERGY TRANSFORMING ASSEMBLY

As described above the assemblies of five TF_1-subunits exhibited pump and gate function, and TFo containing only two components exhibited channel activity of H^+-translocating ATPase. On the other hand all these seven subunits are necessary for reconstitution of the energy transducing activity to convert chemical energy to chemiosmotic energy. Table VI shows that all five subunits are necessary for the reconstitution of ^{32}Pi-ATP exchange and ATP-dependent fluorescence enhancement of 8-anilinonaphthalene sulfonate with TFo-vesicles (exp.1). Vesicles containing CM-TFo which deleted band 4 protein (see Section VI) were as active as vesicles containing TFo. Therefore band 4 protein may not be an essential component of TFo.

From the results presented here, we propose a tentative model of $TFo \cdot F_1$ (Fig.1). Subunit δ and ϵ constitute a pathway of H^+ from TFo to catalytic core of TF_1. Subunit γ have a gate-like function by coupling H^+ flow to ATP hydrolysis (or synthesis). Subunit β is essential for ATP hydrolysis and α seems to be required for coupling activity. Band 6 protein is TF_1-binding protein of the hydrophobic moiety of $TFo \cdot F_1$ and band 8 protein is H^+-channel through membrane, to which DCCD interacts. Experiments with chemical modifiers to hydrophilic amino acid residues in TFo strongly suggest that this channel for H^+ is made of an oligomer of band 8 protein and that protnation and deprotonation of the polar groups faced to the channel are important for H^+ movement.

ACKNOWLEDGEMENT

The authors thank Misses Keiko Ikeba and Toshiko Kambe for excellent technical assistance.

REFERENCES

1. Boyer, P.D., Chance, B., Ernster, L., Mitchell, P., Racker, E., and Slater, E.C. (1977) Ann. Rev. Biochem. 46:955.
2. Sone, N., Yoshida, M., Hirata, H., and Kagawa, Y. (1977) J. Biochem. 81: 519.
3. Sone, N., Yoshida, M., Hirata, H., Okamoto, H., and Kagawa, Y. (1976) J. Membr. Biol. 30:121.
4. Sone, N., Yoshida, M., Hirata, H., and Kagawa, Y. (1977) J. Biol. Chem. 252:2956.
5. Sone, N., Takeuchi, Y., Yoshida, M., and Ohno, K. (1977) J. Biochem 82, 1751-1758.
6. Sone, N., Yoshida, M., Hirata, H., and Kagawa, Y. (1975) J. Biol. Chem. 250:7917.
7. Yoshida, M., Okamoto, H., Sone, N., Hirata, H., and Kagawa, Y. (1977) Proc. Natl. Acad. Sci. USA.74:936.
8. Yoshida, M., Sone, N., Hirata, H., and Kagawa, Y. (1977) J. Biol. Chem. 252:3480.
9. Okamoto, H., Sone, N., Hirata, H., Yoshida, M., and Kagawa, Y. (1977) J. Biol. Chem. 252:6125.
10. Sone, N., Yoshida, M., Hirata, H., and Kagawa, Y. (1978) Proc. Natl. Acad. Sci. USA. 75:4219.
11. Kagawa, Y., Sone, N., Yoshida, M., Hirata, H., and Okamoto, H. (1976) J. Biochem. 80:141.
12. Wakabayashi, T., Kubota, M., Yoshida, M., and Kagawa, Y (1977) J. Mol. Biol. 117:515.
13. Yoshida, M., Sone, N., Hirata, H., and Kagawa, Y (1978) Biochem. Biophys. Res. Commun. in press.
14. Capaldi, R.A., and Vanderkooi, G. (1972) Proc. Natl. Acad. Sci. USA. 69:930.
15. Sierra, M.F., and Tzagoloff, A. (1973) Proc. Natl. Acad. Sci. USA. 70:3155.
16. Fillingame, R.H. (1976) J. Biol. Chem. 251:6630.
17. Sebald, W. (1977) Biochim. Biophys. Acta 463:1.
18. Nelson, N., Eytan, E., Notsani, B., Sigrist, H., Sigrist-Nelson, K., and Gitler, C. (1977) Proc. Natl. Acad. Sci. USA. 74:936.
19. Criddle, R.S., Packer, L., and Shieh, P.(1977) Proc. Natl. Acad. Sci. USA. 74:4306
20. Sone, N., Yoshida, M., Hirata, H., and Kagawa, Y. (1978) J. Biochem. in press.

DISSOCIATION AND RECONSTITUTION OF ESCHERICHIA COLI F_1-ATPase:
ANALYSIS OF THE DEFECT IN UNCOUPLED MUTANTS

Masamitsu Futai
Hiroshi Kanazawa

Department of Microbiology
Okayama University
Okayama, Japan

I. INTRODUCTION

The coupling factors(F_1-F_0) of oxidative phosphorylation that are found in chloroplasts, mitochondria, or bacterial membranes have similar structures and functions(for review, see ref. 1-3). They are responsible for the synthesis and hydrolysis of ATP in reactions coupled to the electrochemical proton gradient. The catalytic portion(F_1-ATPase) of the complex consists of five subunits α, β, γ, δ, and ε, and is extrinsic to membranes.

Studies of F_1-ATPase from E. coli have the distinct advantage that this organism is amenable to genetic analysis. Uncoupled mutants with defective F_1 have been isolated, and mapped around 83 min of the current linkage map(4)(for review, see ref. 5,6). However, the identity of the mutationally altered polypeptides had not been determined at the time we started this work. We were interested in identifying defective subunits in mutant F_1(no ATPase activity), because such information would be of value in understanding the catalytic function of the complex. Identification of the structural genes of polypeptides in F_1 is also important in understanding the regulation of gene expression and assembly of this complex membrane protein. One of the most decisive approaches for determining defective subunits is to assay which wild type subunit complements the mutant F_1 and pro

---

[1] Supported by a grant from Ministry of Education, Science, and Culture of Japan(#311909)

duce activity. Dissociation of wild type F_1 and reconstitution of F_1-ATPase from isolated subunits have been described recently by Futai(7). This development permitted us to establish the complementation system for assaying the defect in mutant F_1.

In this article we review briefly the isolation and properties of pure subunits from E. coli F_1 and reconstitution of ATPase from isolated α, β, and γ subunits. Finally, we discuss the detailed results of the analysis of mutant F_1 by complementation-type assay.

II. DISSOCIATION AND RECONSTITUTION OF E. COLI F_1

A. Isolation of Pure Subunits from E. coli F_1

Dissociation and reconstitution of F_1 were performed to understand the role of each subunit and the assembly of this complex molecule. The two low molecular weight subunits were obtained in pure form(8) and it was shown that both δ and ε are essential for binding of major subunits assembly(α,β,γ) to F_1-depleted membrane(9). Recently Yoshida et al.(10) reconstituted F_1 from isolated subunits of a thermophilic bacterium. However, their procedure for complete resolution of F_1 is applicable only to very stable enzyme, and could not be used for E. coli F_1. Previous workers have shown that β subunit(11), a mixture of α and β(12) and a fraction relatively rich in γ(12) can be obtained in reconstitutively active form after fractionation of cold dissociated E. coli F_1. We have isolated α, β, and γ in practically homogeneous form as discussed below.

E. coli F_1 was dissociated into subunits after dialysis at 4° against high-salt buffer(0.05 M succinate-Tris buffer pH 6.0 containing 1.0 M KCl, 0.1 M KNO_3 and 0.1 mM dithiothreitol). Three subunits were purified by hydrophobic column chromatography(7): α and β were purified by Phenyl Sepharose(Pharmacia Co., Sweden) and γ by butyl agarose(Miles-Yeda Co., Israel). Dissociated F_1 of different batches had to be used for the separation of γ and other subunits, because γ bound to the Phenyl Sepharose and could not be eluted. Two subunits(α,β) could not be purified by butyl agarose column. In an alternative procedure developped recently(13), the three subunits could be isolated from dissociated F_1 of the same batch by chromatography through hydroxylapatite column followed by DEAE-Separose column. Both procedures yielded the three subunits in practically homogeneous forms(7,13).

B. Reconstitution of F_1-ATPase from Isolated Subunits

The ATPase activity was reconstituted by dialyzing a mixture of α, β, and γ against reconstitution buffer(0.05 M succinate-Tris buffer pH 6.0 containing ATP and Mg^{2+} ion(both 1-2 mM) and 10% glycerol) at 22-25°(7). The maximal reconstitution (30-40 units/mg protein) was obtained when the three subunits were mixed in the ratio of 3:3:1(α:β:γ, molar ratio) at a final protein concentration of 0.2-1.0 mg/ml. ATP could not be replaced by ADP, adenylylimidodiphosphate, or AMP. No activity was obtained by dialyzing the subunits against buffer of pH above 8. Individual subunits alone or combinations of any two did not give significant activity except that a mixture of α + β gave about 10% of the maximal activity. Entire F_1 complex was reconstituted by mixing this ATPase(α,β,γ) with two minor subunits (δ,ε) isolated as described previously(8). This F_1 could bind to membranes previously depleted of F_1 and restored ATP-driven transhydrogenase and oxidative phosphorylation(13). These results suggest that F_1 molecules assembled <u>in vitro</u> were active in energy-driven reaction.

C. Properties of Isolated Subunits

It was of interest to study the properties of isolated subunits, because they are active in forming F_1 as discussed above. Isolated α subunit had a high affinity nucleotide binding site, as studied by equilibrium dialysis. The dissociation constants for ATP and ADP were 0.10 and 0.83 μM, respectively. Each nucleotide seemed to have one binding site in the subunit(0.95 and 0.90 site, respectively, for ATP and ADP obtained from Scatchard-type plot). ADP competitively inhibited the binding of ATP with Ki of 0.70 μM. Adenylylimidodiphosphate inhibited the binding of ADP and ATP competitively. These results suggest that isolated α subunit has a single site to which ATP or ADP can bind. Binding of these nucleotides to other subunits could not be detected by equilibrium dialysis, suggesting that they have no high affinity binding sites for nucleotides.

The isolated β subunit bound aurovertin with an approximate Kd of 3.1 μM(Experiment of Stanley Dunn, Cornell University), confirming the results of isolated β subunit from beef heart (14) and yeast(15) mitochondria. Details of these results(section II, C) will be published elsewhere(13).

III. RECONSTITUTION OF F_1-ATPase FROM COMBINATIONS OF ISOLATED SUBUNITS(WILD TYPE) AND INACTIVE F_1(MUTANT)

A. Reconstitution of F_1-ATPase in Crude Extract from Mutant Cells.

It would be convenient if the defect of mutant F_1 in crude fraction could be determined. This can be done using EDTA(ethylenediaminetetraacetate) extract of membranes as discussed below.

As shown previously by immunochemical studies(16), uncoupled strain AN120(uncA401)(17) and DL54(18) have inactive F_1 molecule(no ATPase activity) attached to membranes. About 85 to 90% of these inactive molecules could be solubilized by washing membranes with 0.5 mM EDTA solution containing 1 mM Tris-HCl pH 8.0 and 10% glycerol. The EDTA-extract thus obtained was concentrated by ultrafiltration through an Amicon Filter(UM10) and the concentrated extract was dialyzed against high-salt buffer for 8 hours as described for active F_1. The extract was then frozen stored at -80° until use. The F_1 molecules are known to be dissociated in this condition.

The treated-extract was thawed and mixed with isolated subunits from wild type F_1. As shown in Table 1, ATPase activity was reconstituted by dialyzing the mixture of AN120 extract and α subunit agaist reconstitution buffer described above. No other subunits from the wild type restored ATPase activity. This result suggests that α is defective in F_1 molecule from AN120. ATPase thus reconstituted could bind to membranes which had been previously depleted of F_1 and could restore ATP-dependent formation of proton-motive force which was estimated by quenching of quinacrine fluorescence as described previously(19). These results suggest that active F_1 molecule was reconstituted in crude extract by the addition of a single subunit.

As a control, similar extract was made from other mutant NR 70(20) which has only a small amount of inactive F_1 in membrane. Essentially no reconstitution of ATPase activity was restored by the combination of this extract with isolated single subunit or a mixture of any of the two subunits. This result suggest that the amount of reconstitutively active subunits was low in the extract of this strain, confirming the previous study(16).

These results suggest that the assay using EDTA-extract is a convenient system for the preliminary characterization of the F_1 mutant. However, the following results suggest that the experiments using crude extract should be interpreted carefully; final conclusions being drawn only after the use of highly purified inactive F_1 molecules from each strain. Isolated γ subunit(wild-type) stimulated reconstitution of ATPase from α and AN120 extract(Table 1), whereas it had no effect on the reconstitution of α and purified inactive F_1 as discussed below(Table 2). This may be due to the loss of substantial amounts of γ subunit during experiment, because treatment to dissociate and reconstitute F_1 included incubation of the extract for a considerable time at 4 or 25°. Traces of protease, if present, may act on the subunit during treatment. In this regard it was reported that three subunits of lower molecular weight were

Table 1. Reconstitution of ATPase Activity in EDTA-extract from Various Mutants with Isolated Subunits

Mutant extract (strain)	Wild-type subunit added (μg)			ATPase activity reconstituted (units/mg protein)
	α	β	γ	
AN120[a] (0.56mg protein per assay)	13			0.11
		15		0.00
			2.2	0.00
	13	15		0.11
	13		2.2	0.29
	13		4.4	0.29
		15	2.2	0.00
DL54 (0.54mg protein per assay)	13			0.00
		15		0.02
			2.2	0.01
	13	15		0.03
	13		2.2	0.00
		15	2.2	0.16
NR70[b] (0.64mg protein per assay)	13			0.00
		15		0.00
			2.2	0.00

The EDTA-extract from each mutant was treated with high-salt buffer in the cold and mixed with isolated subunits. The mixture was dialysed against reconstitution buffer described in the text. The ATPase activity reconstituted was expressed as units/mg extract protein. One unit of ATPase was defined as the amount of enzyme required for the hydrolysis of one μ mole of ATP per min(24). No reconstitution of ATPase could be observed when mutant extract was dialyzed against reconstitution buffer.

[a] Other details for this strain was described previously(19).
[b] Addition of a mixture of any of the two subunits did not give ATPase activity.

sensitive to trypsin, whereas α and β were resistant(21).
   The result was more complicated in the case of DL54. For the restoration of ATPase, α and β needed to be mixed with the extract of this strain(Table 1). No single subunit could restore substantial activity of ATPase in the extract; the β subunit, however, restored about 10% of the maximal activity. This result suggest that the α subunit in this F_1 is active and that the β and γ are inactive. The reason for the defects in these two subunits remained to be determined. Isolation of inactive F_1 has so far been unsuccessful, because the amount of inactive F_1 is low in this strain(about one third of the wild type)(16).

B. Reconstitution of ATPase Activity from Combination of Inactive F_1(AN120) and Isolated α Subunit.

   The defect in AN120 was studied further using purified inactive F_1. The inactive F_1 was purified as described previously (24) using immunodiffusion as an assay. As shown previously(19) purified material was apparently homogeneous judging from the result of gel electrophoresis in the presence of sodium dodecyl sulfate. It is noteworthy that no difference was observed between the molecular weights of wild type and mutant α subunit, although this subunit was suggested to be defective in the mutant as discussed above.
   As shown in Table 2, ATPase activity could be reconstituted by dialysis of a mixture of dissociated F_1(AN120) and α subunit from the wild type. This confirmed the experiment with EDTA extract as discussed above. Other subunits from the wild type had no effect on this reconstitution. It is noteworthy that the γ subunit which stimulated reconstitution in the crude system, had no effect on reconstitution in pure inactive F_1. Recently similar results were obtained independently by Dunn(22), although the details of his work were different from ours as discussed previously(19).
   ATPase could be reconstituted from three major subunits(α, β,γ) from wild type as discussed above. Therefore the present results suggest that α is inactive and β and γ are intact in F_1 from AN120. It has already been suggested that the two minor subunits(δ,ε) are active in this strain(19). Thus both defective and intact subunits were shown in F_1 from the mutant by a series of experiments. This clearly indicates the advantage of the complementation-type assay discussed in this article.
   This study also indicates that uncA401 gene which was shown to be defective in AN120(17) is the structural gene for α subunit. Recently, uncD409 is shown to be a structural gene for β subunit(23). Studies on the defects of other mutants are in progress to identify structural genes for other subunits.

Mutation in uncD409 locus also gave inactive F_1 molecules (no ATPase activity)(24). Thus α and β seem to play essential roles in ATPase reaction of F_1 molecule, because mutations in both subunits gave inactive complex.

Table 2. Reconstitution of ATPase Activity from Combinations of Inactive F_1(Strain AN120) and Isolated α Subunit.

Wild-type subunit added(μg)			ATPase activity reconstituted (units/mg protein)
α	β	γ	
6			5.04
10			18.4
20			27.5
	30		0.00
		4.0	0.00
20		4.0	27.5

Inactive F_1 from strain AN120 was treated with high-salt buffer in the cold, and 67 μg of the dissociated material was mixed with isolated subunits. The mixture was dialyzed against reconstitution buffer and ATPase activity reconstituted was expressed as described in the legend of Table 1. For details see text and ref. 19.

REFERENCES

1. Senior, A. E.(1973) Biochim. Biophys. Acta 301 249-277.
2. Harold, F. M.(1977) Curr. Top. Bioenerg. 6 83-149.
3. Racker, E.(1976) A New Look at Mechanisms in Bioenergetics, Academic Press, New York, N. Y.
4. Backman, B., Low, K. B., and Taylor, A. L.(1976) Bacteriol. Rev. 40 116-167.
5. Simoni, R. D. and Postma, P. W.(1975) Annu. Rev. Biochem. 44 523-554.
6. Cox, G. B. and Gibson, F. (1974) Biochim. Biophys. Acta 346 1-25.
7. Futai, M.(1977) Biochem. Biophys. Res. Comm. 70 1231-1237.
8. Smith, J. B. and Sternweis, P. C.(1977) Biochemistry 16 306-377.
9. Sternweis, P. C.(1978) J. Biol. Chem. 253 3123-3128.
10. Yoshida, M., Sone, N., Hirata, H., and Kagawa, Y.(1977) J. Biol. Chem. 252 3480-3485.
11. Vogel, G., and Steinhart, R.(1976) Biochemistry 15 208-216.
12. Larson, R. J. and Smith, J. B.(1977) Biochemistry 16 4266-4270.
13. Dunn, S. D. and Futai, M., in preparation.
14. Verschoor, G. J., van der Sluis, P. R., and Slater, E. C. (1977) Biochim. Biophys. Acta 462 438-449.
15. Gouglas, M. G., Koh, Y., Dockter, M. E., and Schatz, G. (1977) J. Biol. Chem. 252 8333-8335.
16. Maeda, M., Futai, M., and Anraku, Y.(1977) Biochem. Biophys. Res. Comm. 76 331-338.
17. Butlin, J. D., Cox, G. B., and Gibson, F.(1971) Biochem. J. 124 75-81.
18. Simoni, R. D. and Shallenberger, M. K.(1972) Proc. Natl. Acad. Sci. U.S.A. 69 2633-2667.
19. Kanazawa, H., Saito, S., and Futai, M.(1978) J. Biochem. in press.
20. Rosen, B. P.(1973) J. Bacteriol. 116 1124-1129.
21. Nelson, N., Kanner, B. I., and Gutnick, D. L. (1974) Proc. Natl. Acad. Sci. U.S.A. 71 2720-2724.
22. Dunn, S. D.(1978) Biochem. Biophys. Res. Comm. 82 596-602.
23. Fayle, D. R. H., Downie, J. A., Cox, G. B., Gibson, F., and Radik, J.(1978) Biochem. J. 172 523-531.
24. Futai, M., Sternweis, P. C., and Heppel, L. A.(1974) Proc. Natl. Acad. Sci. U.S.A. 71 2725-2729.

ANTIPORT SYSTEMS FOR CALCIUM/PROTON AND SODIUM/PROTON IN ESCHERICHIA COLI

Tomofusa Tsuchiya
Keiko Takeda

Department of Microbiology
Faculty of Pharmaceutical Sciences
Okayama University
Tsushima, Okayama

We have reported previously that everted membrane vesicles of Escherichia coli mediated uptake of calcium when an energy donor such as a respiratory substrate or ATP was supplied(1). The energization of the Ca^{2+} transport by a proton-motive force which in everted vesicles has an orientation acid and positive interior suggested a Ca^{2+}/H^+ antiport mechanism. It was found that an artificially imposed chemical gradient of H^+ across the membrane(interior acidic) induced the uptake of Ca^{2+} (2). This observation was consistent with a Ca^{2+}/H^+ antiport system. West and Mitchell(3) suggested that cells of E. coli possessed an antiport system for Na^+/H^+. They demonstrated efflux of H^+ on addition of Na^+ to an anaerobic cell suspension. Recently, Schuldiner and Fishkes(4) have shown that a Na^+/H^+ antiport could be demonstrated by measuring the fluorescence change of 9-aminoacridine(9AA) upon addition of Na^+.

We have investigated a cation/H^+ antiport reaction in E. coli. As shown in Fig. 1, addition of succinate quenched the fluorescence of 9AA, indicating that H^+ uptake and subsequent establishment of ΔpH had taken place in the everted membrane vesicles. Significant reversal of the fluorescence quenching was observed on addition of Ca^{2+}, implying that the influx of Ca^{2+} elicited the efflux of H^+, and dissipating the ΔpH to some extent. This is supporting evidence for Ca^{2+}/H^+ antiport mechanism. Similar results were obtained with Na^+. Na^+ did not give a significant effect on the Ca^{2+} transport system and vice versa. H^+ efflux due to Ca^{2+}(or Na^+) influx was not observed if vesicles were preincubated with Ca^{2+}(or Na^+). Sr^{2+} and Mn^{2+} inhibited the Ca^{2+}/H^+ antiport, but did not affect

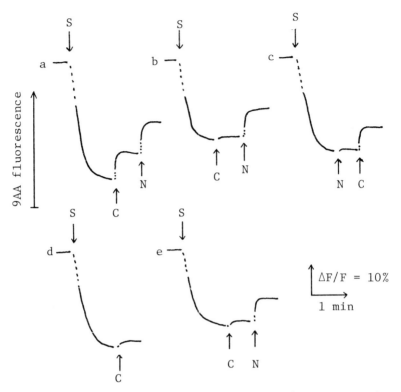

FIGURE 1. Fluorescence change due to Ca^{2+}/H^+ and Na^+/H^+ antiports. Fluorescence quenching of 9AA and partial reversal were measured with everted membrane vesicles. Succinate(K^+ salt)(S), $CaCl_2$(C) and NaCl(N) were added at the points indicated by arrows, to 10 mM, 0.5 mM and 10 mM, respectively. a, control; b, c, d and e, vesicles were preincubated with 0.5 mM $CaCl_2$, 10 mM NaCl, 2 mM $SrCl_2$ or 2 mM $MnCl_2$, respectively.

Na^+/H^+ antiport. H^+ efflux was also detected with a H^+ electrode on addition of Ca^{2+}(or Na^+) to anaerobic cell suspension.

REFERENCES

Tsuchiya, T., & Rosen, B. P. (1975) J. Biol. Chem. 250:7687
Tsuchiya, T., & Rosen, B. P. (1976) J. Biol. Chem. 251:962
West, I. C., & Mitchell, P. (1974) Biochem. J. 144:87
Schuldiner, S., & Fishkes, H. (1978) Biochemistry 17:706

ATP-LINKED CATION FLUX IN
STREPTOCOCCUS FAECALIS

Hiroshi Kobayashi

Research Institute for Chemobiodynamics
Chiba University
Chiba, Japan

Jennifer Van Brunt
Franklin M. Harold

Division of Molecular and Cellular Biology
National Jewish Hospital and Research Center
and
Departments of Biochemistry and Microbiology
University of Colorado Medical Center
Denver, Colorado, U.S.A.

It is now generally accepted that the accumulation of many metabolites and ions by bacteria is linked to the circulation of protons(1). Calcium and sodium transport systems have been studied and it is proposed that these cations are expelled from cells by Ca^{2+}/H^+ or Na^+/H^+ antiporters(2-7). On the other hand, there have been many reports on the presence of bacterial transport systems that require ATP rather than the proton motive force(1). The direct involvement of ATP in these systems remains to be demonstrated. As an approach to this problem we are studying cation transport systems of Streptococcus faecalis using intact cells and everted membrane vesicles and we discovered that the calcium and sodium transport systems of S. faecalis both required ATP(8).

In S. faecalis, calcium uptake by everted membrane vesicles required specifically ATP plus Mg^{2+}. ATP-dependent calcium uptake was not markedly inhibited by either proton conductors or by an inhibitor of the proton-translocating ATPase. $^{45}Ca^{2+}/Ca^{2+}$ exchange reaction also required ATP. We conclude that calcium uptake by everted membrane vesicles is not energized by the proton motive force but by ATP itself. Calcium uptake by intact cells was driven by the membrane potential and did not require ATP. Moreover, the rate of calcium uptake by intact cells did not saturate at high calcium concentration, indicating that calcium uptake by intact cells was not mediated by a specific carrier. On the other hand, calcium extrusion from intact cells required ATP rather than the proton motive force. Other results obtained by studies of calcium extrusion from intact cells were consistent with that of calcium uptake by everted membrane vesicles. Therefore, we conclude that calcium extrusion from intact cells(calcium uptake by everted membrane vesicles) is mediated by an ATP-linked pump. However, it is still unclear whether ATP hydrolysis is essential for calcium transport or not. We were unable to detect either calcium stimulated ATPase activity or phosphorylation of the membrane.

We have also studied the sodium transport system in both everted membrane vesicles and intact cells of S. faecalis. Sodium uptake by everted membrane vesicles required ATP and sodium extrusion from intact cells was driven by ATP rather than the proton motive force. These results suggest that sodium transport of S. faecalis is also mediated by an ATP-linked pump.

REFERENCES

(1). Harold, F. M. (1977). Curr. Top. Bioenerg. 6:83.
(2). Tsuchiya, T., and Rosen, B. P. (1976). J. Biol. Chem. 251:962.
(3). Bhattacharyya, P., and Barnes, E. M., Jr. (1976). J. Biol. Chem. 251:5614.
(4). West, I. C., and Mithell, P. (1974). Biochem. J. 144:87.
(5). Lanyi, J. K., and MacDonald, R. E. (1976). Biochemistry 15:4608.
(6). Bhattacharyya, P., and Barnes, E. M., Jr. (1978). J. Biol. Chem. 253:3848.
(7). Schuldiner, S., and Fishkes, H. (1978). Biochemistry 17:706.
(8). Kobayashi, H., Van Brunt, J., and Harold, F. M. (1978). J. Biol. Chem. 253:2085.

Na^+/K^+ GRADIENT AS AN ENERGY RESERVOIR IN BACTERIA

Vladimir P. Skulachev

Department of Bioenergetics
A. N. Belozersky Laboratory of
Molecular Biology and
Bioorganic Chemistry
Moscow State University
Moscow, U. S. S. R.

I. INTRODUCTION

A. The Question of Biological Function of
Na^+/K^+ Gradient in the Bacterial Cell

It is a general rule that the living cell maintains a gradient of univalent cations between the cytoplasm and environment. For excitable animal tissues, it is well known that Na^+/K^+ gradient is used to support the excitation process. This can, at least partially, explain the biological role of Na^+/K^+ antiport. In nonexcitable animal cells, the function of Na^+/K^+ gradient remained obscure until it was found that a widespread type of osmotic work performed by the outer animal cell membrane is organized as a symport of Na^+ and of a compound accumulated by the cell.

In bacteria, accumulation of solutes is usually carried out by mechanisms other than Na^+ symport (1) and direct coupling of a metabolite and Na^+ translocations is apparently a rather rare case. An exclusion is such a peculiar microorganism as halobacteria (2-5). They are interesting because their intracellular Na^+ concentration can reach 2 M, K^+ concentration being as high as 4 M (6). This important precedent indicates that high level of Na^+ ions inside the cell is not incompatible with life. The fact that some enzyme systems work better in the presence of K^+ than of Na^+ may be regarded as secondary adaptation of enzymes to the K^+-rich and Na^+-poor conditions in cytosol. Considering this problem, we must take into account that energy expenditures required for Na^+/K^+ gradient formation, are very large, i. e. as high as 1/3 of the energy utilized by an average "resting" cell (7). This is apparently too high a price to pay for creating optimal conditions for the functioning of certain enzyme systems. So, the question why bacteria substitute K^+ for Na^+ in their interior, is awaiting solution.

B. The problem of Buffering of $\Delta\bar{\mu}H^+$ Generated Across Bacterial Cytoplasmic Membrane

Recent progress in bioenergetics resulted in the electrochemical potential of H^+ ions ($\Delta\bar{\mu}H^+$) being disclosed in all types of coupling membranes, in accordance with the chemiosmotic principle of energy transduction (8-10). In bacteria, $\Delta\bar{\mu}H^+$ is formed across cytoplasmic membrane. It is produced by thousands of indivudual $\Delta\bar{\mu}H^+$ generators, namely respiratory (or photosynthetic) redox chain enzymes, H^+-ATPase, or H^+-pyrophosphatase. In the same membrane, $\Delta\bar{\mu}H^+$ can be utilized by the above mentioned enzymes to reverse the processes of electron transfer or ATP (PP_i) hydrolysis. Besides, $\Delta\bar{\mu}H^+$ was found to support uphill transport of many substrates across cytoplasmic membrane, as well as rotation of bacterial flagellum (for review, see ref. 11). So, one can conclude that $\Delta\bar{\mu}H^+$ is a convertible form of energy for the bacterial cell.

To perform such a function, energy equivalents stored in the form of $\Delta\bar{\mu}H^+$ (or of component(s) easily equilibrated with $\Delta\bar{\mu}H^+$) must be present in the amounts sufficiently large to buffer the fluctuations of rates of the $\Delta\bar{\mu}H^+$-producing and $\Delta\bar{\mu}H^+$-consuming processes.

Operation of any $\Delta\bar{\mu}H^+$-generator results in electrogenic separation of H^+ and OH^- across a membrane. $\Delta\bar{\mu}H^+$ formed due to this process, is composed of electrical ($\Delta\psi$) and chemical (ΔpH) constituents. As calculations show, the amount of H^+ ions, which must be translocated across the membrane to charge its electric capacity (1 $\mu F/cm^2$) is as small as 1 μmol H^+/g protein (10,12). This amount is of the same order of magnitude as the quantity of enzymes in the bacterial membrane.

Recently, these calculations were confirmed in this group by direct measurement of $\Delta\psi$ formation, when two light-dependent $\Delta\bar{\mu}H^+$ generators were studied. Drs. L. A. Drachev, A. D. Kaulen and A. Yu. Semenov incorporated bacteriorhodopsin membrane sheets from <u>Halobacterium halobium</u> or chromatophores from <u>Rhodospirillum rubrum</u> into a membrane filter impregnated with phospholipids. Illumination of such a system was shown to produce an electric potential difference across the filter, which was measured by a voltmeter (for method, see ref. 13). When a 15 nsec laser flash was used to actuate the system, a single turnover of bacteriorhodopsins (or bacteriochlorophyll reaction center complexes) took place. Such an effect was found to give rise to a potential difference formation of ≈ 80 mV. Thus, the electric capacity of the studied systems was so low, that a rather high $\Delta\psi$ was formed even if $\Delta\bar{\mu}H^+$ generators performed a single catalytic cycle.

To store membrane-linked energy in a "substrate" quantity one must discharge the membrane by means of electrophoretic transmembrane movement of any charged species but H^+.

Flow of a charged penetrant through the membrane down an electric gradient discharges the membrane and hence allows an additional portion of H^+ ions to be translocated by $\Delta\bar{\mu}H^+$ generators. If the amount of the penetrant is sufficiently large, the H^+ concentrations in the membrane-separated compartments change, and a ΔpH forms. The appearance of an H^+ concentration gradient accomapnies that of a gradient of the penetrant. As a result, the energy is stored as ΔpH and $\Delta p[penetrant]$.

Now the quantity of the stored energy depends on:

(i) the pH buffer capacity of the system;

(ii) the amount of the penetrant in the compartment which the penetrant leaves to be transported electrophoretically.

For a bacterial cell, the limiting factor should be pH buffer capacity (which is much higher than the electric capacity of the membrane) if the penetrant is a common cation.

Capacity of the system can be further increased if the formed ΔpH is used to extrude another cation from the cytoplasm by means of an H^+/cation antiport.

Comparing two unsolved problems mentioned above, namely, (1) biological significance of Na^+/K^+ gradient and (2) the requirement of an ion antiport to buffer $\Delta\bar{\mu}H^+$, we postulated that the Na^+/K^+ gradient function is $\Delta\bar{\mu}H^+$ buffering (14).

II. HYPOTHESIS ON Na^+/K^+ GRADIENT AS A $\Delta\bar{\mu}H^+$ BUFFER

It was postulated that electrophoretic influx of potassium ions is responsible for $\Delta\psi$ discharge, while Na^+_{in}/H^+_{out} antiport utilizes the ΔpH formed due to K^+_{out} influx. According to this scheme (14) the initial steps of the energy storage in bacterial cytoplasmic membrane can be described by eqs. (1) and (2):

$$\text{energy sources} \xrightarrow{\text{uphill } H^+_{in} \text{ efflux}} \Delta\psi \qquad (1)$$

$$\Delta\psi \xrightarrow{\text{electrophoretic } K^+_{out} \text{ influx}} \Delta\text{pH}, \Delta\text{pK} \qquad (2)$$

This means that energy sources are utilized by $\Delta\bar{\mu}H^+$ generators to extrude H^+ from the cell. This results in generation of $\Delta\psi$ (internal negative) with no ΔpH being formed. Then electrophoresis of extracellular K^+ into the cell occurs discharging $\Delta\psi$ and allowing a new portion of the energy source to be utilized. As a result, a ΔpH (inside alkaline) and a ΔpK (inside high $[K^+]$) are produced.

K^+ extrusion from the cell down K^+ concentration gradient should prevent $\Delta\psi$ from being lowered when $\Delta\bar{\mu}H^+$ generators are switched off for some time. So, ΔpK can function as a $\Delta\psi$ buffer. However, the cell solving one of its problem in this way, will be confronted with two new problems.

1. Since the intracellular volume is negligible compared to the volume of the environment, a ΔpH formation across bacterial membrane means alkalization of the cytoplasms at a practically constant pH level outside the cell. If $\Delta\psi$ formed by $\Delta\bar{\mu}H^+$ generators is completely transduced to ΔpH, the pH level inside the cell (pH_{in}) must be as high as 11 with the pH outside being neutral. Even partial $\Delta\psi \rightarrow \Delta$pH transduction results in pH_{in} shifts which may prove unfavourable for metabolism. So, stabilizing the membrane-linked energy level we destabilize such an important parameter of cytoplasm as pH. To store energy at low pH changes, high capacity of intracellular pH buffers is needed. In this case, K^+ accumulation should entail an increase in the concentration of deprotonated forms of pH buffers rather than H^+ concentration decrease. However, this hardly solves the problem as pH buffers are enzymes, metabolites etc. Deprotonation of these compounds may result in unfavourable changes of their properties important for their biological functions.

2. A large increase in the concentration of a univalent cation inside the cell is another consequence of operation of the above postulated mechanism. This change in intracellular conditions, like changes in pH and in the degree of protonation of cytoplasmic pH buffers of the cell interior, cannot be

without consequences for cell functioning.

Both problems mentioned can be solved if ΔpH, formed due to K^+ accumulation inside the cell, is used to extrude some other cation, say Na^+, by means of a cation/H^+ antiport (eq. 3):

$$\Delta pH \xrightarrow{Na^+_{in}/H^+_{out} \text{ antiport}} \Delta pNa \qquad (3)$$

This system:

(i) Decreases alkalization of the cell interior.

(ii) Results in lowering the intracellular concentration of a univalent cation (Na^+) which compensates for an increase in that of another univalent cation (K^+).

Under conditions unfavourable for the $\Delta \bar{\mu} H^+$ generator activity, a downhill Na^+ influx accompanied by H^+ efflux might temporary prevent ΔpH disappearance.

ΔpH buffering by Na^+/K^+ exchange system has an obvious advantage over non-specialized intracellular pH buffers since the effect is achieved without any changes in protonation, and hence in the properties, of important intracellular compounds.

Interplay of a K^+ transport system as a $\Delta \psi$ buffer and an Na^+/H^+ exchange mechanism as a ΔpH buffer can greatly facilitate utilization of the energy accumulated in gradients of each of these ions. Let us consider conditions when $\Delta \psi$ is formed due to efflux of pre-accumulated K^+. In this case, performance of $\Delta \bar{\mu} H^+$-dependent work, coupled with an H^+ translocation into the cytoplasm down $\Delta \psi$ must result in acidification of the cell interior, and hence in $\Delta \bar{\mu} H^+$ decrease. This does not occur if H^+ ions transported into the cell when work is performed, are pumped out of the cell in exchange for Na^+_{out}.

Respectively, a work supported by Na^+ influex without K^+ efflux must produce positive charging of the cell interior due to electrogenic H^+ uptake down pH gradient formed by means of Na^+_{out}/H^+_{in} antiport. Downhill K^+ extrusion can compensate for this effect, preventing $\Delta \bar{\mu} H^+$ lowering.

In principle, the role of a $\Delta \bar{\mu} H^+$ buffer could be performed by a gradient of any metabolite transported down $\Delta \psi$ and/or ΔpH, e. g. a gradient of lactose which is known to accumulate in bacteria by means of symport with H^+. However, such mechanisms are hardly effective for a large-scale $\Delta\bar{\mu}_{H^+}$ buffering. In fact, the action of $\Delta\bar{\mu}_{H^+}$ buffer in the cell must result in a situation when $\Delta\bar{\mu}_{H^+}$-consuming functions prove independent of fluctuation of the $\Delta\bar{\mu}_{H^+}$ production and utilization processes. If it were lactose gradient that mainly contributes to the $\Delta\bar{\mu}_{H^+}$ buffering, the work of e. g. the $\Delta\bar{\mu}_{H^+}$-supported flagellar motor of a bacterium should result in a loss of pre-accumulated lactose in response to a decrease of the $\Delta\bar{\mu}_{H^+}$ generator activity. It would be dangerous since lactose is an energy source for the cell.

So, $\Delta\bar{\mu}_{H^+}$ buffering should be carried out by compounds specialized in this role, that are not directly involved in metabolism. Ions of univalent metals meet this requirement. Among them, K^+ accumulation in the cytoplasm and Na^+ extrusion to extracellular space seem to be the most convenient system. It is easy to obtain a large Na^+ gradient extruding Na^+ from the cell. To do so, it is sufficient to decrease Na^+ cencentration in the cell interior occupying a very small portion of the medium. As to extracellular Na^+ concentration, it is sufficiently high since Na^+ is the most common univalent cation in the environment. As far as K^+, the second widespread univalent cation, is concerned, it may be used as a component accumulated inside the cell. Its concentration outside is not sufficiently low to hinder the search for this cation in the environment. On the other hand, it is not too high to induce a large osmotic unbalance across the cytoplasmic membrane if K^+ is accumulated in the cell down $\Delta\psi$ and Na^+ is extruded down ΔpH.

It is difficult to think of an ion pair other than K^+ and Na^+ specialized in $\Delta\bar{\mu}_{H^+}$ buffering in bacteria. For example, Ca^{2+}/Mg^{2+} antiport can hardly perform this function because the affinities of these cations to intracellular substances are vastly different. Besides, accumulation of a bivalent cation inside the cell up to 0.1 M concentration, as is the case with K^+, seems to be impossible without dramatic changes in the state of cytoplasm. Cl^-/F^- or Cl^-/Br^- exchanges are not expedient due to extremely low $[F^-]$ and $[Br^-]$ in the environment. Cl^-/SO_4^{2-} pair seems to be bad as many important properties of these two anions are too different.

The above hypothesis assumes that Na^+/K^+ gradient plays the same role for $\Delta\bar{\mu}_{H^+}$ as creatine phosphate for the other convertible form of energy, ATP. In this context, an important difference between these two energy reservoirs should be discussed. Equilibrium in the creatine phosphotransferase reaction is strongly shifted to the formation of ATP and creatine. So, creatine phosphorylation cannot compete with the majority

of the ATP-utilizing reactions. It starts only when ATP/ADP
ratio reaches a very high level, i. e. when the rate of ATP-
producing reactions is much higher than that of ATP-consuming
processes. This means that initiation of ATP production immediately actuates processes of ATP-dependent work in spite of
the presence of creatine.

On the other hand, Na^+/K^+ mechanism <u>per se</u> does not pre-
sume any threshold $\Delta \bar{\mu} H^+$ value to be actuated. K^+ electrophoresis might, in principle, take place at low $\Delta \psi$ and Na^+/K^+
antiport might occur at small ΔpH. Thus, K^+ and Na^+ fluxes,
if always active, can effectively compete with other $\Delta \bar{\mu} H^+$-sup-
ported processes. Hence, Na^+/K^+ gradient formation, in addition to such a favourable effect as $\Delta \bar{\mu} H^+$ stabilization, can
decelerate initiation of $\Delta \bar{\mu} H^+$-utilizating functions when $\Delta \bar{\mu} H^+$-
generators are switched on.

One may overcome such a difficulty assuming that in bacteria fluxes of univalent cations are regulated like those in
animal cells. For example, opening or closing K^+ channel, we
can weitch on or off Na^+/K^+ mechanism.

When K^+ channel is closed, $\Delta \bar{\mu} H^+$ must be predominantly in
the form of $\Delta \psi$ which reaches its maximum very fast, and the
membrane becomes ready to perform work immediately after energy
is furnished. When it is open, the membrane-linked energy flow
is divided between $\Delta \psi$-, ΔpH-, ΔpK- and ΔpNa-producing processes.
In such a case, energization requires more time and a larger
amount of energy. The first regime is fast but unstable, the
second slow but fluctuation-proof. The latter becomes especially important if the membrane performs simultaneously more than
one type of work, or if the level of available energy sources
strongly changes in time.

III. THE STUDY ON E. <u>coli</u> MEMBRANE ENERGIZATION BY K^+/Na^+ GRADIENT

This group, in cooperation with group of Dr. L. Grinius in
Vilnius State University, undertook a study to verify the above
hypothesis (15). In this chapter, some results of this investigation will be reported.

In the majority of experiments, an <u>E. coli</u> strain without
H^+-ATPase was used to avoid side effects linked with fluctuation in the intracellular ATP level.

First, motility of the bacteria was chosen as a $\Delta \bar{\mu} H^+$-linked type of work of the cell since this function can be easily and directly measured.

In H^+-ATPase-deficient E. coli cells, respiration is the only mechanism competent in the de novo production of $\Delta\bar{\mu}H^+$. $\Delta\bar{\mu}H^+$ buffer, if exists, must prolong the motility for some time after transition to anaerobiosis.

The cells were kept with K^+ and without Na^+ in an open vessel. Then the cells were diluted by the incubation medium (see below) and a drop of the diluted cell suspension was placed under a cover slip to measure motility (for method see ref. 16). Since air was not accessible during the motility measurement, oxygen consumption by the bacteria resulted in anaerobiosis in several min. The hypothesis on the Na^+/K^+ energy buffer implies that an incubation medium with Na^+ and without K^+ must support the motility for some time in anaerobic conditions. In such a medium, downhill K^+_{in} efflux and Na^+_{out} influx were assumed to generate a $\Delta\psi$ (negative inside) and a ΔpH (alkaline inside), respectively. Experiments showed that this is the case.

The data of a typical experiment are given in Fig. 1. One can see that incubation of E. coli in the K^+ medium results in fast decrease in the rate of motility which disappears in 7 min (see the curve designated by K^+). Under these conditions, high $[K^+]_{in}$ is counterbalanced by high $[K^+]_{out}$, and Na^+ is absent both inside and outside the cell. So, there are no K^+ or Na^+ gradients, and $\Delta\bar{\mu}H^+$ buffering proves impossible.

In a medium containing $Tris^+$ as a univalent cation, 13 min are necessary to achieve cessation of motility. In this case there is a K^+ gradient across bacterial membrane as $[K^+]_{in}$ is high and $[K^+]_{out}$ is very low.

The maximal prolongation of motility (up to 17 min) was observed in a Na^+ medium. In the latter case, there are gradients of both K^+ and Na^+ of proper directions.

In all three cases, aeration resulted in immediate activation of motility.

Measurements of the respiration rates in the K^+ and $Tris^+$ media showed that they differ by not more than 5%. So, the observed prolongation of motility in the $Tris^+$ medium cannot be due to difference in time required for the bacterial suspension to consume all the oxygen in the incubation medium.

Respiration in a Na^+ medium proved somewhat lower than in the two other cases studied. Nevertheless, the observed differences in the motilities can hardly be explained by the lower rate of respiration in the Na^+ medium, especially if we take into account characteristic differences in the form of the motility-time relationship. As can be seen in Fig. 1, there is a sizable plateau in the Na^+ curve when the motility rate is stabilized at the level of about 25% of the maximal. Further support for the Na^+ influx-induced membrane energization was obtained in the ANS^- experiments (see below, Figs. 3 and 4).

As to the above postulated K^+ efflux in a K^+-deficient medium, it was indeed observed under anaerobic conditions when K^+ and O_2 concentrations were monitored with K^+ and O_2 electrodes, respectively. As seen in Fig. 2, addition of the cells to an aerobic medium with low K^+ level immediately results in a K^+ uptake which gives way to a K^+ release in about 1 min after complete O_2 exhaustion. It should be stressed that in this and other experiments of the study no K^+ ionophore was added.

In the next series of experiments, anilinonaphthalene sulfonate (ANS^-) was used as a probe for membrane energization (for method, see ref. 15). In Fig. 3, an H^+-ATPase-deficient E. coli mutant was used as in Figs. 1 and 2. The cells were added to an anaerobic incubation medium containing alternatively K^+ or Na^+. In a K^+ medium, a biphasic increase in the ANS^- fluorescence took place, the first phase being smaller than the second. Subsequent aeration resulted in strong fluorescence decrease, which indicates membrane energetization (see (17)). The effect of air was inhibited by uncouplers (not shown). In the Na^+ medium, the first phase was of the same magnitude as in the K^+ medium. However, development of the second phase was greatly decelerated as if the membrane de-energization induced by anaerobiosis proceeded much slower.

In Fig. 4 we used a system differing from the previous one in two respects: (i) An E. coli strain possessing H^+-ATPase was used and (ii) the bacteria were placed first to an aerobic medium and then, several min later, O_2 was evacuated. In all samples 0.03 M MOPS was present. Three media were used, i. e. with K^+, with Na^+ and without any univalent metal ion. As an ANS^- probe showed, O_2 evacuation gave rise to fast de-energization in the K^+ medium, which could be reversed by addition of air.

No de-energization took place for at least 8 min in the Na^+ medium, and subsequent aeration was without measurable effect on the ANS^- fluorescence.

In the sample without K^+ and Na^+, during the first two min after O_2 evacuation, no ANS⁻ fluorescence changes were observed. Then a fluorescence increase that accelerated in time occurred.

Thus, the obtained data prove to be in agreement with the predictions of the above hypothesis.

IV. SOME RELATED OBSERVATIONS

There are several lines of evidence of an electrophoretic K^+ transport through the bacterial cytoplasmic membrane (reviewed by Harold (1)). Among them the data of Rhoads and Epstein (18) are especially interesting. It was found that in E. coli there is an electrophoretic system of K^+ accumulation (the so-called TrkF system), which is characterized by a relatively low rate and unsaturable dependence of this rate upon the outer K^+ concentration. It is the TrkF system that is responsible for K^+ transport if (i) the outer K^+ level is not too low; (ii) the ATP and/or $\Delta\bar{\mu}H^+$ levels are below a certain critical value.

If both ATP and $\Delta\bar{\mu}H^+$ are sufficiently high, another K^+ transport system (TrkA) is actuated which directly utilizes the energy of ATP, and probably the $\Delta\psi$ energy as well. This system is saturable (K_m 1.55 mM) and its activity is very high (V_{max} 550 μg ions K^+/g·min). The TrkF and TrkA systems are constitutive. Growth under K^+-deficient conditions induces formation of a third (Kdp) system of a very high affinity for K^+ (K_m 0.002 mM, V_{max} 150 μg ions K^+/g·min). This system uses ATP, requires a periplasmic protein and is operative even in the absence of $\Delta\bar{\mu}H^+$.

In terms of the above concept, the TrkF system is necessary to accumulate K^+ and transduce a part of $\Delta\psi$ to ΔpH at a usual $[K^+]_{out}$ and a suboptimal level of energy production in the cell. Since this system is of low rate and affinity for K^+, it cannot compete with other energy-consuming processes, so that only a small portion of the energy produced is utilized for uphill K^+ influx. Cessation of $\Delta\bar{\mu}H^+$-generating systems may induce efflux of the pre-accumulated K^+ and hence $\Delta\psi$ buffering. When the activity of the energy-producing mechanisms exceeds that of the energy-consuming ones, ATP and $\Delta\bar{\mu}H^+$-levels are maximal and the TrkA system switches on to transport K^+ much faster and up to a higher gradient than TrkF. As a result, a larger amount of energy can be stored as ΔpK. The $\Delta\psi$-generating efflux of K^+ in this case, as well as in the case of the Kdp system, might occur via the TrkF system, since $[K^+]_{in}$ is high, and this is favourable for TrkF having an unsaturable

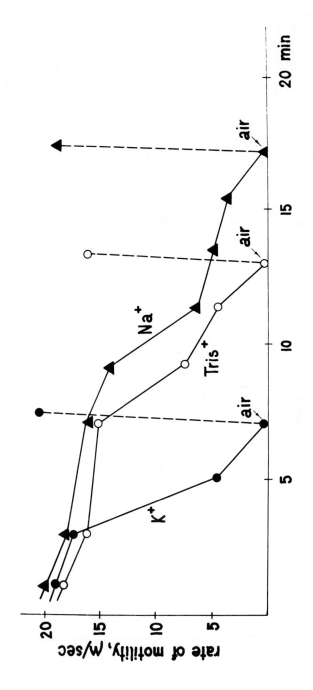

Figure 1. Na$^+$/K$^+$ gradient as an energy source for anaerobic motility of an E. coli strain lacking H$^+$-ATPase (From ref. 15).

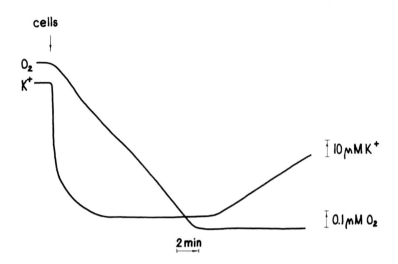

Figure 2. Respiration and K^+ transport in an E. coli strain lacking H^+-ATPase (From ref. 11).

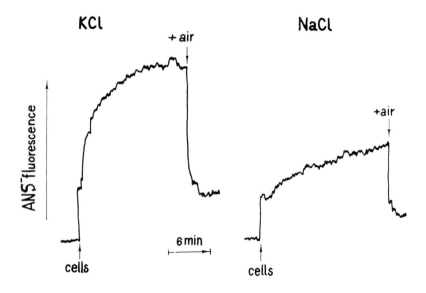

Figure 3. ANS^- responses in an E. coli strain lacking H^+-ATPase (From ref. 15).

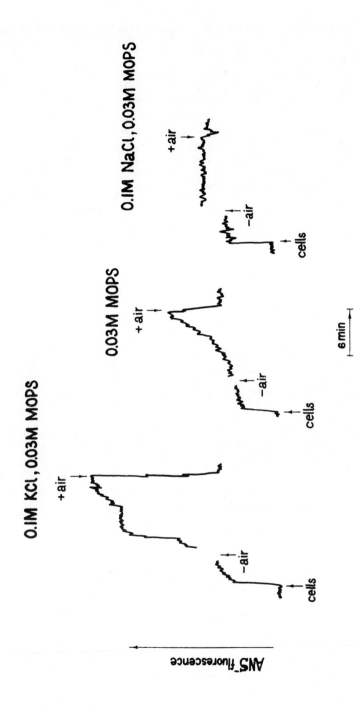

Figure 4. ANS⁻ responses in an E. coli strain possessing H⁺-ATPase (From ref. 15).

[K^+] dependence.

Another example of the above type of regulation was demonstrated by Harold et al. in Streptococcus faecalis (1, 19-21). K^+ accumulation in their S. faecalis strain was found to be electrophoretic, being supported by $\Delta\psi$ formed by H^+-ATPase utilizing glycolytic ATP. It was shown that not only K^+ accumulation but also $^{42}K^+/K^+$ exchange requires glycolysis, and this effect is not abolished by an H^+-ATPase inhibitor, DCCD. The simplest explanation for this fact is that ATP is an allosteric activator of the K^+ carrier. When the ATP level is low, large amounts of energy cannot be spent on K^+ transport since the K^+ carrier is inactive. Only if there is an excess of energy, K^+ transport is actuated, and a large-scale $\Delta\psi$-buffering takes place.

Harold (1) mentioned that Na^+/K^+ antiport in S. faecalis seems to require also ATP as allosteric activator. As to energy source for this exchange, when it occurs against Na^+ concentration gradient, it is ΔpH, so that Na^+ flux is directed from the cell interior to the extracellular medium.

There are observations indicating ΔpK-supported ATP formation in bacteria loaded with K^+ and then suspended in a K^+-free medium containing valinomycin (reviewed by Harold (1)). The only question is whether this effect takes place without valinomycin. A precedent of this kind was apparently furnished by Wagner and Oesterhelt (22) who noted that a downhill K^+ efflux stabilizes the ATP level in Halobacterium halobium cells without valinomycin. The authors concluded that the K^+ gradient in H. halobium can be an energy reserve for ATP synthesis. However, it cannot be ruled out that this property is a feature specific for extremal halophilic bacteria.

In the same paper, capacities of ΔpK and ATP as energy reservoirs were compared. It was found that the switching on of the light, that actuates light-dependent bacteriorhodopsin $\Delta\bar{\mu}H^+$-generators in addition to H^+ pumps of the respiratory chain and ATPase operating in the dark, gives rise to an increase in (i) $\Delta\psi$ by about 20 mV; (ii) ΔpH by 0.5 unit; (iii) [ATP] by 2.8 nmol/kg cell water; (iv) [K^+]$_{in}$ by 1500 nmol/kg cell water.

Assuming H^+-ATPase to transport $2H^+$/ATP, we may conclude that the ratio of the capacity of the ΔpK energy reservoir to that of ATP is 750:2.8, or approx. 250:1. Again, such a high ratio can be specific for halobacteria containing about 4 M KCl inside the cell. However, the above change in $\Delta\psi$ was only about 20%. Higher $\Delta\psi$ changes must induce much higher changes in $[K^+]_{in}$ which exponentially depends upon $\Delta\psi$ (e. g. in the above experiment, 20% increase in $\Delta\psi$ from the original level of about 100 mV resulted in doubling $[K^+]_{in}$).

Wagner et al. (23) calculated that the use of ΔpK instead of $\Delta\psi$ as a form of stored energy increases the capacity of the system by 5 orders. They measured the rate of the dark efflux of K^+ ions pre-accumulated in the light, and concluded that complete exhaustion of the light-dependent K^+ pool took about 24 hr. The authors concluded that ΔpK formed in the light may be used by H. halobium as an energy source during the night.

Wagner and Oesterhelt (22) have also measured the $\Delta\psi$ value across H. halobium membrane as a function of $[K^+]_{out}$. The data confirm the conclusion that K^+ movement is electrogenic. Very high ion specificity of K^+ transport system was demonstrated. Again, as in the cases of E. coli and S. faecalis, a fast K^+ uptake could be observed only when the energy sources were in excess. Namely, the K^+ influx reached its maximum at much higher light intensities than did ATP synthesis (24)

With H. halobium it was shown (2-5) that the ΔpNa energy can be used for uphill transport of many amino acids into the cell, which seems to be coupled with downhill Na^+ influx. In the same bacterium, Na^+/H^+ antiport has been directly demonstrated (5).

Earlier Na^+/H^+ antiport was described by West and Mitchell (25) in E. coli.

An interesting observation has been made by Shuldiner and Fishkes (26) when Na^+ export was studied in E. coli membrane system at neutral and alkaline pH values. It was found that in an alkaline medium, when pH outside is higher than inside and Na^+/H^+ antiport must result in Na^+ being accumulated rather than extruded, this process becomes electrogenic since more than one H^+ is injected into the cell per each Na^+ ejected. In this case, Na^+/H^+ antiport may function as an $\Delta\bar{\mu}H^+$ buffer and, in addition, as a mechanism of the cytoplasm acidification under alkaline external conditions.

V. CONCLUSION

The data of the study undertaken in this group to verify the $\Delta\bar{\mu}H^+$ buffer hypothesis as well as related results of other laboratories testify to validity of the suggestion (14) that Na^+/K^+ gradient across bacterial cytoplasmic membrane plays the role of a $\Delta\bar{\mu}H^+$ buffer.

Further investigation along this line seems to be promising. First of all, it is necessary to study systematically, in different kinds of bacteria, the ΔpK and ΔpNa-supported formation of electric potential and pH differences, correspondingly. The mechanisms of electrophoretic K^+ transport and electroneutral Na^+/H^+ antiport are other very important problems. Regulation of K^+ channel also seems an interesting question, as well as the possible existence of two regimes of cytoplasmic membrane energetics, i. e. fast unbuffered, and slow buffered. Finally, the role of Na^+/K^+ gradient as a reservoir of a membrane-linked energy in eucaryotic membranes appears to be an important problem.

ACKNOWLEDGMENTS

I wish to thank Dr. A. N. Glagolev, Dr. L. Grinium and Professor P. Mitchell for useful discussion, Dr. G. Wagner for sending his manuscript before publication, Ms. N. M. Goreyshina for the help in the preparation of the manuscript, and Ms. T. I. Kheifets for correcting the English version of the paper.

REFERENCES

1. Harold, F. M., Curr. Top. Bioenerg. 6:83 (1977).
2. MacDonald, R. E., and Lanyi, J. K., Biochemistry 14:2882 (1975).
3. Lanyi, J. K., Yearwood-Drayton, V., and MacDonald, R. E., ibid. 15:1595(1976).
4. Lanyi, J. K., Renthal, R., and MacDonald, R. E., ibid. 15:1603 (1976).
5. Lanyi, J. K., and MacDonald, R. E., Biophys. J. 17:32a (1977).
6. Ginzburg, M., Sachs, L., and Ginzburg, B. Z., J. Gen. Physiol. 55:187 (1970).
7. De Witt, W., "Biology of the Cell", W. B. Saunders Co., Phyiladelphia, London, Toronto, 1977.
8. Mitchell, P., Nature 191:144 (1961).
9. Mitchell, P., "Chemiosmotic Coupling in Oxidative and Photosynthetic Phosphorylation", Glynn Research, Bodmin, 1966.
10. Mitchell, P., "Chemiosmotic Coupling and Energy Transduction", Glynn Research, Bodmin, 1968.
11. Skulachev, V. P., FEBS Lett. 74:1 (1977).

12. Mitchell, P., ibid. 78:1 (1977).
13. Drachev, L. A., Kaulen, A. D., and Skulachev, V. P., Biokhimiya (Russian) 41:1478 (1978).
14. Skulachev, V. P., FEBS Lett. 87:171 (1978).
15. Broun, I. I., Glagolev, A. N., Grinius, L. L., and Skulachev, V. P., DAN USSR (Russian) (1979).
16. Belyakova, T. N., Glagolev, A. N., and Skulachev, V. P., Biokhimiya (Russian) 41:1478 (1976).
17. Skulachev, V. P., "Energy Transformation in Biomembranes", Nauka, Moscow (Russian), 1972.
18. Rhoads, D. B., and Epstein, W., J. Biol. Chem. 252:1394 (1977).
19. Harold, F. M., and Papinean, D., J. Membr. Biol. 8:27 (1972).
20. Harold, F. M., and Papinean, D., ibid. 8:45 (1972).
21. Harold, F. M., and Altendorf, K. H., Curr. Top. Membr. Transp. 5:2 (1974).
22. Wagner, G., and Oesterhelt, D., Ber. Deutsch. Bot. Ges. 89:289 (1976).
23. Wagner, G., Hartmann, R., and Oesterhelt, D., Eur. J. Biochem.   (1978).
24. Oesterhelt, D., Gottschlich, R., Hartmann, R., Michel, H., and Wagner, G., Soc. Gen. Microbiol. Symp. 27:333 (1977).
25. West, I., and Mitchell, P., Biochem. J. 144:87 (1974).
26. Schuldiner, S., and Fishkes, H., Biochemistry 17:706 (1978).

ON THE PROTON-PUMPING FUNCTION OF CYTOCHROME c OXIDASE

Mårten Wikström
Klaas Krab

Department of Medical Chemistry
University of Helsinki
Helsinki, Finland

I. INTRODUCTION

The main postulate of Mitchell's chemiosmotic theory, according to which the coupling between mitochondrial respiration and ATP synthesis takes place by means of an electrochemical proton gradient ($\Delta\tilde{\mu}_{H^+}$) generated across the mitochondrial inner membrane (1), is largely accepted today. The molecular mechanisms of generation and utilization of $\Delta\tilde{\mu}_{H^+}$ by the respiratory chain and the mitochondrial H^+-ATPase respectively, are as yet unsolved, and represent central issues of today's bioenergetics.

Mitchell's redox loop hypothesis of the mechanism of generation of $\Delta\tilde{\mu}_{H^+}$ by the respiratory chain (1) withstood most experimental tests until recently, when Brand et al. (2) demonstrated that the $\overleftarrow{H^+}/2e^-$ quotients of redox-linked proton ejection of 4.0 and 6.0 in mitochondria oxidising succinate or NADH-linked substrates respectively (3), were underestimated by 2.0 due to fast influx of H^+ together with P_i into the mitochondria via the H^+/P_i symporter. True $\overleftarrow{H^+}/2e^-$ quotients of 6.0 and 8.0 respectively are clearly very difficult or impossible to reconcile with the redox loop arrangement of the respiratory

---

Abbreviations: DADH$_2$, diaminodurene (2,3,5,6-tetramethylphenylenediamine); EGTA, ethyleneglycolbis(aminoethyl)tetraacetate; E_m, half-reduction (midpoint) redox potential; FCCP, carbonylcyanide trifluoromethoxyphenylhydrazone; $\overleftarrow{H^+}$, proton ejection, ejected protons; HEPES, N-2-hydroxylethylpiperazine-N'-ethanesulfonate; NEM, N-ethylmaleimide; $\Delta\psi$, electrical membrane potential; $\Delta\tilde{\mu}_{H^+}$, electrochemical proton gradient.

chain. More recently, Lehninger et al. (4,5) and Azzone et al. (6) have claimed even higher $\overleftarrow{H}^+/2e^-$ stoichiometries, but we have found it impossible to reconcile these with our own findings (see ref. 7 and below).

The most obvious alternative mechanism to redox loops is that the respiratory chain complexes function as redox-linked proton pumps, the molecular structure and mechanism of which would be expected to differ fundamentally from the vectorially orientated redox reactions (H- and e$^-$ transfer) that characterise redox loops.

The cytochrome c oxidase reaction has usually been taken as the best example of a pure electron-translocating step of a redox loop (see 1,8-10). Critical inspection of the literature inevitably leads to the conclusion that this is the only respiratory chain segment for which some positive evidence for pure electron translocation has been reported, whereas the evidence for other segments is merely consistent with this notion.

We have recently shown evidence suggesting that cytochrome oxidase is a redox-linked proton pump (11-14). Our work with intact mitochondria and submitochondrial particles respectively, has been largely confirmed by Sigel and Carafoli (15) and by Sorgato et al. (16). In addition, we have provided evidence for the proton-pumping function of cytochrome oxidase with artificial phospholipid vesicles into the membranes of which the isolated and purified enzyme was incorporated (12,14).

However, our results and interpretation have also been seriously challenged, on the one hand by Moyle and Mitchell (17-20) who maintain that the oxidase is simply electron-translocating, and on the other, by Lehninger et al. (5,21) and Azzone et al. (6, and see this volume), who have claimed recently that the cytochrome oxidase reaction is linked to ejection of $4H^+/2e^-$ in mitochondria rather than $2H^+/2e^-$ as we originally found in all studied systems.

We have already shown that some of the criticism by Moyle and Mitchell is unfounded (22,23), and in this paper we will proceed by showing that a "classical" oxygen pulse type of experiment (cf. 2,3) with rat-liver mitochondria oxidising added cytochrome c, unequivocally demonstrates that H^+-pumping function of cytochrome oxidase. We will also show that an important piece of evidence, previously thought to prove an electron-translocating function of the enzyme, must be reconsidered in the light of new data.

II. PROTON TRANSLOCATION LINKED TO OXIDATION OF ADDED CYTOCHROME c IN MITOCHONDRIA

Fig. 1 shows that pulses of O_2 added to anaerobic mitochond-

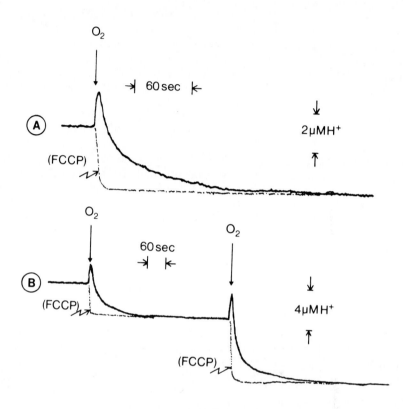

FIGURE 1. Proton translocation linked to oxidation of cytochrome c in rat-liver mitochondria. 120 mM KCl, 10 mM MgSO$_4$, 1 mM EGTA, 1 mM HEPES was supplemented with 15 µM rotenone, 0.53 µg/ml valinomycin, 5.3 µg/ml oligomycin, 0.26 mM NEM and mitochondria (0.46 µM cytochrome aa_3). After saturation with N$_2$, 118 µM ferrocytochrome c and 84 pmoles antimycin per mg protein were added. Volume 3.8 ml, pH 7.04, 24±0.1°C. After 20 min of preincubation known amounts of oxygen were injected as aliquots of medium. pH was recorded with a glass electrode and the scale calibrated in terms of H$^+$ using standard HCl. Acidification = upward deflection. Anaerobiosis followed 1-2 sec after addition of O$_2$. The oxygen addition in A, and the first addition in B are 1.57 µM O. Second addition in B is 3.13 µM O. Dotted traces: In the presence of 0.33 µM FCCP.

ria suspended in KCl-medium in the presence of added cytochrome $\underline{c}$, results unequivocally in ejection of H^+ ($\overleftarrow{H^+}$). The effect is fully abolished by uncoupling agents (FCCP), and requires the presence of valinomycin and N-ethylmaleimide (NEM). It is of utmost importance (cf. below) to note that the extent of the overall consumption of H^+ during the reaction is independent of FCCP and is, whthin experimental error (<5%) equal to the amount of 1/4 O_2 added. This is expected from the stoichiometry of the overall chemical reaction, viz,

(1)   cyt. $\underline{c}^{2+}$ + 1/4 O_2 + H^+ → cyt. $\underline{c}^{3+}$ + 1/2 H_2O

and excludes the operation of any scalar acid-producing reactions under the experimental conditions, for instance, such as the one suggested by Moyle and Mitchell (17-20). Similar controls in the cytochrome oxidase vesicle system (12,14) also excluded artefacts of this kind.

Fig. 1 alone provides firm evidence for the proton-translocating function of cytochrome oxidase. In fact, the evidence is just as strong as that for respiratory chain-linked proton translocation originally reported by Mitchell and Moyle (3), who introduced the O_2-pulse technique. The extrapolated (cf. 3) $\overleftarrow{H^+}/e^-$ quotients of proton ejection in experiments such as that of Fig. 1 are close to 1.0 (a typical series of eight O_2 pulses yielded a mean of 0.78 ± 0.14 S.D.), in agreement with our earlier findings. Experiments have also been performed pulsing aerobic mitochondria with ferrocytochrome $\underline{c}$ with an equally undisputable result (22).

This result contrasts diametrically to the reports by Moyle and Mitchell (17-20), who failed to observe any $\overleftarrow{H^+}$ linked to oxidation of added cytochrome $\underline{c}$. Instead, these workers describe an artefactual (scalar) production of H^+, which they can abolish by adding Mg^{2+} (17,19,20). Inspection of their published results reveals that in all cases where this artefact was suppressed, the experiment was performed either in the absence of added K^+ (18-20) and/or in the absence NEM (20). In the first case, the necessary charge compensation for H^+ translocation in the form of K^+ uptake catalysed by valinomycin is not properly fulfilled (see e.g. ref. 3), so that no $\overleftarrow{H^+}$ is detected extramitochondrially. Under such conditions there was also no effect of NEM (18), which is hardly surprising due to the limiting effect of $\Delta\psi$ on net ejection of H^+. In the experiments performed in KCl medium with appropriate suppression of the acidification artefact (20) no NEM was present. This reagent is particularly important in O_2-pulse experiments as the long anaerobic incubation of the mitochondria leads to extensive efflux of P_i, which is then rapidly taken up with H^+ following the oxygen pulse unless the H^+/P_i symporter is blocked by NEM (2,23). We should emphasise that in the experiment of Fig. 1

the conditions were such that Moyle's and Mitchell's acidification artefact is suppressed (10 mM Mg^{2+}, see ref. 20). The absence of such an artefact can also be directly excluded as discussed above.

In most aerobic reductant pulse experiments, as opposed to O_2-pulse experiments, NEM is not required for demonstration of H^+ translocation by cytochrome oxidase. This is because the very slow rotenone- and antimycin-insensitive endogenous respiration is nevertheless sufficient to retain most of the P_i due to the tightness of the membrane. However, in cases where the mitochondria, for one reason or another, are less tightly coupled, P_i will leak out also aerobically and then NEM will be required for optimal observation of H^+ ejection (cf. refs. 11,12,22). This variability in the effect of NEM can easily be demonstrated experimentally by intentional partial uncoupling of the mitochondria (unpublished).

III. A NOTE ON THE OXIDATION OF DIAMINODURENE

The experiments by Moyle and Mitchell using diaminodurene as substrate (18) were also performed in a sucrose medium, which, in part, explains the low $\overleftarrow{H}^+/O$ quotients reported as we have shown directly (23). In addition, these authors assumed that oxidation of $DADH_2$ at pH 7 gives rise to $2H^+$. We have shown that much less than $2H^+$ are released from the redox mediator on oxidation at this pH (23). This was confirmed by Mitchell and Moyle (20), but they claimed that the release of 2 $H^+/DADH_2$ oxidised is achieved in the presence of mitochondria. However, under the conditions of a typical O_2-pulse experiment, the formed oxidised DAD is rapidly rereduced by endogenous mitochondrial hydrogen donors. This can be shown directly by monitoring spectrophotometrically the reduction of ferricyanide in a mitochondrial suspension containing diaminodurene (unpublished). We previously showed (23) that the H^+/O quotient of the cytochrome oxidase reaction (after appropriate correction for H^+ released from the redox mediator) is close to 2.0 with $DADH_2$ as substrate in a KCl-medium supplemented with NEM.

IV. THE DEPENDENCE OF THE REDOX POTENTIAL OF CYTOCHROME a ON MEMBRANE POTENTIAL

Hinkle and Mitchell (24) showed that the redox potential of cytochrome $\underline{a}$ is shifted relative to that of cytochrome $\underline{c}$ when a diffusion potential of either K^+ or H^+ ions is applied

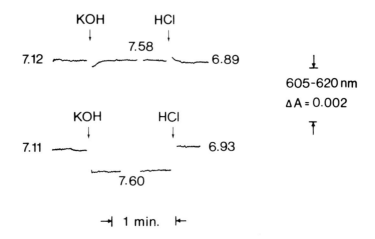

FIGURE 2. Dependence of E_m of cytochrome a on intramitochondrial pH. Experimental conditions were similar to those described in ref. 25, but with an additional 40 nmol/mg protein of NEM. Ferro-/ferricyanide redox buffer is added to poise cytochrome c at a constant potential at which cytochrome a is approx. half-reduced. Alkali or acid pulses are given in the absence (upper trace) or presence (lower trace) of FCCP and valinomycin. The corresponding pH changes in the suspension are indicated in the figure. The mitochondrial suspension is aerobic in the rpesence of rotenone, antimycin and cyanide (cf. ref. 25). Oxidation of cytochrome a results in downward deflection of the trace at the employed wavelength couple. Control experiments showed that no change occurred in the redox state of cytochrome c.

across the membrane of CO-inhibited mitochondria. Since the apparent E_m of cytochrome a was independent of extramitochondrial pH, it was concluded on the basis of these findings that cytochromes c and a interact by vectorial electron transfer orientated perpendicular to the membrane so that the redox equilibrium is affected by the electrostatic field across the membrane. Mitchell and Moyle (20) recently made the conclusive statement that this result •••"excludes any energetic role of the redox change of cytochrome a in proton pumping through cytochrome oxidase". However, Hinkle and Mitchell did not exclude the possibility that the electrical field would interact with cytochrome a through perturbation of the proton activity in a proton well connecting the cytochrome with the matrix (M-) phase. In fact, Artzatbanov et al. (25) have recently shown that the E_m of cytochrome a is dependent of pH in the M-phase in cyanide-inhibited mitochondria. As shown in Fig. 2, we have confirmed this finding. A pH jump in the mitochondrial suspension has little or no effect on the E_m of cytochrome a unless either FCCP (not shown) or FCCP plus valinomycin are present. The latter condition shows that the effect cannot be due to a membrane potential, but must result from the change in pH in the M-phase. The absence of effect of extramitochondrial pH is more dramatic if the H^+/P_i symporter is blocked, for instance by NEM (cf. Fig. 2 and ref. 25).

Thus, in cyanide-inhibited mitochondria, the reduction of cytochrome a by cytochrome c is linked to uptake of H^+ from the M-phase. This is, however, not the case in CO-inhibited mitochondria (unpublished), where the E_m of cytochrome a is independent of pH also in the M-phase. It must be concluded that the behaviour of cytochrome a with respect to redox-linked proton uptake is a function of the redox and/or ligand state of cytochrome a_3, and therefore, that the categorical statement cited above is premature.

V. CONCLUSION AND FINAL REMARKS

In contrast to other coupling regions of the respiratory chain, the span between cytochrome c and O_2 encompasses no classical hydrogen carriers. An unequivocal demonstration of proton pumping catalysed by cytochrome c oxidase therefore provides strong evidence for the notion that respiratory chain complexes in general function as redox-linked proton pumps rather than being organised as redox loops.

As shown in this and two previous papers (22,23), we have carefully considered the criticisms raised by Moyle and Mitchell (17-20) against our proposal. In each case a likely reason for the apparent discrepancy has been pinpointed, and proton pumping has been demonstrated under appropriate experimental conditions. Moreover, we have published direct experimental controls that rule out the specific artefacts proposed by Mitchell and Moyle (20) to be the basis for our viewpoint.

Lehninger et al. (5,21) have recently claimed that the stoichiometry of proton pumping by cytochrome oxidase is 4 $\tilde{H}^+/2e^-$, which is twice the quotient that we originally reported. They suggest that the high stoichiometry is obtained at high concentrations of ferrocyanide in disagreement with our original paper (12), where experiments were reported at ferrocyanide concentrations up to 7.2 mM. We have more recently rechecked these data under conditions as close as possible to those used by

Lehninger (sucrose-based media and ferrocyanide up to 10 mM), but we have been unable to obtain $\overleftarrow{H^+}/2e^-$ quotients higher than 2.0. Lehningers data with TMPD plus ascorbate as substrate also disagree with the recent findings by Sigel and Carafoli (15), who reported a $\overleftarrow{H^+}/2e^-$ stoichiometry of nearly 2 also in this system in full agreement with our findings.

We conclude that cytochrome c oxidase is the first clearcut case of a proton pumping respiratory chain complex. Since, as concluded in section IV, transmembraneous electron transfer in cytochrome oxidase must be considered doubtful, there remains no strong evidence for electron transfer orientated in this fashion in mitochondria. It is possible that respiratory chain complexes are organised in much the same fashion as the mitochondrial H^+-ATPase, viz. with the catalytic (redox) portion of the protein protruding into either the aqueous C- or M-phase, and with a proton-conducting portion (see 26) residing within the hydrophobic membrane proper.

REFERENCES

1. Mitchell, P., "Chemiosmotic Coupling in Oxidative and Photosynthetic Phosphorylation", Glynn Research Ltd., Bodmin, 1966.
2. Brand, M. D., Reynafarje, B., and Lehninger, A. L., J. Biol. Chem. 251:5670 (1976).
3. Mitchell, P., and Moyle, J., Biochem. J. 105:1147 (1967).
4. Reynafarje, B., Brand, M. D., and Lehninger, A. L., J. Biol. Chem. 251:7442 (1976)
5. Lehninger, A. L., Reynafarje, B., and Alexandre, A., this volume, P.
6. Azzone, G. F., Pozzan, T., Di Virgilio, F., and Miconi, V., in "Frontiers of Biological Energetics: From Electrons to Tissues" (P. L. Dutton et al. eds.), Academic Press, New York, in press.
7. Wikström, M., and Krab, K., in "Proc. 29th Mosbach Coll." (G. Schäfer and M. Klingenberg, eds.), Springer Verlag, in press.
8. Hinkle, P. C., Fed. Proc. 32:1988 (1973).
9. Skulachev, V. P., Ann. N. Y. Acad. Sci. 227:188 (1974).
10. Papa, S., Biochim. Biophys. Acta 456:39 (1976).
11. Wikström, M. K. F., Nature 266:271 (1977).
12. Wikström, M. K. F., and Saari, H. T., Biochim. Biophys. Acta 462:347 (1977).
13. Wikström, M., in "The Proton and Calcium Pumps" (G. F. Azzone et al. eds.), p. 215. Elsevier/North-Holland Biomedical Press, 1978.
14. Krab, K., and Wikström, M., Biochim. Biophys. Acta 504:200

(1978).
15. Sigel, E., and Carafoli, E., Eur. J. Biochem. 89:119 (1978).
16. Sorgato, M. C., Ferguson, S. J., Kell, D. B., and John, P., Biochem. J. 174:237 (1978).
17. Moyle, J., and Mitchell, P., FEBS Lett. 88:268 (1978).
18. Moyle, J., and Mitchell, P., FEBS Lett. 90:361 (1978).
19. Mitchell, P., and Moyle, J., in "Frontiers of Biological Energetics: From Electrons to Tissues" (P. L. Dutton et al. eds.), Academic Press, New York, in press.
20. Mitchell, P., and Moyle, J., in "Proc. Japanese-American Conference on Cytochrome Oxidase, Kobe, Japan" (T. E. King et al. eds), in press.
21. Alexandre, A., Reynafarje, B., and Lehninger, A. L., Proc. Nat. Acad. Sci. U.S. in press.
22. Wikström, M. K. F., and Krab, K., FEBS Lett. 91:8 (1978).
23. Wikström, M., and Krab, K., in "Frontiers of Biological Energetics: From Electrons to Tissues" (P. L. Dutton et al. eds.), Academic Press, New York, in press.
24. Hinkle, P., and Mitchell, P., J. Bioenergetics 1:45 (1970).
25. Artzatbanov, V. Yu., Konstantinov, A. A., and Skulachev, V. P., FEBS Lett. 87:180 (1978).
26. Wikström, M., Saari, H., Penttilä, T., and Saraste, M., in "Fed. Eur. Biochem. Soc. Symp., Vol. 45" (P. Nicholls et al. eds.), p.85. Pergamon Press, Oxford and New York, 1978.

MECHANISM OF H$^+$ PUMP IN RAT LIVER MITOCHONDRIA

G. F. Azzone
T. Pozzan
F. Di Virgilio
M. Miconi

C.N.R. Unit for the Study of Physiology of Mitochondria and Institute of General Pathology, University of Padova, Padova, Italy

I. INTRODUCTION

A fundamental achievement of bioenergetics in the last 20 years is the recognition that pumping of H$^+$ is the basis for energy conservation. This is true for all energy transducing membranes whether mitochondria, bacteria and chloroplasts. In spite of its fundamental importance the nature of H$^+$ transport is completely unknown. Mitchell (1) has pioneered a "direct" mechanism based on the idea that H$^+$ and e$^-$ do not move separately across the membrane but combined through a hydrogen carrier. The subsequent movement of e$^-$ through an e$^-$ carrier gives rise to release of H$^+$ and charge separation. The whole respiratory chain is then seen as an alternation of H and e$^-$ carriers arranged in transmembrane loops. If the "direct" mechanism is correct, the H$^+$/e$^-$ ratio should be 1 at each coupling sites in all energy transducing membranes. The indirect mechanism (2-4) on the other hand assumes that H$^+$ and e$^-$ move independently, i.e. the transfer of e$^-$ causes conformational changes in the proteins of each respiratory chain complex whereby H$^+$ are displaced along specific channels. In the "indirect" mechanism H$^+$/e$^-$ ratios higher than 1 may be envisaged. Determination of the H$^+$/e$^-$ stoichiometry appears the most direct means to distinguish between "direct" and "indirect" mechanism of H$^+$ transport. The number of controversial reports on the H$^+$/e$^-$ ratios indicates however this determination as a difficult task.

Mitchell and Moyle (5,6) introduced the oxygen pulse technique to determine the H$^+$/e$^-$ ratio and obtained H$^+$/e$^-$ ratios exactly as predicted by the loop scheme of the chemiosmotic mechanism. With oxygen as electron acceptor, the H^+/oxygen ratio were 6 and 4 with NAD linked and succinate as substrates, respectively. With Fe''' as electron acceptor the $H^+/2\ e^-$ ratio were 4 and 2 after correction for the FCCP insensitive release, with NAD-linked and succinate as substrates, respectively.

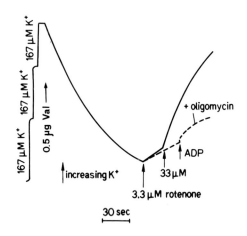

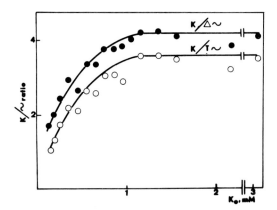

Fig. 1 - <u>Kinetics and stoichiometry of the H^+ pump.</u>
The upper experiment is after Rossi and Azzone (10) and the lower after Azzone and Massari (8). Rat liver mitochondria oxidizing aerobically endogenous substrates or β-hydroxybutyrate.

Ten years ago our laboratory (7-9) suggested to determine the H^+/e^- ratio by measuring: a) the rates of the ion fluxes in steady state and b) K^+ instead of H^+. The steady state technique has the great advantage of permitting determination of the stoichiometry on an amount of ion translocation 2 orders of magnitude larger that involved in the oxygen pulse technique. This renders the measurements not only more precise but also independent of side reactions. Given that in steady state macroscopic electroneutrality necessarily holds, replacement of H^+ with K^+ renders the determination independent of the reuptake of weak acids (which mask part of the H^+ extrusion).

Fig. 1 shows the experimental system used by our laboratory (10). Mitochondria were incubated aerobically in the presence of KCl and Pi, and K^+ uptake initiated by the addition of valinomycin. A phase of K^+ uptake, paralleled by a stimulation of the respiration, occurs. This phase lasts several minutes. K^+ uptake and respiration then level off. Addition of rotenone initiates a K^+ efflux which is coupled to ATP synthesis. The important feature of the experiment is the length and dimension of the phase of K^+ uptake. Fig. 1 shows the $K^+/\sim$ ratios determined on this experimental system with β-hydroxybutyrate as substrate. The ratios approached a value of 4 at about 1 mM KCl (8).

The experiments of Fig. 1 were taken to indicate that the mitochondrial respiratory chain extrudes 4 H^+ per coupling site, i.e. 12 with NAD linked substrates and 8 with succinate. This was in agreement with the ratios previously found on Ca^{++} transport by Chance (11) and by Rossi and Lehninger (12). At the time the Ca^{++}/oxygen ratios were measured their relevance as to the H^+ pump stoichiometry was not understood also in view of the fact that the H^+/Ca^{++} ratio was found to be 1.

Mitchell (13) has criticized the $K^+/\sim$ ratios on the basis of two main arguments: a) that it is measured on K^+ and not on H^+ and b) that there may be a K^+ uptake independent of respiration. Both arguments would be valid were the stoichiometry measured on an amount of K^+ uptake lower than the amount of charge separation required to build up a $\Delta\psi$ of 200 mV (this being 1 nmol H^+ x mg $prot^{-1}$). However in the experiment described in Fig. 1 the amount of K^+ transport in steady state is more than 2 orders of magnitude greater than that required to charge the membrane. Thus H^+ and K^+ must be equivalent and any preexisting $\Delta\psi$ cannot affect the steady state stoichiometries. The criticism of Mitchell has met a large consensus with two consequences. First, the low stoichiometry has become extensively accepted while the high stoichiometry has been mostly ignored. Second, the oxygen pulse technique has maintained a privileged consideration as compared with the steady state method.

In 1975 Lehninger and coworkers (14,15) started a systematic investigation of the H^+ pump stoichiometry. They measured again the stoichiometry of aerobic Ca^{++} uptake, under conditions where there is a stoichiometric uptake of weak acids, and obtained a ratio of 4 charges/site. Furthermore they measured the H^+/oxygen ratios both by oxygen pulses and steady state methods and found a substantial increase of the H^+/oxygen ratio (say from 2 to 3-4) when the reuptake of endogenous Pi was inhibited by N-ethylmaleimide.

A more recent development concerns the dissection of the stoichiometry of the H^+ pump at each of the three respiratory chain coupling sites. Wikström (4) first reported, in contrast with Mitchell and Moyle (5), that addition of Fe" to antimycin inhibited mitochondria results in a H^+ extrusion with a H^+/e^- ratio of 1. While Wikström observations were under attack by Moyle and Mitchell (16), an extensive investigation was carried out both in the laboratory of Lehninger (17) and in ours on the stoichiometry of the H^+ pump at the three coupling sites. Some of the basic finding are reported below.

II. METHODS

Preparation of rat liver mitochondria, measurements of H^+, K^+ and Ca^{++} transport and of oxygen uptake were carried out according to standard procedures (18).

III. RESULTS AND DISCUSSION

Stoichiometry of the H^+ pump at site III.

In the experiment of Fig. 2, mitochondria were incubated in 0.12 M KCl and supplemented with 1 µM rotenone + 60 pmols antimycin A x mg protein^{-1}. The reaction was initiated by the addition of 15 pmols valinomycin x mg protein^{-1}, 30 seconds after Fe". It is seen that the rate of H^+ extrusion increased with the increase of Fe" concentration, reaching a plateau at about 10 mM Fe". Parallel to the increase of H^+ extrusion there was also an increase of the rates of Fe" oxidation and of oxygen uptake. At all Fe" concentrations both $H^+/2e^-$ and H^+/oxygen ratios were higher than 3 and often approached the value of 4. Fig. 2 shows also that the number of e^- moving through coupling site III was about 20% less when calculated on the Fe" oxidation than when calculated on the oxygen uptake.

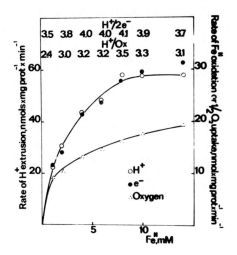

Fig. 2 - $H^+/2e^-$ and H^+/oxygen stoichiometry at coupling site III. The medium contained 0.12 M KCl, 2 mM Tris-Cl pH 7.4, 0.1 mM EGTA, 200 μM NEM, 8 mM $MgCl_2$, 1 μM rotenone, 60 pmol antimycin A x mg protein^{-1}, and 2 mg mitochondrial protein/ml. Temperature 14°C. After 3 minutes of preincubation, variable amounts of Fe" and then 45 pmol valinomycin x mg prot^{-1}. Rates of Fe" oxidation were corrected for the antimycin insensitive Fe"' reduction.

Consequently the H^+/oxygen ratio were 20% lower when calculated on the oxygen uptake than when calculated on the Fe" oxidation data. Moyle and Mitchell (16) have suggested that some of the Fe"' formed during Fe" oxidation be reduced by other hydrogenated component of the respiratory chain. The slight discrepancy between oxygen and Fe"' measurements may however be due to other reasons. One is the mitochondrial swelling due to K^+ uptake. Since swelling causes decrease of absorbance, this results in an apparent decrease of the rate of Fe"' formation. When the Fe" oxidation data were corrected for the swelling induced decrease of absorbance (as occurring during H^+ extrusion with ascorbate + TMPD at equal respiratory rates) the discrepancy between oxygen electrode and spectrophotometric data was less than 10%. In the present work both polarographic and spectrophotometric measurements were used.

In Fig. 3 mitochondria were incubated in a sucrose medium and then supplemented with 2 mM Fe". This initiated a short phase of Fe" oxidation (upper trace) which rapidly levelled off. Addition of 100 μM Ca^{++} induced a new phase of Fe" oxidation which levelled off in about 2 minutes. The lower trace shows the process of Ca^{++} uptake. The uptake of 100 µM Ca^{++} was completed in about 2 minutes parallel to the enhancement of the Fe" oxidation. In the Figure are also reported the charge/site ratios as measured after 30, 60, 90 and 120 seconds of incubation. In the case of Fig. 3, Ca^{++} uptake leads to the precipitation of $Ca_3(PO_4)_2$ in the matrix and thus no swelling occurs.

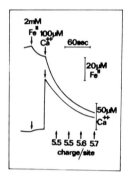

Fig. 3 - <u>Charge/site ratio at coupling site III</u>. The medium contained 0.25 M sucrose, 10 mM Tris-Cl, pH 7.8, 0.2 mM Pi-Tris, 1 µM rotenone, 1 mM $MgCl_2$, 60 pmols antimycin A x mg $prot^{-1}$, 500 pmol oligomycin x mg $prot^{-1}$, and 4 mg prot/ml. Temperature 14°C. The upper trace indicates Fe" oxidation and the lower Ca^{++} uptake, in presence of 100 µM murexide. Stoichiometries are calculated on total Fe" oxidation and Ca^{++} uptake.

The question arises as to why the H^+/site ratios of Fig. 2 and the charge/site ratios of Fig. 3 are higher than those reported by Wikström (4,19). In the case of the H^+/site ratio, where the H^+ extrusion lasts only few seconds an important difference lies in the way the reaction is initiated. Higher ratios are obtained by adding valinomycin to mitochondria supplemented with Fe" as compared to adding Fe" to mitochondria treated with valinomycin. The differences between the two experimental conditions are that, in the former case, complex IV is already reduced and $\Delta\psi$ present when H^+ extrusion is initiated.

The discrepancy is open to several interpretations. The simplest is that that when the reaction is initiated with Fe" there is an overlap of scalar redox reactions with the vectorial H^+ extrusion. Alternatively, the higher rate of H^+ extrusion may be due to being some components of the H^+ pump at coupling site III already in reduced form when H^+ extrusion is initiated. Another possibility is that initiation of H^+ pumping in state 4 mitochondria has a higher efficiency than in state 5 mitochondria. This would imply a sort of "activation" of the H^+ pump by electrical or conformational energy. The way the reaction is initiated may also affect the charge/site ratios measured on the initial rates of Ca^{++} transport, (cf Wikström (19)) in NEM supplemented mitochondria. On the other hand the determination of the stoichiometry over a steady state rate of $Ca_3(PO_4)_2$ uptake lasting 2 minutes (Fig. 3) is independent of this factor.

Addition of tetramethyl-p-phenylendiamine (TMPD) is known to by-pass the antimycin A inhibition of succinate oxidation. The TMPD-dependent succinate oxidation in antimycin A inhibited mitochondria is coupled with ATP synthesis with a P/O ratio of 1. This is due to operation of coupling site III while coupling site II is inhibited by antimycin A. Addition of valinomycin to antimycin A inhibited mitochondria supplemented with various amounts of TMPD gives rise to a large H^+ extrusion the rate of which increased with the concentration of TMPD. Table I shows the effect of pH on the rate of H^+ extrusion during aerobic succinate oxidation + 400 μM TMPD by antimycin A inhibited mitochondria. The rate of H^+ extrusion increased from 54.6 at pH 6.5 to 100.1 nmol x mg prot^{-1} x min^{-1} at pH 7.5. Although coupling site II is completely inhibited by antimycin A and no energy conservation takes place, it may be conceived that H^+ release still occurs if Wurster Blue$^+$ oxidizes an hydrogen carrier in the $b-c_1$ complex. In this case the aerobic H^+ extrusion in the presence of succinate + TMPD would not simply be due to operation of the H^+ pump at site III but would include also an antimycin A insensitive H^+ release. Table I shows that the rate of antimycin A insensitive H^+ release increased from 7.4 at pH 6.5 to 47.5 nmols x mg prot^{-1} x min^{-1} at pH 7.5. The rate of e^- transport also increased from 20.8 at pH 6.5 to 36.6 nmols e^- x mg prot^{-1} x min^{-1} at pH 7.5. Three types of H^+/e^- stoichiometries can be calculated. First, the ratio between the rates of total H^+ extrusion and oxygen reduction: a/c, which reflects the sum of site II + site III stoichiometries and is between 2.60 and 2.79. This is in accord with an extrusion of 2 H^+/e^- at site III and with 1 H^+/e^- through the antimycin A inhibited $b-c_1$ complex. Second the ratio between the rates of H^+ extrusion + KCN and oxygen reduction: a-b/c, which should reflect the pure site III stoichiometry and is between 2.27 and 1.49. This is also in accord with the extrusion of 2 H^+/e^- at site III. Third, the ratio

TABLE I
H^+/e^- ratio with succinate + TMPD in antimycin A inhibited mitochondria

pH	V_H^+ suc→TMPD→O_2 (a)	V_H^+ suc→WB^+ (b)	V_e^- suc→TMPD→O_2 (c)	H^+/e^- a/c	a-b/c	b/c
	nmols x mg prot^{-1} x min^{-1}					
6.5	54.6	7.4	20.8	2.63	2.27	.36
6.9	78.0	15.6	30.0	2.60	2.07	.52
7.2	87.5	32.6	35.0	2.50	1.57	.93
7.5	100.1	47.5	36.6	2.79	1.49	1.30

The medium contained: 0.12 M sucrose, 60 mM KCl, 2 mM Tris-Cl, 1 µM rotenone 0.1 mM EGTA, 50 pmol antimycin A x mg prot^{-1} and 3 mg prot/ml. Temperature 18°C. After 5 minutes of aerobic pre incubation in presence of 400 µM TMPD were added 0.2 mM succinate, 100 µM NEM and 100 pmol valinomycin x mg prot^{-1}. The H^+ release in the span succinate Wurster Blue$^+$ (WB^+) was measured in the presence of 1 mM KCN.

---

between the rates of KCN insensitive H^+ release and oxygen reduction: b/c. Since the rate of oxygen reduction reflects that of e^- transfer in the b-c_1 complex the ratio reflects the site II stoichiometry in the presence of antimycin A and is between 0.36 and 1.30. The variation of the H^+/e^- ratio at site II as a function of pH may have significant implications. The low H^+/e^- ratio at acidic pH may be taken as an evidence for dissociation between H^+ and e^- transport in the b-c_1 complex.

Stoichiometries at sites I and II.

Table II shows the H^+/site ratios as obtained by adding Fe''' to KCN and valinomycin treated mitochondria oxidizing either endogenous substrates or succinate (+ rotenone). The amount of H^+ extrusion following the addition of 10 µM Fe''' was, with endogenous substrates, 27 and 38 µM in the absence and presence of NEM, respectively. Since during oxidation of endogenous substrates both sites I and II are operating this results in a H^+/site ratio of 2.7 and 3.8 without and with NEM, respectively. Table II shows also the charge/site ratio as determined by adding Fe''' to KCN treated mitochondria in the presence of Ca^{++} and oxidizing either glutamate + malate

TABLE II

Stoichiometries at coupling site I and II

H^+/site ratio

Substrate	Fe''' added µM	H extruded µM		H^+/site ratio	
		+ NEM	− NEM	+ NEM	− NEM
Endogenous	10	38+2	27+4	3.8+0.2	2.7+0.4
Succinate	20	37+2	30+1	3.7+0.1	3.0+0.1

Charge/site ratio

Substrate	Fe^{3+} reduction nmol/mg/min	Ca uptake nmol/mg/min	Charge/site
Glutamate + malate	141+8	114+6	2.48+0.18
Succinate	168+17	74+6	1.76+0.09

ATP/site ratio^x

Substrate	Fe^{3+} reduction nmol/mg	ATP synthesis nmol/mg	ATP/site ratio
Glutamate + malate	62+3	42+4	0.68+0.02
Succinate	108+5	33+3	0.60+0.01

Medium for H^+/site ratios: 120 mM KCl, 2 mM Tris Cl pH 7.3, 2 mM KCN, 200 µM EGTA, 2 mM $MgCl_2$, 200 pmol val/mg protein. Succinate 1 mM + 3 µM Rotenone. NEM 50 nmol/mg protein. 4 mg protein/ml. T°22. Reaction initiated by Fe''' H^+ extrusion calculated by extrapolation.

Medium for charge/site ratios: 200 mM Sucrose, 20 mM Tris Cl, pH 7.3 2 mM NaCN, 2 mM $MgCl_2$, 0.1% BSA, 0.2 mM Pi Tris, 200 µM $CaCl_2$. Charge/site measured on Ca^{++} uptake. Reaction initiated by 200 µM Fe'''. T° 26. Glutamate and malate were 4 and 2 mM respectively. Succinate 1 mM + 3 µM Rotenone. 1 mg protein/ml. Medium for ATP/site ratios was identical. Ca^{++} was replaced with 200 µM EGTA, 2 mM Pi. Reaction initiated by 100 µM ADP.

---

x
The ATP/oxygen ratio was also measured at coupling site III with Ferrocyanide + ascorbate as substrate (+ 60 pmols antimycin A x mg prot^{-1}). The ATP/oxygen ratio in this system was 30% lower than that with succinate, i.e. 1.35 with Fe'' vs 1.98 with succinate.

or succinate (+ rotenone). The rates of Fe''' reduction were 101 and 168 nmols Fe''' x mg prot^{-1} x min^{-1} and those of Ca^{++} uptake 119 and 74 nmols Ca^{++} x mg prot^{-1} x min^{-1} with glutamate + malate and succinate, respectively. The Ca^{++}/site ratio were 2.28 during operation of sites I + II (glutamate + malate) and 1.76 during operation of site II.

The results of the charge/site ratios are at first site in contrast with those of the H^+/site ratios. The question arises as to whether the decrease of the charge/site ratio be due to the drain of e^- by Fe''' at the C side of the membrane. The e^- drain would account for a decrease of the amount of charges separated by the H^+ pumps and the extent of decrease should be higher with succinate (one pump) than with glutamate + malate (two pumps). It may be recalled that the ATP/2 e^- ratios found with Fe''' as e^- acceptors were lower than predicted (20). If the decrease of the charge/site ratio is due to a Fe''' dependent e^- drain, this should affect to the same extent also the ATP/site ratio. Table II indicates that this is, indeed, the case. The rates of Fe''' reduction were 124 and 216, and those of ATP synthesis 85 and 66 nmols x mg prot^{-1} x min^{-1}, with glutamate + malate and succinate, respectively. Accordingly, the ATP/site ratios were 0.68 and 0.60 with glutamate + malate and succinate, respectively. If the charge/site and ATP/site ratios of Table II are corrected for an e^- leak amounting to 25% and 50% the rate of Fe''' reduction, in the case of glutamate + malate and succinate respectively, the charge/site ratios become close to 4/site and the ATP/site ratio close to 1/site.

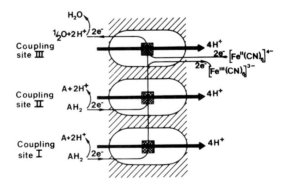

Fig. 4 - The three H^+ pumps of the mitochondrial respiratory chain.

IV. CONCLUSION

The nature of the H^+ pump.

Fig. 4 is a schematic representation of the respiratory chain H^+ pumps. Each complex is seen as possessing a H^+ pump the operation of which leads to the extrusion of $4H^+$ across the membrane and to the separation of 4 charges. The scneme is based on the assumptions that: a) NADH and succinate donate e^- at the M side, b) oxygen is reduced by cytochrome oxidase also at the M side; and c) Fe''' and Fe'' interact with cytochrome C at the C side. Being e^- donated and accepted at the same side of the membrane, charge separation during oxidation of NADH linked substrates or succinate results only through the operation of the H^+ pumps. In addition to the charges separated by the pumps, there is charge separation and consumption when e^- acceptors and donors drain e^- or feed e^- in the middle of the chain at one side of the membrane. This is the case with Fe''' which takes out 2 e^- or of Fe'' which donates 2 e^- at the C side, and explains why use of Fe''' results in the consumption of 2 charges while use of Fe'' in the separation of 2 charges. The charge separation due to e^- transfer through the complex is therefore seen as independent from that due to operation of the H^+ pumps.

The coupling mechanism is indicated as a "black box" since little is known about the process by which the transfer of 2 e^- through the complex leads to the extrusion of 4 H^+. Several alternatives may be envisaged as to the molecular nature of the H^+ transport. Each pump should contain a specific H^+ channel for the transmembrane movement of H^+. However the channel may be either identical or different for each of the three complexes. The H^+ channel may transport either 4 or only 2 H^+, being the other 2 translocated by other mechanisms for example via H carrier together with e^-.

REFERENCES

1. Mitchell, P. (1966) Biol. Rev. 41, 445-499
2. Massari, S. and Azzone, G.F. (1970) Europ. J. Biochem. 12, 300-309
3. Chance, B., Crofts, A.R., Nishimura, M. and Price, B. (1970) Europ. J. Biochem. 13, 364-374
4. Wikström, M.F. (1977) Nature 266, 271-273
5. Mitchell, P. and Moyle, J. (1966) in The Biochemistry of Mitochondria, eds. E.C. Slater et al., Academic Press, London 53-74
6. Mitchell, P. and Moyle, J. (1967) Biochem. J. 104, 588-600
7. Rossi, E. and Azzone, G.F. (1969) Europ. J. Biochem. 7, 418-426
8. Azzone, G.F. and Massari, S. (1971) Europ. J. Biochem. 19, 97-107
9. Azzone, G.F. and Massari, S. (1973) Biochim. Biophys. Acta 301, 195-226
10. Rossi, E. and Azzone, G.F. (1970) Europ. J. Biochem. 12, 319-327
11. Chance, B. (1975) J. Biol. Chem. 240, 2729-2748
12. Rossi, C.S. and Lehninger, A.L. (1974) J. Biol. Chem. 239, 3971-3980
13. Mitchell, P. (1973) VIII FEBS Meeting Amsterdam (North-Holland, Elsevier) pp. 352-370
14. Brand, M.D., Chen, C.H. and Lehninger, A.L. (1976) J. Biol. Chem. 251, 968-974
15. Brand, M.D., Reynafarje, B. and Lehninger, A.L. (1976) J. Biol. Chem. 251, 5670-5679
16. Moyle, J. and Mitchell, P. (1978) FEBS Letters 88, 268-272
17. Alexandre, A. and Lehninger, A.L. (1978) Abstract XII FEBS Meeting
18. Azzone, G.F., Pozzan, T., Bragadin, M. and Dell'Antone, P. (1977) Biochim. Biophys. Acta 459, 96-109
19. Wikström, M.F. (1978) in The Proton and Calcium Pumps, eds. G.F. Azzone et al., Elsevier-North Holland, Amsterdam pp. 227-238
20. Copenhaver, J.H. Jr. and Lardy, H.A. (1952) J. Biol. Chem. 195, 225-238

THE STOICHIOMETRY OF VECTORIAL H⁺ MOVEMENTS COUPLED TO
ELECTRON TRANSPORT AND ATP SYNTHESIS
IN MITOCHONDRIA

Albert L. Lehninger[1]
Baltazar Reynafarje
Adolfo Alexandre

Department of Physiological Chemistry
Johns Hopkins University School of Medicine
Baltimore, Maryland

I. INTRODUCTION

It has long been appreciated that proton movements across, within, or on the surface of the mitochondrial inner membrane are intimately related to the mechanism(s) of energy transduction during electron transport. However, it is remarkable that there has been significant uncertainty for over a decade regarding the precise stoichiometric relationships between electron transport and H⁺ movements, and also between H⁺ movements and ATP synthesis. Although Mitchell and Moyle had concluded in the period 1965-1967 that outward movement of 2 H⁺ accompanies passage of an electron pair through each of the three energy-conserving sites of the respiratory chain, other investigators have from time to time observed higher values or have noted experimental inconsistencies (review in 1). Since it is axiomatic that understanding of mechanism can come only after knowledge of the reaction stoichiometry, we have undertaken a comprehensive re-evaluation of the H⁺/site and H⁺/ATP stoichiometry of mitochondrial electron transport and ATP hydrolysis and synthesis.

Earlier reports from this laboratory have shown that close to 12 H⁺ are ejected from the matrix to the medium as a pair of electrons pass through the three energy-conserving sites of

---

[1] Supported by NIH grant GM05919, NSF grant PCM 75-21923, NCI contract N01-CP-45610.

the mitochondrial respiratory chain to molecular oxygen (1-8). Thus the <u>average</u> H^+/site ratio, the number of H^+ ejected per pair of electrons per energy-conserving site, approaches 4. This value of the H^+/site ratio has been observed with four different procedures applied to rat liver mitochondria (2,3,8,9) and has also been observed in mitochondria isolated from rat heart (3,6), rat brain, and Ehrlich ascites tumor cells (A. Villalobo, J. Benavides, and A.L. Lehninger, unpublished data). The value of 4 for the average H^+/site ratio provides better agreement with thermodynamic data on electron transport, ATP synthesis, and observed transmembrane electrochemical gradients (4,5,10) than the H^+/site ratio of 2 proposed by Mitchell and Moyle on the basis of oxygen pulse experiments (11), which are subject to serious underestimation (5,9,12).

This paper considers two major questions regarding the H^+ stoichiometry of electron transport and oxidative phosphorylation. The first is posed by discrepancies between the average H^+/site ratio of 4 observed with natural substrates, such as pyruvate or succinate, and available reports on the $H^+/2e^-$ ratio for energy-conserving site 3 of the respiratory chain, i.e. the cytochrome oxidase reaction, when tested with artificial electron donors. Mitchell (13-16), as well as Papa (17), has concluded that no H^+ ejection is coupled to the cytochrome oxidase reaction (13-16); in the "loop" concept proposed in the chemiosmotic hypothesis the cytochrome oxidase reaction is assumed to transport only electrons (14-16). On the other hand, Wikström (18-20) has provided evidence that 2 H^+ are ejected per pair of electrons transported from cytochrome c to oxygen in rat liver mitochondria and in cytochrome oxidase vesicles. Moreover, Sorgato and Ferguson (21) have recently demonstrated that electron transport through site 3 develops a transmembrane gradient of both $\Delta\Psi$ and ΔpH in inverted mitochondrial vesicles. There is other evidence that the three sites of the respiratory chain are energetically equivalent, since each yields one molecule of ATP or drives uptake of 2 Ca^{2+}.

The second point at issue is the value of the H^+/ATP ratio. The available data (4,22,23) indicate that the H^+/ATP ratio for hydrolysis of extramitochondrial ATP is close to 2, although higher K^+/ATP uptake ratios have been reported (24,25). To be compatible with an average H^+/site ratio of 4, the H^+/ATP ratio is expected to be 3, as pointed out before (4,5,10); the fourth H^+ is required to bring about the energy-requiring electrogenic exchange of ADP^{3-} (out) for ATP^{4-} (in), via the electroneutral H^+-$H_2PO_4^-$ symport process (4,26,27). Although an earlier communication from this laboratory reported a determination of the H^+/ATP with a new rate method giving values close to 3 (5), some unexpected observations subsequently made on the transmembrane movements of ATP, ADP, and phosphate, required resolution before this value could be accepted. These questions have since been resolved, following the discovery of a hitherto unreported membrane transport system for ATP, ADP, and phosphate that is operative during ATP hydrolysis (28).

In this communication we describe determinations of the H^+/site ratio of sites 2 and 3 of the respiratory chain and of the H^+/ATP ratio by a new rate procedure (10,28,29). The results, which are compatible with each other and with data reported earlier, fully support the value 4 for the H^+/site ratio and assign value 3 for the H^+/ATP ratio.

II. THE H^+/SITE RATIO FOR SITE 2

Energy-conserving site 2 of the respiratory chain of rat liver mitochondria was isolated by use of succinate as electron donor in the presence of rotenone to prevent oxidation of endogenous NADH; ferricyanide was used as electron acceptor in the presence of cyanide, to block electron flow to oxygen. K^+ (+ valinomycin) was used as the mobile charge-compensating cation to enter the matrix in exchange for the ejected H^+. N-ethylmaleimide was added to suppress interfering H^+ uptake on the H^+-$H_2PO_4^-$ carrier (3,5,8), and EGTA was added to prevent Ca^{2+} movements. H^+ ejection was followed with a glass electrode and ferricyanide reduction was measured spectrophotometrically with the wavelength pair 420-500 nm. Under these conditions ferricyanide, which does not penetrate the membrane, accepts electrons from mitochondrial cytochrome c, which is located on the outer surface of the inner membrane (32).

Figure 1 shows traces for the ejection of H^+ and the reduction of ferricyanide by succinate. Both H^+ formation and ferricyanide disappearance proceeded for a sufficiently long steady-state period to allow accurate determination of their rates (italic numbers). The ratio <u>rate of H^+ ejection/rate of ferrocyanide reduction</u> (in ng-ions or nmoles per min per mg protein) was 157/84.5 = 1.86, equivalent to an $H^+/2e^-$ ratio of 2 x 1.86 = 3.72 for the segment of the respiratory chain between succinate and cytochrome c, i.e. site 2.

Appropriate control experiments showed that addition of antimycin A, which blocks electron flow through site 2, inhibited both ferricyanide reduction and H^+ ejection over 96 per cent. When ferrocyanide was added to the test system instead of ferricyanide no net H^+ ejection occurred. When 100 μM Ca^{2+} was the permeant cation (with valinomycin and EGTA omitted), nearly identical results to those in Figure 2 were observed, with an $H^+/2e^-$ ratio of 3.86. Thus either K^+ (+ valinomycin) or Ca^{2+} can serve as the permeant charge-compensating cation.

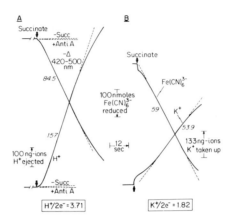

FIGURE 1. The $H^+/2e^-$ and $K^+/2e^-$ ratios at site 2 with succinate as electron donor and ferricyanide as electron acceptor. The test system (2.6 ml, 25°) contained 125 mM sucrose, 60 mM KCl, 2 mM Li-Hepes pH 7.15, 3 µM rotenone, 0.15 mM Tris-EGTA, 40 nmoles N-ethylmaleimide per mg protein and 10 mg mitochondrial protein. Valinomycin (0.1 nmoles per mg protein) was added at 2 min and 0.5 mM KCN at 4 min. After addition of 600 nmoles ferricyanide for calibration the reaction was initiated with 0.3 mM lithium succinate. In B the medium was the same except that 12 mg mitochondrial protein was used and 60 mM LiCl plus 1.7 mM KCl replaced the 60 mM KCl. Ferricyanide reduction was monitored at 420-500 nm, pH changes with a glass electrode, and K^+ uptake with K^+-selective electrode.

To verify these measurements, we have also measured the rate of the simultaneous uptake of the charge-compensating cation, using selective electrodes. Figure 1B shows the results of such an experiment in which K^+ was the mobile cation; the ratio rate of K^+ uptake/rate of H^+ ejection was 53.9/59 = 0.91, equivalent to a $K^+/2e^-$ uptake ratio of 1.82, close to the expected value 2.0. Since succinate is dehydrogenated on the inner side of the inner membrane and delivers 2 H^+ into the matrix compartment, where they provide charge-compensation for 2 of the 4 vectorial H^+ ejected from the matrix to the medium. Entry of 2 K^+ from the medium via valinomycin compensates for the 2 other H^+ ejected. We conclude that as each pair of electrons passes through site 2, 4 H^+ are ejected from the matrix to the medium.

III. THE $H^+/2e^-$ RATIO AT SITE 3

Since the oxidation of succinate by molecular oxygen via sites 2 + 3 has already been shown to have an H^+/O ratio of 8 and since the experiment in Figure 1 showed that site 2 tested directly has an $H^+/2e^-$ ratio of 4, by difference site 3 must also have an $H^+/2e^-$ ratio of 4, rather than the value 0 proposed by Mitchell or the value 2 proposed by Wikström and colleagues (18-20,33,34).

Shown in Figure 2 are results from a typical experiment in which the $H^+/2e^-$ ratio at site 3 was measured by a steady-state rate method. The electron donor was 6 mM ferrocyanide and oxygen was electron acceptor. The rat liver mitochondria were supplemented with rotenone and antimycin A to prevent electron flow from endogenous substrates into site 3. Ferrocyanide was rapidly oxidized to ferricyanide, whose appearance was monitored spectrophotometrically. Concurrently H^+ was recorded spectrophotometrically in a parallel vessel containing phenol red indicator. Both ferrocyanide oxidation and H^+ ejection showed nearly linear steady-state rates. The ratio rate of H^+ ejection/rate of ferrocyanide oxidation (in ng-ions or nmoles per min per mg) was found to be 293/160 = 1.83 for a one-electron change, equivalent to an $H^+/2e^-$ ratio of 3.66 for site 3.

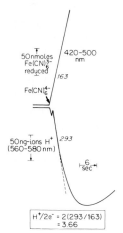

FIGURE 2. The $H^+/2e^-$ ratio at site 3 with ferrocyanide as substrate. The test system (2.3 ml, 25°) contained 200 mM sucrose, 22 mM KCl, 1.5 mM K Hepes pH 7.3, 3.5 µM rotenone, 0.2 nmoles antimycin A per mg protein, 40 nmoles N-ethylmaleimide per mg, and 8.0 mg mitochondrial protein added at zero time. Valinomycin (0.1 nmoles per mg) was added at 30s and the reaction started 4 min later with 6.0 mM potassium ferrocyanide.

Control experiments showed that in the absence of ferrocyanide no significant absorption changes occurred at either wavelength. When cyanide was added no significant electron flow or H^+ ejection took place. Replacement of ferrocyanide with ferricyanide (3 mM) resulted in negligible H^+ production. No evidence for significant "extra" H^+ ejection, as described by Moyle and Mitchell (16), could be found; in any case, the rates of "extra" H^+ ejection reported by them are trivially small compared to the overall rates of ferrocyanide oxidation observed.

We have also observed $H^+/2e^-$ values approaching 4 for site 3 with the use of ascorbate + ferrocyanide (6-8 mM) and ascorbate + 200 μM N,N,N',N'-tetramethylphenylenediamine (TMPD) as electron donor systems. In the latter cases the rate of electron flow was measured with an oxygen electrode, and H^+ output with a glass electrode. Scalar H^+ changes consequent to the oxidation of ascorbate to dehydroascorbate (1 H^+ per ascorbate oxidized) were deducted to obtain the vectorial H^+ ejection. Although the response time of the oxygen electrode (∼ 1.4s) is longer than that of the glass electrode (∼ 0.4s) the steady-state periods were sufficiently long (12-20s) to cause little or no error in determination of the slopes.

The most important parameters leading to the higher values close to 4.0 for the $H^+/2e^-$ ratio at site 3 in this study, as compared to the findings of Wikström et al., were found to be (1) use of a suspending medium containing sucrose at 120-200 mM and (2) the use of much higher concentrations of ferrocyanide. We have found (9, also A. Alexandre and A.L. Lehninger, unpublished data) that in the medium of 120 mM KCl employed by Wikström (18,19) initiation of electron flow in the presence of K^+ and valinomycin causes rapid osmotic swelling of mitochondria, due to increased uptake of Cl^-, together with K^+ (35,36), accompanied by an increased leak of H^+. Such effects are prevented by the nonpermeant sucrose or LiCl (3,9).

The other and more important difference is the use of relatively high concentrations (6-8 mM) of ferrocyanide as electron donor, compared to the relatively low concentrations (< 1 mM) employed in the experiments of Wikström (18,19). At 1 mM ferrocyanide the rate of oxidation and H^+ ejection is very low, in comparison with the rate of H^+ back-leakage, whereas at 6-8 mM ferrocyanide the rate of H^+ ejection is many-fold higher, some 300 ng-ions per min per mg. We found the $H^+/2e^-$ ratio to be about 2.0 at 1 mM ferrocyanide, but increased to about 4 at 6 mM (10,29).

The $H^+/2e^-$ ratio of close to 4 at site 3 was verified by measurements of the uptake of the charge-compensating cation Ca^{2+}. The ratio <u>rate of Ca^{2+} uptake/rate of oxygen consumption</u> (in ng-ions or ng-atoms per min per mg) was found to be 2.72, or 2 x 2.72 = 5.44 positive charges per site, close to the 6 positive charges expected for site 3 under the conditions described. Four positive charges must enter to replace the 4 H^+ ejected from the matrix to the medium and two additional positive charges are required to compensate for the 2 H^+ absorbed from the matrix as an atom of oxygen is reduced to yield H_2O. While this accounting of the charge separations indicates that a total of 6 charges are separated as a pair of electrons passes from the artificial electron donor via mitochondrial cytochrome c and cytochrome oxidase to oxygen. The intrinsic $H^+/2e^-$ ratio for site 3 is 4, as it is for the other two sites. Normally, when natural substrates such as pyruvate or succinate furnish electrons to the respiratory chain, each substrate molecule donates 2 H^+ to the matrix on dehydrogenation and 2 H^+ are subsequently absorbed from the matrix during reduction of 1/2 O_2 to form H_2O; thus the scalar H^+ formation and absorption in the matrix normally cancels out. However, separation of 6 electric charges when ferrocyanide is the donor is a matter of some interest. Tests are under way to determine whether the $ATP/2e^-$ ratio for oxidative phosphorylation at site 3 will exceed the value 1.0, as might be expected if all the electrochemical potential generated on oxidation of ferrocyanide can be utilized for ATP synthesis.

IV. THE H^+/ATP STOICHIOMETRIC RATIO FOR THE HYDROLYSIS AND SYNTHESIS OF EXTRAMITOCHONDRIAL ATP

Earlier measurements of the H^+/ATP (out) hydrolysis ratio (4,22,23) employed the ATP pulse technique (22), which gave values close to 2.0. However, estimation of the H^+/ATP ratio by pulse methods is unsatisfactory, since the concentration of ATP added in such experiments is well below K_M for ATPase activity; thus the ATPase reaction approaches completion at an increasingly lower rate at a time when the rate of ΔH^+ back-decay is relatively high. After unsuccessful attempts to resolve this and other technical difficulties in the pulse method, we developed a much simpler and more direct steady-state rate method (28). In this procedure a relatively high concentration of ATP, well above K_M for the ATPase activity, is added to respiration-inhibited, de-energized rat liver mitochondria, supplemented with diadenosine pentaphosphate to inhibit adenylate kinase activity (37) and with K^+ (plus valinomycin) as the charge-compensating cation. Under these conditions the initial rate of H^+ ejection in relation to ΔH^+ back-decay is very high. A rapid sampling method was used to obtain colorimetric measurements of the P_i formed. The ratio rate of H^+ ejection/rate of ATP hydrolysis (ng-ions or nmoles per min per mg) gives the number of H^+ ejected per molecule of

ATP hydrolyzed.

Figure 3 shows a typical experiment carried out on rat liver mitochondria at 15°. From the steady-state rates of H^+ ejection and ATP hydrolysis the H^+/ATP ratio was 72.8/23.2 = 3.1. Many such experiments have given an average value of 2.90, with little deviation.

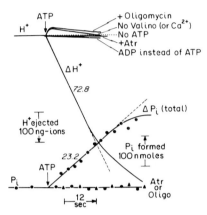

FIGURE 3. The H^+/ATP ratio from steady-state rates of H^+ ejection and ATP hydrolysis. The test system (8.0 ml) contained 125 mM sucrose, 40 mM LiCl, 20 mM KCl, 3.0 mM K^+ Hepes pH 7.05, 0.5 mM EGTA, 20 μM diadenosine pentaphosphate, 2 μM rotenone, 40 nmoles N-ethylmaleimide per mg protein, rat liver mitochondria (2.0 mg protein per ml) at time zero, valinomycin (100 ng mg^{-1}) at 2 min, and ATP (200 μM) at 6 min. The incubation was carried out at 15°.

Under the conditions outlined in Figure 3 ATP is hydrolyzed only to ADP and phosphate. ATP hydrolysis and coupled H^+ ejection are completely inhibited by atractyloside and by oligomycin. No significant endogenous changes occur in the absence of added ATP; added ADP is not hydrolyzed. In the absence of valinomycin, little or no hydrolysis ensues. Under the experimental conditions employed the scalar H^+ is retained in the alkalinized matrix (cf. 4,23) and does not appear in the medium; thus no correction is required for its formation. Either K^+ or Ca^{2+} can be used as charge-compensating cation. N-ethylmaleimide is not required in the medium; the H^+/ATP rate ratio is 3.0 in its absence or presence.

These and other data on K^+ uptake and on the membrane transport processes involving ATP, ADP, and phosphate during ATP hydrolysis, which are electroneutral and thus do not require charge-compensation by K^+ (28), thus verify that the H^+/ATP ratio is 3.

V. DISCUSSION

The observed H^+/site ratios of close to 4.0 obtained for both sites 2 and 3, each measured individually, are in full agreement with our earlier measurements indicating that the average H^+/site ratio is close to 4 (1-8) and are also supported by recent measurements in Azzone's laboratory (38). Several measurement techniques have been employed in these studies, with fully consistent results, thus eliminating the possibility of systematic measurement artifacts.

Perhaps the most crucial point is the finding that the $H^+/2e^-$ ratio of site 3 is close to 4, determined with 3 different substrate systems. This value contrasts sharply with the view of Mitchell that no H^+ ejection is associated with the cytochrome oxidase reaction (13-15), based on pulse-type experiments (13), which are subject to underestimation, particularly due to very rapid interfering uptake of H^+ together with phosphate on the H^+-$H_2PO_4^-$ symporter (cf. 1,3,5), which is much too fast to be corrected for by the extrapolation procedure employed by Mitchell and Moyle (11). Although the latter have recently denied that uptake of H^+ plus phosphate is a factor leading to underestimation and have concluded that N-ethylmaleimide increases the observed H^+/site ratios spuriously through its capacity to evoke H^+ ejection coupled to oxidation of endogenous substrates (39), these arguments had already been considered and rejected in our earlier communications (3, 5,12). In any case, we have already shown that N-ethylmaleimide or other inhibitors of the $H^+/H_2PO_4^-$ symporter are not required for the observation of H^+/site ratios close to 4; simple removal of phosphate from the test medium (3) or measurement of $K^+/2e^-$ uptake ratios (8) allow observation of average $H^+/2e^-$ ratios close to 4 in the absence of N-ethylmaleimide.

Of greater importance is the difference between the $H^+/2e^-$ ratio of 2 for site 3 observed by Wikström and colleagues (18-20,33,34) and the value of close to 4 reported here. However, the considerations and data reviewed here (cf. 10,29) strongly suggest that these differences are a reflection of the precise experimental conditions employed, as indicated above.

The new value for the H^+/ATP ratio of 3.0 briefly reported earlier (5), has been verified by data in this paper and by quantitative examination of the transport pathways and stoichiometry involved in the transmembrane movements of ATP, ADP and phosphate during ATP hydrolysis (28). The·newer stoichiometric data may be placed into a revised "proton cycle" (Figure 4, left) for the formation and utilization of electrochemical H^+ energy units; this scheme is an updating of that described earlier (4). It proposes that an electrochemical

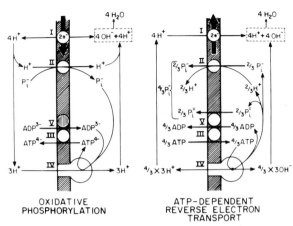

FIGURE 4. "Proton cycles" for the formation and utilization of H^+ energy equivalents in oxidative phosphorylation (left) and ATP-dependent reverse electron flow (right).

gradient equivalent to 4 H^+ per pair of electrons is generated by electron flow through each site. The electrochemical equivalent of 3 H^+ is used to bring about synthesis of ATP from ADP and phosphate in the matrix, and the fourth H^+ equivalent is used to drive the electrogenic expulsion of ATP^{4-} from the mitochondria on the atractyloside-sensitive ADP-ATP carrier in exchange for incoming ADP^{3-}, via the action of the electroneutral H^+-$H_2PO_4^-$ symporter, which is known to participate in oxidative phosphorylation.

The newer stoichiometric data reported here must be accounted for by any valid theory for the mechanism of oxidative phosphorylation. Some aspects of the chemiosmotic hypothesis, particularly the stoichiometric and thermodynamic relationships, are not supported by our data. Moreover, the $H^+/2e^-$ ratio of 4 per energy-conserving site cannot be accounted for solely by the loop mechanism postulated by Mitchell; in particular the ejection of 4 H^+ by site 3 suggests participation of a conformationally-driven H^+ transport mechanism. These and other theoretical and biological implications of the proton stoichiometry have been further developed in earlier communications from this (1,4,5,10) and other laboratories (18-20,38).

Finally, we have recently reported (28) a hitherto undescribed membrane transport process operating during hydrolysis of ATP (out) by mitochondria, which promotes the transmembrane exchange

$$ATP^{4-} \text{ (out)} \xrightarrow{M} ADP^{3-} + 0.5\ P_i^{2-}$$

This exchange process is electroneutral, atractyloside-sensitive, and mersalyl (and NEM)-insensitive. Phosphate efflux is obligatorily linked to the ATP/ADP exchange and cannot be dissociated. This exchange process operates only in the direction written and is believed to function normally in ATP-dependent reverse electron transport, as is indicated by the vectorial diagram in Figure 4 (right). Thus two different sets of pathways presumably independently regulated, are employed for the transport of adenine nucleotides and phosphate. The pathway in Figure 4 (right) accounts for the recent observation that reversal of electron flow through site 1 requires 1.33 ATPs per 2e$^-$ (40).

Other details of the experiments reviewed here are now in press (41,42).

REFERENCES

1. Brand, M. D., Reynafarje, B., and Lehninger, A. L. (1976) *Proc. Natl. Acad. Sci. USA* 73:437.
2. Brand, M. D., Chen, C-H., and Lehninger, A. L. (1976) *J. Biol. Chem.* 251:968.
3. Reynafarje, B., Brand, M. D., and Lehninger, A. L. (1976) *J. Biol. Chem.* 251:7442.
4. Brand, M. D. and Lehninger, A. L. (1977) *Proc. Natl. Acad. Sci. USA* 74:1955.
5. Lehninger, A. L., Reynafarje, B., and Alexandre, A. (1977). In *"Structure and Function of Energy-Transducing Membranes"* (K. van Dam and B. V. van Gelder, eds.), p. 95. Elsevier, Amsterdam.
6. Vercesi, A., Reynafarje, B., and Lehninger, A. L. (1978) *J. Biol. Chem.* 253:0000.
7. Reynafarje, B. and Lehninger, A. L. (1977) *Biochem. Biophys. Res. Communs.* 77:1273.
8. Reynafarje, B. and Lehninger, A. L. (1978) *J. Biol. Chem.* 253:0000.
9. Alexandre, A. (1977) *Fed. Proc.* 36:814.
10. Lehninger, A. L. and Reynafarje, B. (1978) *12th FEBS Meeting*, Dresden, July, *Abstracts, Vol. 1*, #1732.
11. Mitchell, P. and Moyle, J. (1967) *Biochem. J.* 105:1147.
12. Brand, M. D., Reynafarje, B., and Lehninger, A. L. (1976) *J. Biol. Chem.* 251:5670.
13. Mitchell, P. (1969). In *"The Molecular Basis of Membrane Functions"* (D. C. Tosteson, ed.), p. 483. Prentice-Hall, Englewood Cliffs, N.J.
14. Mitchell, P. (1966) *"Chemiosmotic Coupling in Oxidative and Photosynthetic Phosphorylation"*. Glynn Research Ltd., Bodmin.

15. Mitchell, P. and Moyle, J. (1970). In *"Electron Transport and Energy Conservation"*. (J. M. Tager et al., eds.), p. 575. Adriatica Editrice, Bari.
16. Moyle, J. and Mitchell, P. (1978) *FEBS Letts. 88*:268.
17. Papa, S. (1976) *Biochim. Biophys. Acta 456*:39.
18. Wikström, M. K. F. (1977) *Nature 266*:271.
19. Wikström, M. K. F. and Saari, H. T. (1977) *Biochim. Biophys. Acta 462*:347.
20. Wikström, M. K. F. (1978). In *"The Proton and Calcium Pumps"*. (G. F. Azzone et al., eds.), p. 215. Elsevier, Amsterdam.
21. Sorgato, M. C. and Ferguson, S. J. (1978) *FEBS Letts. 90*:178.
22. Mitchell, P. and Moyle, J. (1968) *Eur. J. Biochem. 4*:530.
23. Moyle, J. and Mitchell, P. (1973) *FEBS Letts. 30*:317.
24. Cockrell, R. S., Harris, E. J., and Pressman, B. (1966) *Biochemistry 6*:3902.
25. Azzone, G. F. and Massari, S. (1971) *Eur. J. Biochem. 19*:97.
26. Wulf, R., Kaltstein, A., and Klingenberg, M. (1978) *Eur. J. Biochem. 82*:585.
27. LaNoue, K., Mizani, S. M., and Klingenberg, M. (1978) *J. Biol. Chem. 253*:191.
28. Reynafarje, B. and Lehninger, A. L. (1978) *Proc. Natl. Acad. Sci. USA 75*:0000.
29. Alexandre, A. and Lehninger, A. L. (1978) *Fed. Proc. 37*:1326.
30. Reynafarje, B. (1978) *Fed. Proc. 37*:1753.
31. Alexandre, A. and Lehninger, A. L. (1978) *12th FEBS Meeting*, Dresden, July, *Abstracts, Vol. 1,* #1733.
32. Jacobs, E. E. and Sanadi, D. R. (1960) *Biochim. Biophys. Acta 38*:12.
33. Wikström, M. K. F. and Krab, K. (1978) *FEBS Letts. 91*:8.
34. Krab, K. and Wikström, M. K. F. (1978) *Biochim. Biophys. Acta,* in press.
35. Azzi, A. and Azzone, G. F. (1967) *Biochim. Biophys. Acta 131*:468.
36. Brierley, G. P. (1970) *Biochemistry 9*:697.
37. Lüstorff, J. and Schlimme, E. (1976) *Experientia 32*:298.
38. Pozzan, T., Di Virgilio, F., Bragadin, M., and Azzone, G. F. *Proc. Nat. Acad. Sci. USA,* submitted.
39. Moyle, J. and Mitchell, P. (1978) *FEBS Letts. 90*:361.
40. Rottenberg, H. and Gutman, M. (1977) *Biochemistry 16*:3220.
41. Alexandre, A., Reynafarje, B., and Lehninger, A. L. (1978) *Proc. Natl. Acad. Sci.,* in press.
42. Alexandre, A. and Lehninger, A. L. (1978) *J. Biol. Chem.,* in press.

ANISOTROPIC CHARGE MODEL FOR H^+-EJECTION FROM MITOCHONDRIA

Tomihiko Higuti, Naokatu Arakaki, Makoto Yokota,
Akimasa Hattori, Singo Niimi, and Setuko Nakasima

Faculty of Pharmaceutical Sciences, University of
Tokushima, Tokushima, Japan

According to the anisotropic charge model [1,2] for energy transduction in oxidative phosphorylation, the first step in energy-transduction is an anisotropic distribution of electric charges in the mitochondrial inner membrane (negative charges are created on the surface of the C-side of the membrane, and positive charges on the M-side). H^+ and OH^- are drawn electrostatically to these negative and positive charges, respectively. This model is interpretated in Fig. 1.

In good accordance with this model, it has been shown[1-4] that amphipathic cations (ethidium, acriflavine, tetraphenyl-arsonium, etc.) inhibit energy transduction in oxidative phosphorylation in intact mitochondria by binding to, and neutralizing, negative charges created on the C-side, but that these cations have no inhibitory activity in submitochondrial particles, which are inside-out relative to the membranes of intact mitochondria. Conversely amphipathic anions (anilino-naphthalene sulfonate and tropaeolin 00) inhibit energy transduction in oxidative phosphorylation in submitochondrial particles, but these anions have no appreciable inhibitory activity in intact mitochondria. It has been proposed [1] that this new type of inhibitor should be called an "anisotropic inhibitor of energy transduction".

The present study shows that: (1) The binding of the positively charged anisotropic inhibitors to intact mitochondria energized by succinate (or ATP) caused ejection of H^+ into the suspending medium and this ejection of H^+ increased to the level of valinomycin (+KCl)-dependent H^+-ejection with increase in the concentration of inhibitors added. (2) The total amounts of H^+ ejected on addition of these inhibitors and on subsequent addition of valinomycin (+KCl) to energized mitochondria were constant, irrespective of the amount of the inhibitors added. (3) The maximum amount of ethidium-induced H^+-ejection from the mitochondria energized with ATP at infinite dye concentration decreased with increase in the amount of ATPase system ($F_1 \cdot F_0$) inactivated by oligomycin. (4) The negatively charged anisotropic inhibitors of energy transduction caused OH^--ejection from submitochondrial particles energized with succinate. (5) Oligomycin (3 μg/mg protein) did not inhibit antimycin A-induced reuptake of H^+ by intact mitochondria respiring with succinate in the presence or absence of Pi and ADP.

These findings suggest that the inhibitors- and valinomycin-dependent H^+-ejection and OH^--ejection from the energized mitochondrial inner membrane occurs with the appearance of anisotropic charges: the positively charged inhibitors bind electrostatically to the negative charges on the C-side and then the protons drawn electrostatically to the negative charges are released into the suspending medium to maintain the electroneutrality of the medium, as shown in Fig. 1.

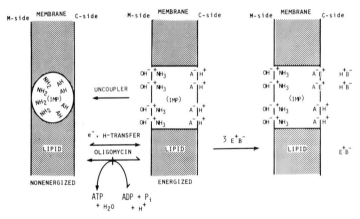

Fig. 1 Anisotropic charge model for ATP synthesis and H^+-translocation in mitochondria. IMP, inner mitochondrial membrane protein (F_0, electron transport complex or other protein); A^-, protein anionic groups; NH_2, basic amino acids in protein; E^+, ethidium cation; B^-, bromine anion; H^+, proton; OH^-, hydroxyl ion.

1 Higuti, T., Yokota, M., Arakaki, N., Hattori, A. and Tani, I. (1978) Biochim. Biophys. Acta 503, 211
2 Higuti, T., Yokota, M., Arakaki, N., Hattori, A. and Tani, I. (1977) Japan Bioenerg. Group, Abst., 3, 60
3 Higuti, T., Arakaki, N., Yokota, M., Hattori, A., and Tani, I. (1978) FEBS Lett., 87, 87
4 Higuti, T., Ohnishi, M., Arakaki, N., Nakasima, S. and Yokota, M. Anal. Biochem., in press

LIPOPHILIC ALKYLAMINES AS INHIBITORS OF COUPLING AND UNCOUPLING IN BEEF HEART MITOCHONDRIA

Edmund Bäuerlein
Heinz Trasch

Abteilung für Naturstoffchemie
Max-Planck-Institut für medizinische Forschung
Heidelberg, Fed. Rep. Germany

I. INTRODUCTION

Inhibition of coupled respiration was obtained by lipophilic N-alkyl-imidazole-2-thiols (Ia) or N-alkyl-benzimidazole-2-thiols (IIa) only, if glutamate + malate were the

Ia: X = SH
Ib: X = H
$C_{10}H_{21}$

IIa: X = SH
IIb: X = H
$C_{10}H_{21}$

substrates and the chain lengths were at least nine carbon atoms long (Trasch and Bäuerlein, 1978a). Because this inhibition could not be released by uncouplers, the action of these agents was located near to the electron transport chain. They were introduced among others (Bäuerlein and Kiehl, 1976; Kiehl and Bäuerlein, 1976, 1977) with the intention of modifying assumed functional groups of the ATP synthase (Bäuerlein, 1978), for example the disulfide group (R-S-S-R → RS^+ + RS^- + P_i → R $SOPO_3H_2$ + RS^-) as intermediate of a proton-driven ATP synthesis. The sulfenyl group (RS^+) was thought to be trapped by the thiol compounds as

---

Abbreviations: DNP = 2,4-Dinitrophenol; SF 6847 = p-hydroxy-di-tert butyl-benzyliden-malodinitrile; TMPD = tetramethyl-p-phenylenediamine.

mixed disulfides. We were very surprised that the corresponding sulfur free compounds, Ib and IIb, were much more reactive in the inhibition reaction than the thiol compounds Ia and IIa, though they were unable to form the assumed disulfides. Because the reactivity from Ia to Ib and from IIa to IIb increased with increasing basicity, it appeared more probable that these compounds interacted with the proposed proton pump of the NADH dehydrogenase than with its electron transport segment. This assumed inhibition of a proton pump provided a very simple interpretation of both, coupling and uncoupling, by these basic lipophilic substances. To study these effects by the most simple combination of a lipophilic and basic group, primary unbranched alkylamines

$$CH_3-(CH_2)_{n=1}NH_2$$

were used instead of the imidazole compounds (Trasch and Bäuerlein, 1978b).

II. RESULTS AND DISCUSSION

Coupled respiration of beef heart mitochondria was inhibited, if we used primary alkylamines with a chain length of at least eleven carbon atoms and if glutamate + malate were used as substrates (Fig. 1). The inhibition was accompanied by a 10-35% stimulation of state 4 and could not be released by the uncouplers DNP or SF 6874 (Fig. 2). With ascorbate + TMPD as the substrate a similar inhibition was found using similar concentrations of the corresponding alkylamines (Fig. 2). With succinate as substrate, however, state 4 was stimulated completely to state 3 respiration by the above described inhibitory concentrations (Fig. 1 and 2). With the corresponding hydrocarbon n-tetradecane no effect could be detected up to 300 nmoles/mg protein with any of the substrates.

The action of a representative compound, tetradecylamine, on state 4 respiration was studied to determine, if the inhibition of coupled respiration may not be but an accidental value on the way to complete electron transport inhibition. The most striking result was that with ascorbate + TMPD as the substrate state 4 respiration could not be inhibited more than 35% by high concentrations of the amine (Fig. 3).

To distinguish now more precisely between the inhibition of a proton pump and of the electron transport, we used the most sensitive analytical method for this problem, the changes of the membrane potential $\Delta\psi$ and the proton gradient

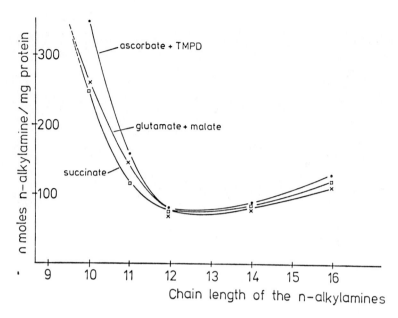

Fig. 1. The concentrations of primary alkylamines required for either complete inhibition of coupled respiration with glutamate + malate (x) or ascorbate + TMPD (o) as the substrates or for complete stimulation of state 4 to state 3 respiration with succinate (□). The conditions are the same as described in Figure 2.

ΔpH of steady state mitochondria. We found that both remained unchanged up to the inhibitory concentration of the amine and decreased around it with different ratios for each of the substrates used (Fig. 4).

Equimolar amounts of picric acid together with the inhibitory concentration of tetradecylamine stimulated state 4 to 50% of state 3 respiration with glutamate + malate and to 160% with ascorbate + TMPD as substrates. Inhibition of coupled respiration by tetradecylamine, however, could not be released by subsequent addition of picric acid.

III. DISCUSSION

Primary unbranched alkylamines inhibited coupled respiration, if glutamate + malate or ascorbate + TMPD were the substrates and if the alkyl chain of the amines were at least 11 carbon atoms long. Because these inhibition reactions could not be released by various uncouplers, it was

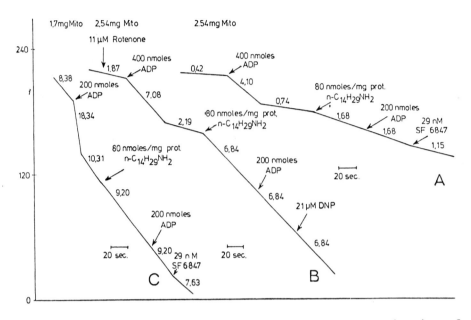

Fig. 2. Effect of tetradecylamine on coupled respiration of beef heart mitochondria. The traces represent the output from an oxygen electrode. The numbers on the traces are respiration rates, µmoles of oxygen mg^{-1} protein h^{-1} at 25°C. Experiment A: Beef heart mitochondria (2.54 mg) were added to 2.4 ml of a reaction mixture which consisted of 0.25 M sucrose, 2.5 mM glutamate, 2.5 mM D,L-malate, 5 mM malonate, 20 mM KCl, 5 mM MgCl$_2$, 10 mM phosphate and 20 mM Tris·HCl, pH 7.3. Experiment B: Beef heart mitochondria (2.54 mg) were added to 2.4 ml of a reaction mixture consisting of 0.25 mM sucrose, 10 mM succinate, 20 mM KCl, 5 mM MgCl$_2$, 10 mM phosphate and 20 mM Tris·HCl, pH 7.3. Experiment C: Beef heart mitochondria (1.7 mg) were added to 2.4 ml of a reaction mixture containing 0.25 M sucrose, 5 mM ascorbate, 0.25 mM TMPD, 20 mM KCl, 5 mM MgCl$_2$, 10 mM phosphate and 20 mM Tris·HCl, pH 7.3.

very probable, that they were located near to the electron transport chain. The usual interpretation would be that these inhibition reactions, which could not be uncoupled, may be ascribed, rather, to the action of the amines on the electron transport than to a coupling phenomenon. Yet it is not generally considered that the production of a proton gradient as part of the proton motive force is coupled to the electron transport; thus, some inhibitors of the electron transport may act primarily as inhibitors of the proton translocation reaction on the individual electron transport complex I, III

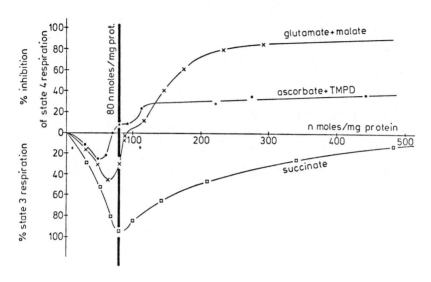

Fig. 3. The action of tetradecylamine on state 4 respiration. The conditions are the same as described in Figure 2.

or IV.

In the case of the alkylamines it was not necessary to completely prevent electron transport for the inhibition of ATP synthesis, but only to lower it from state 3 to state 4 respiration (Fig. 2). Up to this inhibitory concentration of the amines the membrane potential $\Delta\psi$ as well as the proton gradient ΔpH remained unchanged; above it both, $\Delta\psi$ and ΔpH decreased with varying degrees with the different substrates (Fig. 4). It was concluded that this inhibition may be ascribed to the interaction of the basic amino group (linked at least to the undecyl group) with two of the proton translocating units of the electron transport chain. It was probable, that the lipophilic amine was tightly bound both by the alkyl chain and by an ionic linkage, formed by an acid-base reaction of an undissociated carboxyl group in a channel or pump with the amine, by which proton transport, and concomitantly coupling and uncoupling, were prevented. This idea was supported by the action of picric acid, which could not reverse the alkylamine inhibition by subsequent addition, however, when acid and amine were added simultaneously, the mitochondria became uncoupled. Because the inhibition of state 3 respiration was obviously coupled directly to the inhibition of proton translocation (Fig. 4), state 4 respiration may be an artefact and without any function for the proton

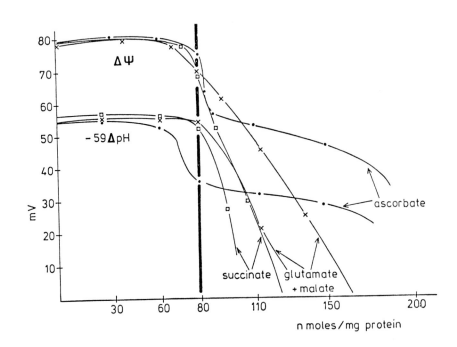

Fig. 4. The changes of proton concentration gradient and the membrane potential $\Delta\psi$ in steady state beef heart mitochondria by increasing amounts of tetradecylamine in the presence of the different substrates. The matrix volume was determined with ^3H_2O and $[^{14}C]$ sucrose according to E. Pandan and H. Rottenberg (1973); the medium contained 2.11 ml of 0.2 M sucrose. 10 mM Tris·HCl, pH 7.2, 2.7 mg protein and 1 mM of the respective substrate(s). ΔpH was determined by measuring the distributions of $[^{14}C]$-acetate according to D. G. Nicholls (1974); $[^3H]$-sucrose was added to each sample to follow the changes in the extramatrix volume; the medium contained 2.16 ml of 0.2 M sucrose, 10 mM Tris·HCl, 25 mM Tris-acetate, pH 7.2, 2.7 mg protein and 1 mM of the respective substrate(s). $\Delta\psi$ was determined by a similar procedure using $[^{86}RbCl]$ and $[^3H]$ sucrose (Mark III liquid scintillation spectrometer: autoisotope programm no. 9); the medium contained 2.18 ml of 0.2 M sucrose, 10 mM Tris·HCl, pH 7.2, 10 mM RbCl, 20 nM valinomycin, 2.7 mg protein and 1 mM of the respective substrate(s) (K-salts). Each sample was incubated for 2 min with the respective substrate(s). The reaction was started by the addition of n-tetradecylamine and terminated after 2 min by centrifugation.

motive force, if the inhibition of coupling and uncoupling reflected, in fact, the inhibition of a proton pump. Finally, it should be mentioned that the minimal chain length of eleven carbon atoms for the specific inhibition may be taken also as evidence for R.J.P.Williams' proposal of localized protons buried in the lipid of the membrane.

REFERENCES

Bäuerlein, E. (1978) Chemical Mechanisms in Energy
    Coupling, in "Energy Conservation in Biological
    Membranes" (G. Schäfer and M. Klingenberg, eds.),
    Springer Verlag, Heidelberg, in press.
Bäuerlein, E. and Kiehl, R. (1976) FEBS Lett. 61:68.
Kiehl, R. and Bäuerlein, E. (1976) ibid. 72:24.
Kiehl, R. and Bäuerlein, E. (1977) ibid. 83:311.
Trasch, H. and Bäuerlein, E. (1978a) Hoppe-Seyler's Z.
    physiol. Chem. 359:1157.
Trasch, H. and Bäuerlein, E. (1978b) FEBS Lett. in press.

A CYANINE DYE:
PHOSPHATE-DEPENDENT CATIONIC UNCOUPLER
IN MITOCHONDRIA

Hiroshi Terada[1]

Faculty of Pharmaceutical Sciences,
University of Tokushima
Tokushima

I. INTRODUCTION

Many organic compounds with various chemical structures, such as phenols(1), salicylanilides(2), benzimidazoles(3) and phenylhydrazones(4) are known to uncouple oxidative phosphorylation in mitochondria. The chemical structures of the

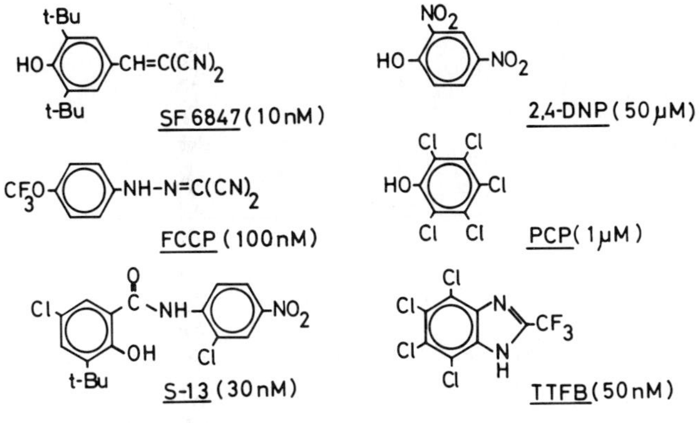

Fig. 1. Chemical structures of currently used weakly acidic uncouplers and their activities in mitochondria.

---

[1]Supported by a grant from the Ministry of Education, Japan.

currently used potent uncouplers and their activities expressed as the concentration required for maximal release of the respiration in State 4 rat-liver mitochondria are shown in Fig. 1. Since potent uncouplers are weakly acidic, and they increases proton permeability across membranes in mitochondria(5,6) and model membrane systems(7,8), dissipation of the proton gradient across the mitochondrial membrane is regarded as essential for exhibiting their activities, although this mechanism is not fully accepted(9,10).

However, little is known about the mechanism of action of cationic uncouplers in mitochondria, partly because there are few cationic uncouplers that are potent on mitochondria. Recently, the cyanine dye, NK-19(4,4'-dimethyl-3,3'-di-n-heptyl-8-[2-(4-methyl-3-n-heptylthiazole)]-2,2'-dicarbocyanine diiodide) was found to be an effective uncoupler in mitochondria, releasing State 4 respiration and the oligomycin-inhibited respiration and activating ATPase(11,12). This paper deals with the action of NK-19, which was kindly supplied from Nihon Kankoh Shikiso Research Laboratory, Okayama, on mitochondrial function in more detail.

NK-19

II. EFFECT OF NK-19 ON STATE 4 MITOCHONDRIA

Fig. 2 shows the effect of NK-19 on State 4 mitochondria in inorganic phosphate(Pi) containing medium using succinate as substrate. The addition of 20 µM of NK-19 increases the rate of respiration to about 5-fold that in State 4. However, the effect is transient and slight inhibition of the respiration takes place about 1 min after the addition of NK-19. The typical uncoupler, SF 6847(1), cannot release this inhibition. Transient stimulation of the respiration was

A Cyanine Dye

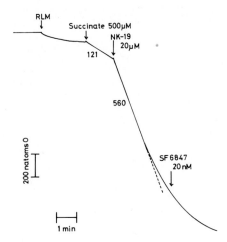

Fig. 2. Effect of NK-19 on the respiration of State 4 mitochondria. Rat-liver mitochondria (1 mg/ml) were suspended in medium consisting of 150 mM sucrose, 20 mM KCl, 5 mM $MgCl_2$, 2 mM EDTA and 1 mM potassium phosphate, pH 7.2, in a total volume of 4.35 ml at 25° C. Succinate (with 1 μg rotenone) was used as substrate.

consistently observed at all concentrations of NK-19, and the degree of the inhibition became greater as the concentration of NK-19 increased. Fig. 3 shows the relation between the initial respiration rate induced by NK-19 and the concentration of NK-19 added to the reaction medium containing Pi. Less than 10 μM NK-19 is enough for inducing the maximal release of the respiration when succinate is used as substrate.

III. EFFECT OF INORGANIC PHOSPHATE ON THE ACTIVITY

Fig. 4 shows the effect of Pi on the uncoupling activity of NK-19. In the absence of Pi, 9.2 μM NK-19 scarcely stimulates the respiration using succinate as substrate. The relation between the concentration of NK-19 and the rate of respiration is also shown in Fig. 3, where less than two-fold stimulation of respiration is observed in the absence of Pi. Addition of Br^-, Cl^- or I^- instead of Pi did not enhance the respiration, and the permeant anion SCN^- also failed to

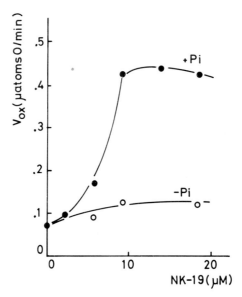

Fig. 3. Changes in the respiratory rate (V_{ox}) with the concentration of NK-19 in the presence and absence of Pi. The medium for +Pi contained 200 mM sucrose, 2 mM $MgCl_2$, 1 mM EDTA and 10 mM potassium phosphate, pH 7.2, and that for -Pi contained 10 mM Tris·Cl instead of phosphate. State 4 respiration of rat-liver mitochondria (3.1 mg/4.35 ml) was measured using succinate with 1 μg rotenone as substrate.

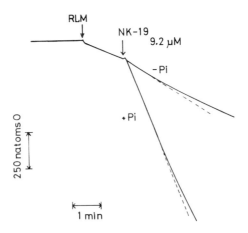

Fig. 4. Phosphate-dependent effect of NK-19 on respiration of State 4 mitochondria. Experimental conditions, see Fig. 3.

stimulate the respiration. Stimulation by NK-19 was induced with CH_3COO^-, but the degree of the stimulation was about half that induced with Pi. These results would indicate that for exhibiting its effect, NK-19 must penetrate to the inner side of the mitochondria. The process of penetration does not seem simply to be due to formation of an ion-pair complex with the permeant ion, and may involve the Pi-translocating system in mitochondrial membrane.

IV. PROTON EJECTION BY NK-19

As shown in Fig. 5, the addition of 20 µM NK-19 causes instant ejection of H^+ from mitochondria into the reaction medium in the presence of Pi. The amount of H^+ ejected attains about 50 nequiv. per mg protein. The H^+-ejection is reversed on addition of SF 6847.

It is interesting to compare the result in Fig. 5 with that in Fig. 2, the two experimental conditions being almost the same. The extent of H^+-ejection is not influenced by the inhibitory effect of NK-19 on the respiration, and SF 6847, which is ineffective in stimulating NK-19-inhibited respiration, reverses the ejection of H^+. Furthermore, the ejection of H^+, though slower, was also observed on adding 20 µM NK-19 to the mitochondrial suspension in the absence of

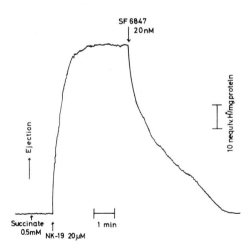

Fig. 5. Ejection of H^+ by NK-19 and its reversal by SF 6847. Rat-liver mitochondria(1 mg/ml) were suspended in the medium described in Fig. 2 in a total volume of 3.0 ml.

Pi, where NK-19 induced slight respiratory stimulation. The extent of the ejection was about 50 nequiv. per mg protein, which is the same as that observed in the presence of Pi. These results indicate that the uncoupling action of NK-19 is not directly correlated with change in the H^+ movement across mitochondrial membrane. Thus it is important for understanding the mechanism of action of NK-19 to measure the membrane potential across mitochondrial membrane during the action of NK-19.

From the above results, NK-19 was shown to be a potent phosphate-dependent cationic uncoupler in oxidative phosphorylation in mitochondria. The mechanism of action of NK-19 is very complex and requires further study.

REFERENCES

1. Muraoka, S., and Terada, H. (1972). *Biochim. Biophys. Acta* 275:271.
2. Williamson, R.L., and Metcalf, R.L. (1967). *Scinece* 158:1694.
3. Beechey, R.B. (1966). *Biochem. J.* 98:284.
4. Heytler, P.G., and Prichard, W.W. (1962). *Biochem. Biophys. Res. Commun.* 7:272.
5. Cunarro, J., and Weiner, M.W. (1975). *Biochim. Biophys. Acta* 387:234.
6. Terada, H., Uda, M., Kametani, F., and Kubota, S. (1978). *Biochim. Biophys. Acta* in press.
7. Skulachev, V.P., Sharaf, A.A., and Liberman, E.A. (1967). *Nature* 216:718.
8. Yamaguchi, A., and Anraku, Y. (1978). *Biochim. Biophys. Acta* 501:136.
9. Hanstein, W.G., and Hatefi, Y. (1974). *J. Biol. Chem.* 249:1356.
10. Kessler, R.J., Zande, H.V., Tyson, C.A., Blondin, G.A., Fairfield, J., Glasser, P., and Green, D.E. (1977). *Proc. Natl. Acad. Sci. USA* 74:2241.
11. Kanemasa, Y. (1969). *Acta Med. Okayama* 23:337.
12. Terada, H. (1979). *Kankohshikiso* in press.

ND THE MECHANISM OF Ca^{2+}-INDUCED SWELLING
OF RAT LIVER MITOCHONDRIA

Masanao Kobayashi
Yoshiharu Shimomura
Masashi Tanaka
Kazuo Katsumata
Harold Baum[1]
Takayuki Ozawa

Department of Biomedical Chemistry
Faculty of Medicine
University of Nagoya
Nagoya, Japan

I. INTRODUCTION

The precise mechanism of Ca^{2+}-transport across the mitochondrial membrane has not been fully understood, despite its potential physiological importance in the regulation of Ca^{2+} concentration within the cell and hence Ca^{2+}-controlled intracellular biological activities (Rasmussen, 1970; Borle, 1975; Bygrave, 1977). Up to this time, it has been assumed that Ca^{2+}-uptake into mitochondria takes place in an energy-dependent process involving a specific transport system, or carrier, which is located on the inner mitochondrial membrane and sensitive to specific inhibitors such as La^{3+} or ruthenium red (see Bygrave, 1977; Lehninger, et al., 1978).

We have found that several compounds including 1-butyl-3-(p-tolylsulfonyl)urea (tolbutamide), a widely used oral bypoglycemic agent, coenzyme Q_2 and Q_3 induce a rapid swelling of rat liver mitochondria in the presence of Ca^{2+}. On the other hand, uncoupling effect of tolbutamide on liver mitochondria has been reported by one of the authors (Katsumata and Hagiha-

---

[1]Visiting professor from University of London under Japan-UK Faculty exchange program.

ra, 1973). Both phenomena, swelling and uncoupling, appear to be based on a common underlying mechanism, having some relevance to Ca^{2+}-transport. Present report describes the effect of these agents on the rate of respiration and swelling of rat liver mitochondria in the presence of Ca^{2+}, as well as on mitochondrial Ca^{2+}-transport; the results suggest the involvement of a cyclic flux of accumulated Ca^{2+} across the inner membrane induced by these agents in the swelling and apparent uncoupling described above.

II. RESULTS

A. Effect of Tolbutamide and Ubiquinones on Mitochondrial Respiration

Oxygen consumption by mitochondria was recorded using Beckman 39550 oxygen electrode. The medium used for prepara-

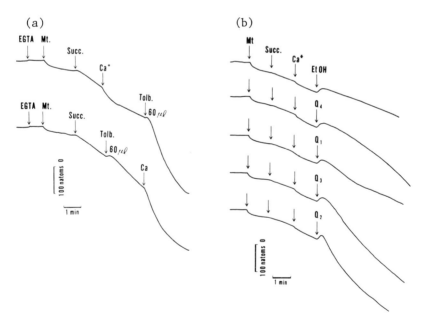

FIGURE 1. Effect of tolbutamide and ubiquinones on mitochondrial respiration. Oxygen consumption was measured in a total volume of 6 ml. (a) Tolbutamide was added in 60 μl ethanol to give a final concentration of 1 mM. (b) Ubiquinones were added in 60 μl ethanol to give a final concentration of 0.1 mM.

tion of mitochondria and throughout this and following analyses was 0.25 M sucrose - 10 mM TrisHCl (pH 7.2). Succinate (10 mM) was used as substrate and 0.1 mM EGTA was added to prevent the possible interference by endogenous Ca^{2+}.

As shown in FIGURE 1a, 1 mM tolbutamide induced distinct and continuous release of the controlled respiration, only in the presence of Ca^{2+} (0.3 mM) and *vice versa*. This effect of tolbutamide was dose-dependent and was detected at a concentration as low as 0.1 mM. In the absence of Ca^{2+}, 1mM tolbutamide had little effect; at a higher concentration, 2 mM, however, a distinct release of controlled respiration was observed as reported previously (not shown). A derivative of tolbutamide in which a carboxyl group was introduced at the terminal of the butyl group (tolb-COOH) was of no effect (not shown).

Similar release of the controlled respiration was also observed with 0.1 mM coenzyme Q_2 and Q_3 (FIGURE 1b), again only in the presence of Ca^{2+} (0.3 mM); their shorter (coenzyme Q_1) or longer (coenzyme Q_4; coenzyme Q_{8-10} not shown) homologs, however, had no or only slightly stimulating effect on the respiration rate.

B. Effect of Tolbutamide and Ubiquinones on Swelling of Mitochondria

Swelling of mitochondria was followed by decrease in absorbance at 540 nm using Shimadzu Multipurpose Recording Spectrophotometer MPS-50L. The medium and additions were essentially the same as those described in the preceding section.

As shown in FIGURE 2a, in the presence of Ca^{2+} (0.3 mM), tolbutamide induced a rapid swelling of mitochondria in a dose-dependent manner; at a concentration as low as 0.1 mM a definite potentiation of swelling was observed compared with the control (EtOH). Addition of ATP or ADP (1 mM) prior to tolbutamide resulted in a lag period before the onset of the swelling (not shown). Ruthenium red (10 μM). when added before Ca^{2+} or tolbutamide, but not after both, completely blocked the swelling (not shown). Tolb-COOH showed no swelling effect.

Similarly, as shown in FIGURE 2b, coenzyme Q_2 and Q_3 (0.1 mM) also induced, only in the presence of Ca^{2+}, a rapid swelling. In accordance with the results on respiration (FIGURE 1b), shorter or longer homologs were of no effect (results with coenzyme Q_{8-10} not shown) compared with the control (EtOH). Ruthenium red (10 μM) added prior to Ca^{2+} completely prevented the swelling induced by coenzyme Q_3 (not shown).

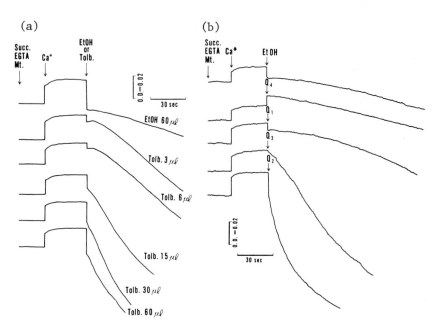

FIGURE 2. Effect of tolbutamide and ubiquinones on swelling of mitochondria. Absorbance at 540 nm was followed in a total volume of 3 ml. (a) Aliquots of tolbutamide solution in ethanol were added as indicated, 30 μl giving a final concentration of 1 mM. (b) Ubiquinones were added in 30 μl ethanol to give each a final concentration of 0.1 mM.

C. Effect of Tolbutamide and Ubiquinones on Ca^{2+}-transport in Mitochondria

Since the observations described above suggested the involvement of Ca^{2+}-transport as a common underlying mechanism, the effect of these agents on Ca^{2+}-uptake into mitochondria was studied. Ca^{2+}-uptake was computed from the decrease of $^{45}Ca^{2+}$ (initial concentration: 0.3 mM) from the supernatant obtained by centrifugation of the reaction mixture after 2 min incubation with mitochondria. Other conditions were essentially the same as those for respiration and swelling except omission of EGTA. The agents to be tested were added before the addition of $^{45}Ca^{2+}$ or after its incorporation into mitochondria, followed by further 2 min incubation. The results in TABLE I clearly indicate that tolbutamide, in a dose-dependent manner, and coenzyme Q_2 induced a efflux of the incorporated $^{45}Ca^{2+}$. The effect was much less with coenzyme Q_3 and Q_1, and was not observed with coenzyme Q_4 and tolb-COOH.

TABLE I $^{45}Ca^{2+}$-uptake by Mitochondria in the presence of Tolbutamide or Ubiquinones

Additions	(mM)	$^{45}Ca^{2+}$-uptake[a]	
		Before[b]	After[c]
None		39.2	
Tolbutamide	0.25	34.8	29.5
	0.5	19.8	21.4
	1.0	9.2	20.5
Tolb-COOH	1.0	35.8	39.3
Coenzyme Q_1	0.1	28.4	34.7
Coenzyme Q_2	0.1	17.9	17.6
Coenzyme Q_3	0.1	27.5	34.0
Coenzyme Q_4	0.1	38.3	37.0

[a] Values are in nmoles/mg protein.
[b] Additions first, then mitochondria and $^{45}Ca^{2+}$.
[c] Mitochondria and $^{45}Ca^{2+}$ first, then additions.

III. DISCUSSION

The swelling and the release of controlled respiration of mitochondria induced by tolbutamide, coenzyme Q_2 or Q_3 described above showed a common feature that both were observed only in the presence of Ca^{2+} and were inhibited by ruthenium red, a specific inhibitor of Ca^{2+}-transport system. This suggested the involvement of Ca^{2+}-transport in the observed swelling and release of respiration. ^{45}Ca-uptake study indicated that these agents induced a efflux of the incorporated Ca^{2+} from mitochondria. It is conceivable that a cyclic energy-dissipating flux of Ca^{2+} thus induced, expending energy for its continuous recycling, results in accelerated electron transport, an apparent uncoupled respiration of mitochondria, eventually accompanied by collapse of the structural integrity of mitochondria. Such a mechanism, Ca^{2+}-cycling, has been proposed by others in order to explain the uncoupling effect of ionophore A23187 (Reed and Lardy, 1972) or of prostaglandins (Carafoli, et al., 1975), and the Pi- or diamide-induced efflux of Mg^{2+} and K^+ from liver mitochondria (Siliprandi, et al., 1978). Similar experiments with triiodothyronine (data not shown) suggested that the wellknown swelling and uncoupling effect of this and thyroxine on mitochondria may also be based on this mechanism.

The mechanism by which the agents described in this report induce the efflux of intramitochondrial Ca^{2+} is not clear. However, some structural similarities between tolbutamide and coenzyme Q_2, requirement for a certain extent of hydrophobicity of the effective reagents, as shown by ineffectiveness of tolb-COOH and shorter or longer coenzyme Q homologs, together with inhibition of swelling by specific agents such as ATP and ADP, may suggest that these agents have not acted as mere ionophores, but through some specific binding site on the inner mitochondrial membrane. With regard to the latter observation, Sordahl and Asimakis (1978) have recently reported results suggesting the involvement of intramitochondrial adenine nucleotides bound to membrane, probably to adenine translocase, in retention of accumulated Ca^{2+}.

REFERENCES

Borle, A.B. (1975). In "Calcium Transport in Contraction and Secretion" (E. Carafoli, F. Clementi, W. Drabikowski, and A. Margreth, eds.), p. 77. North-Holland Publishing Co., Amsterdam.
Bygrave, F.L. (1977). Current Topics in Bioenergetics 6: 259.
Carafoli, E., Malström, K., Capano, M., Sigel, E., and Crompton, M. (1975). In "Calcium Transport in Contraction and Secretion" (E. Carafoli, F. Clementi, W. Drabikowski, and E. Margreth, eds.), p. 53. North-Holland Publishing Co., Amsterdam.
Katsumata, K. and Hagihara, M. (1973). Nagoya J. Med. Sci. 35: 69.
Lehninger, A.L., Reynafarje, B., Vercesi, A., and Tew, W.P. (1978). In "The Proton and Calcium Pumps" (G.F. Azzone, M. Avron, J.C. Metcalfe, E. Quagliariello, and N. Siliprandi, eds.), p. 203. Elsevier/North-Holland Biomedical Press, Amsterdam.
Rasmussen, H. (1970). Science 170: 404.
Reed, P.W., and Lardy, H.A. (1972). J. Biol. Chem. 247: 6970.
Siliprandi, N., Siliprandi, D., Toniello, A., Rugolo, M. and Zoccarato, F. (1978). In "The Proton and Calcium Pumps" (G.F. Azzone, M. Avron, J.C. Metcalfe, E. Quagliariello, and N. Siliprandi, eds.), p. 263. Elsevier/North-Holland Biomedical Press, Amsterdam.
Sordahl, S.A., and Asimakis, G.K. (1978). ibid., p. 273.

THE KINETICS OF THE REACTION OF CALCIUM WITH MITOCHONDRIA AT SUBZERO TEMPERATURES[1]

Britton Chance
Yuzo Nakase[2]
Fanny Itsak[3]

Johnson Research Foundation
University of Pennsylvania
Philadelphia, PA

INTRODUCTION

Calcium evokes the most rapid and effective transition from the resting state 4 to the active state 3 in mitochondria (1). The state 4 to 3 transition occurs in 70 msec at 23°, and the maximal respiratory rates can be achieved in the calcium stimulated state 3. The calcium reaction is remarkable in that it can be observed at temperatures well below +4°, at which temperature ADP ceases to stimulate respiratory activity (2).

Studies of mitochondrial kinetics at subzero temperatures are afforded by the stop flow apparatus using ethylene glycol as an aprotic solvent down to -30° (3). Temperatures down to -125° can be reached in the triple trap solid state, and a variety of intermediate compounds in cytochrome oxidase and oxygen have been identified (4). Energy coupling reactions, usually inhibited by aprotic solvents, are still active in 15% ethylene glycol, and in a quaternary mixture of ethylene glycol, dimethyl sulfoxide (DMSO), methanol and water of a freezing point of -22° (5,6,7). This communication describes how such a reaction

---

[1]This work has been supported by NIH Grants GM 12202 and HL 18708.
[2]Present address: Department of Physiology, Wakayama Medical School, Wakayama-city, JAPAN.
[3]Present address: Department of Biochemistry, The Weizmann Institute of Science, Rehovot, ISRAEL.

medium can be used to explore calcium kinetics and membrane activation and deactivation processes under these remarkable conditions.

EXPERIMENTAL METHODS

The experimental methods are described in detail elsewhere (7), together with the observation that the pH is little affected by the temperature when the protein concentration is as high as in these particular experiments [approximately 10 mg mitochondrial protein per ml (8)]. The optical assays are made with the wavelength scanning dual wavelength spectrophotometer and with a time sharing interference filter dual wavelength spectrophotometer.

RESULTS

<u>Observation of Calcium Stimulated Respiration at Low Temperatures</u>. By employing an aprotic colvent in the case of Figure 1, including 15% DMSO, the oxygen electrode trace (time

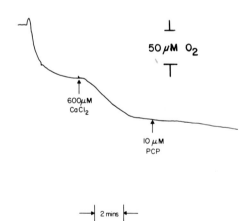

FIGURE 1. An illustration of respiratory control in the accumulation of calcium in rat liver mitochondria at -4°. The points of addition of 600 µM calcium and 10 µM PCP are indicated by the arrows. The reaction medium contains 8 mg mitochondrial protein per ml, 6 mM succinate and glutamate, 10 mM phosphate buffer, pH 7.4, 225 mM mannitol, 75 mM sucrose, 50 mM morpholinopropanesulfonate (MOPS), and 15% DMSO.

moving from left to right and oxygen concentration decreasing downwards) shows that addition of anaerobic substrate-supplemented mitochondria to the reaction mixture causes a decrease in oxygen concentration proportional to the fractions of the volume they occupy. Thereafter, the addition of 600 µM calcium, indicated by an arrow, shows a delay characteristic of the Clark electrode followed by a state 4 to 3 transition. Subsequently, the addition of 10 µM PCP causes no detectable stimulation of respiration. This result identifies calcium as one of the few stimulants (if not the only one) of mitochondrial respiration in the subzero temperature range.

In other experiments the distinction between the uncoupler and calcium is underlined where the second addition of calcium readily stimulates the respiratory rate (1). In the figures shown in this paper the effect of respiratory stimulation for the two calcium additions, shown in separate graphs, is clearly defined and distinctly different.

In order to study bioenergetic reactions effectively at low temperatures, it is necessary to explore the effects of different aprotic solvents. Figures 2 and 3 show characteristically different effects of two useful aprotic solvents, ethylene glycol (Figure 2) and DMSO (Figure 3). The experiments are conveniently carried out at 0° in order that the system be tested under controlled conditions with no aprotic solvent present.

In Figure 2A in the case of varied ethylene glycol concentration with NADH as substrate, the results for the first and second calcium additions show characteristic increased respiratory inhibition with increased concentration of ethylene glycol. This inhibition is more severe for the second addition of calcium, since the calcium loading has increased. At 30% ethylene glycol, the decrease of respiratory rate appears to be small for the first calcium addition and is in fact 43% at the extreme. The state 4 data are of interest — there is no rise in rate, as would be expected if uncoupling had occurred.

Figure 2B illustrates the effect of ethylene glycol concentration upon respiratory control ratios with succinate as substrate. It is apparent that the inhibition of electron transport is greater than with NADH as substrate, although the level reached in the inhibited state is roughly the same in the two cases and the decrease of activity is to 56% of the initial value.

The second addition of calcium evokes a more marked inhibition of respiration, and at the same time a rise of the state 4 rate characteristic of a mild uncoupling effect of ethylene glycol. The end result is that at 30% ethylene glycol no respiratory control can be demonstrated.

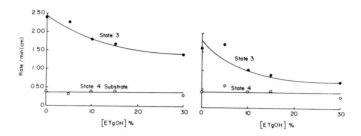

FIGURE 2A. The effect of ethylene glycol concentration upon the respiratory control ratio in rat liver mitochondria at 0° with NADH as substrate. The respiration is activated by two sequential additions of 600 μM calcium, as shown in the two separate graphs. The state 3 and 4 rates are plotted as a function of the total volume of the reaction medium occupied by ethylene glycol. Other conditions similar to those in Figure 1.

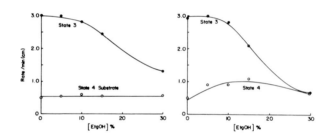

FIGURE 2B. The effect of ethylene glycol concentration on the respiratory control ratio at 0° with succinate as substrate (6 mM). Reaction medium as in Figure 1; calcium additions as in Figure 2A.

By way of contrast, the effect of DMSO shown in Figure 3 is scarcely perceptable up to approximately 18%; thereafter, an abrupt inhibition of the state 3 rate occurs, giving an inhibition at 30% DMSO of 40% of the initial rate. Again the state 4 rate is unaffected — at least for the first calcium addition. For the second calcium addition, at 10% DMSO there begins an abrupt uncoupling event which raises the state 4 rate nearly to the state 3 rate. This type of uncoupling event is characteristic of some of the barbiturates (9) and represents a second type of effect of the aprotic solvent.

Multicomponent Aprotic Solvent Mixtures. In order to enhance the freezing point depression and at the same time to maintain a low mole fraction of the aprotic solvents, mixtures based upon

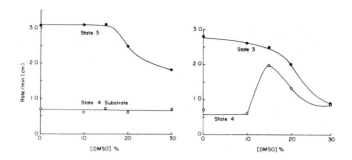

FIGURE 3. The effect of DMSO concentration upon the respiratory control ratio in rat liver mitochondria at 0° with succinate as substrate. Other conditions as in Figures 1 (reaction medium) and 2 (calcium additions).

the results of Figures 2 and 3 and other data as well are illustrated in Figures 4A and B. In this case, fixed amounts of the two preceding aprotic solvents, 15% ethylene glycol and 5% DMSO, are supplemented by concentrations of methanol indicated on the abscissa. In Figure 4A, for the first addition of calcium and with NADH as substrate, the effects of added methanol on respiration are characteristic of those observed previously with ethylene glycol. There is a depression of respiration so that, for example, at 5% methanol, 34% depression is obtained. The effect is similar for the second addition of calcium.

In Figure 4B the effect is much more striking because of the higher respiratory rate with succinate. No stimulation over the state 4 rate can be observed for the second addition of calcium.

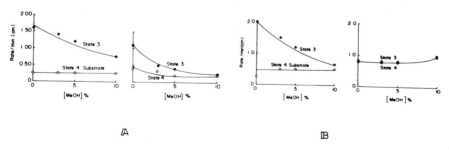

FIGURE 4. The effect of methanol concentration upon the respiratory control ratio in rat liver mitochondria at 0°, A, with NADH as substrate, and B, with succinate as substrate. The reaction medium contains 15% EtgOH and 5% DMSO in water. Other conditions as in Figures 1 (reaction medium) and 2 (calcium additions).

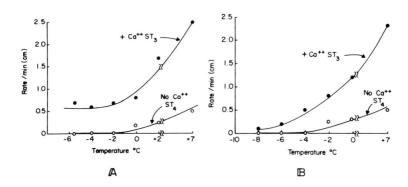

FIGURE 5. The effect of temperature upon the respiratory control ratio in calcium stimulated respiration. Conditions as in Figure 4, except that the methanol concentration is maintained at 5%; A, with NADH as substrate, and B, with succinate as substrate.

Effect of Temperature. For the same conditions as in Figure 4, we show in Figures 5A and B the temperature profile for the first calcium addition with NADH as substrate (Figure 5A) and with succinate as substrate (Figure 5B). As expected from the ethylene glycol profiles of the previous charts, no respiratory control is evidenced with succinate as substrate at the lowest temperature, while significant respiratory control can be observed with NADH as substrate. Considering then the latter case, the respiratory control ratio becomes very large as the temperature approaches -6°. In fact, the state 4 rate only becomes measurable above 0°.

There is an extremely low temperature coefficient for NADH respiration in the temperature interval from 0° to -6°. In the presence of succinate, however, the temperature dependence on calcium stimulated respiration is high and the inhibition by the aprotic solvent is continuous down to -8°, where the state 3 and 4 rates cannot be distinguished.

Spectroscopic Studies at Low Temperatures. Chance and Williams found that a low concentration of azide is useful in magnifying "crossover" responses of mitochondria in the state 4 to 3 transition (10). Figure 6 illustrates the spectroscopic changes obtainable at -6° in the presence of aprotic solvents upon adding calcium to the mitochondrial suspension. The recordings are taken by the dual wavelength spectrophotometer with a computer-corrected baseline. The state 4 baseline prior to the addition of calcium is so labelled. Trace 1 is obtained immediately upon adding calcium; traces 2 and 3 are 50 and 150 sec thereafter. Trace 1 shows increased reduction of cytochrome

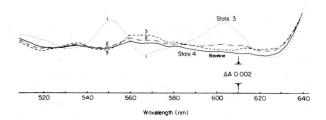

FIGURE 6. Effect of calcium upon the crossover response of mitochondria. Conditions similar to those in Figure 1, except the aprotic solvent mixture of Figure 5B is employed; temperature -6°, 54 mM azide. Trace 1 is taken immediately after 50 μM calcium addition; trace 2, 50 sec after 50 μM calcium addition; trace 3, 150 sec after 50 μM calcium addition. The solid curve represents the baseline trace prior to any addition.

$a + a_3$ and c, and an oxidation of cytochrome b (presumably the long-wave component), a typical crossover response for these conditions and one that reverts very nearly to the initial conditions (baseline) after the uptake of calcium (traces 2 and 3).

<u>Kinetic Studies at Low Temperatures</u>. The kinetics of the state 4 to 3 transitions under these conditions are indicated by Figure 7, which identifies the responses of the cytochromes, but in this case without the addition of azide in order that

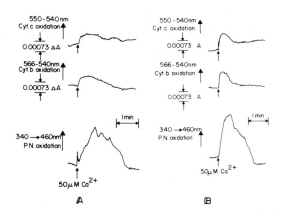

FIGURE 7. The kinetics of the crossover responses per calcium addition; A, at -8.3°, and B, at -4.8°.

the respiratory chain exhibit its full functional capability in electron transport. The traces in Figure 7A, at -8.3°, show measurable rise times for cytochrome c and cytochrome b oxidation (in the absence of azide both these components are altered in the direction of oxidation) with half-times of approximately 4 sec. Thereafter, the traces show cycling back, in a more pronounced way for cytochrome b than cytochrome c, but generally in accordance with the traces of Figure 6. The half-times for the response of cytochrome c and cytochrome b are indistinguishable under these conditions. However, the half-time for NADH oxidation (measured fluorometrically) is much slower, being approximately 20 sec.

The temperature coefficient for the effect is rather large; if the experiment is repeated at -4.8° (shown in Figure 7B), then the rise times of the cytochrome traces are unresolved (<1 sec) while the half-time for the NADH change is approximately 7 sec.

DISCUSSION

These experiments demonstrate the feasibility of maintaining highly significant respiratory control ratios at low temperature, particularly in the case of NADH oxidation where in the region of -4° the state 4 rate is undetectable. At the same time, responses of respiratory carriers are significant and can be time resolved along the respiratory chain, showing large differences between the speed response of cytochrome carriers and the NADH pool.

Under these conditions, previous work has shown that there is no stimulation of respiration by ADP; this lack of response is presumably due to a failure of the translocase to provide an increased matrix space concentration of ADP. Nevertheless, the calcium transport system works effectively and rapidly at the low temperatures, affecting all of the energy coupling sites of the respiratory chain in apparently the same manner as at room temperature and permitting a kinetic analysis of the response of the carriers by the ordinary manual mixing method [the room temperature half-time is 70 msec (1)].

With any technological advance, often more questions can be raised than can be answered in the initial study, as is true in this case. It is obvious that the method points to the possibility of measuring hydrogen ion and potassium ion movements under the same conditions as calcium uptake, thereby identifying their roles in the state 4 to 3 transition in a way not readily possible at the higher temperatures.

SUMMARY

A new method is described for studying the state 4 to 3 transition in aprotic solvents and under conditions where kinetic resolution of the transition is readily obtained. This method will be useful in the study of ion uptake, membrane potential, hydrogen ion gradients and kinetics in mitochondrial suspensions.

REFERENCES

1. Chance, B. (1965) J. Biol. Chem. 240:2729-2748.
2. Chance, B., Takeda, H., Nakase, Y., and Itsak, F. (1978) FED. PROC. 37:851a.
3. Erecińska, M., and Chance, B. (1972) Arch. Biochem. Biophys. 151:304-315.
4. Chance, B., Graham, N., and Legallias, V. (1975) Anal. Biochem. 67:552-579.
5. Douzou, P., Sireix, R., and Travers, F. (1970) Proc. Natl. Acad. Sci. 66:787-792.
6. Douzou, P., Hui Bon Hoa, G., Maurel, P., and Travers, F., in CRC Handbook of Biochemistry and Molecular Biology, 3rd Ed., Vol. 1, (Fasman, G.C., ed).
7. Chance, B., and Itsak, F., "Effects of Aprotic Solvents on Mitochondrial Electron Transport and Respiratory Control", submitted to J. Biol. Chem., August 1978.
8. Williams-Smith, D.L., Bray, R.C., Barter, M.J., Tsopanakis, A.D., and Vincent, S.P. (1977) Biochem. J. 167:593-600.
9. Chance, B., and Hollunger, G. (1961) J. Biol. Chem. 236:1534-1543.
10. Chance, B., and Williams, G.R. (1955) J. Biol. Chem. 217:383-393.

ADP,ATP TRANSPORT IN MITOCHONDRIA AS ANION TRANSLOCATION

Martin Klingenberg

Institut für Physikalische Biochemie
der Universität München

It is now generally established since our first formulations (Pfaff and Klingenberg, 1968; Klingenberg et al., 1969a; Klingenberg et al., 1969b) that the ADP,ATP transport in the hetero mode (ATP against ADP) carries one negative charge through the membrane. The first evidence came from the influence of membrane energization in making the exchange reaction asymmetric such that the uptake of ATP was preferred to that of ADP, whereas in uncoupled mitochondria this difference was largely abolished (Klingenberg and Pfaff, 1966). A similar effect has valinomycin plus K^+, indicating that the collaps of membrane potential is responsible for the equalizing effect (Klingenberg, 1972). Under these conditions the uptake of ATP was shown to be accompanied by equimolar influx of either H^+ (with uncoupler) or K^+ (with valinomycin) (Klingenberg et al., 1969b; Wulf et al., 1978; LaNoue et al., 1978). As a result, under equilibrium or steady-state conditions the difference in the ATP/ADP ratio outside was found to be higher than inside (Klingenberg et al., 1969a; Heldt et al., 1972), and a linear correlation of the double ratio outside to inside to the membrane potential was determined (Klingenberg and Rottenberg, 1977).

These are the three major steps in defining the electrical ATP,ADP transport to which a large variety of further results can be added. Some of these are important for the understanding of how the membrane potential affects the translocation step making it an electrophoretic process in the ATP synthesis.

This work was supported by a grant from the Deutsche Forschungsgemeinschaft (SFB 51).

In the present contribution I shall dwell on some further data concerning asymmetry of ATP,ADP exchange, electrophoretic and electrogenic ATP,ADP transport as reflected in the respiratory chain and present some results concerning the membrane potential influence on translocation mechanism.

The Energy Control of the ATP Extrusion from Mitochondria

Although it has been known that the "energization" differentiates between ATP and ADP uptake in mitochondria, it remained experimentally unclear whether also the efflux of ATP and ADP is influenced by the energization of mitochondria. In our earlier reports we concluded (Pfaff and Klingenberg, 1968; Heldt and Pfaff, 1969) that the energization influences only the forward rate but not the efflux such that the ADP and ATP efflux remains about equal in the energized state. In later studies, however, it was shown that the ADP influx is dependent on the intramitochondrial ratio ATP/ADP such that it is considerably accelerated when there is more ATP present (Klingenberg, 1975). Therefore, it was concluded that also ATP is preferably extruded as compared to ADP in exchange for external ADP.

However, real proof could be only given if it would be possible to follow directly the efflux of endogenous ADP and ATP, both in the external and internal space. This required measurement of nucleotide and radioactive label distribution in both the internal and external space over the three species ATP, ADP and AMP. Furthermore, it was desirable to follow the kinetics of the exchange in both these reaction spaces. For this purpose a new device was developed ("RAMPRESA"), involving rapid mixing, sampling and separation of the mitochondria from the medium by pressure Millipore filtration and quenching by acid addition. The internal and external volume were differentiated by taking at each time a pair of samples, one with the total volume and one with only the external volume after separating mitochondria by rapid pressure filtration. The difference was used to calculate the internal nucleotide. The total process was fully automated and computer controlled by a preset time program.

The experiment in Figure 1 shows the exchange of ADP against the endogenous nucleotides under conditions where the ATPase is inhibited by oligomycin and the mitochondria are energized or uncoupled. Prior to addition of oligomycin, in the intramitochondrial pool about equivalent amounts of ADP and ATP were established in order to differentiate between the efflux of ATP versus ADP.

The results demonstrate that in the energized state (Fig. 1A) (without FCCP) ATP is extruded from the internal pool in preference to ADP. This is verified by measuring the disappearance of ATP from the endogenous pool and the appearance of the nucleotides in the external space. In the internal space the nucleotides are replaced in exchange for the exogenous ADP. ATP formation is blocked by oligomycin. In the uncoupled state (Fig. 1B) the efflux of ADP is preferred to ATP, again as shown by the movements of nucleotides in the internal and external space.

These results show that the energization does not only regulate the uptake portion of the ADP,ATP exchange but also the efflux portion. Among the four possible exchange modes, such as the two homo-types of exchange (ADP_e against ADP_i and ATP_e against ATP_i) and the two types of hetero exchange (ATP_e against ADP_i and ADP_e against ATP_i), the last one is strongly preferred by the energization of the membrane. The over-all reaction of the ADP,ATP exchange is therefore regulated both for the entrance and exit portions such as to prefer that mode which would be productive for the generation of external ATP in oxidative phosphorylation. The homo-mode is futile, or the ATP against ADP exchange even counter productive. The shift of the exchange reaction towards the ADP against ATP exchange is important in view of the limited over-all capacity of the ATP,ADP exchange which then is used to better advantage for oxidative phosphorylation.

The problem of the exit rates needs further interpretation in view of the high Mg^{++} content in mitochondria, which may cause most of the ATP and also of the ADP to be present as the Mg^{++}-complex. With an about equal total concentration of ADP and ATP present, the ratio of the free forms should be $(ATP/ADP)_{free} \simeq 10^{-1}$, provided an excess Mg^{++}-concentration ($Mg^{++} \simeq 10$ mM). This makes understandable that in the uncoupled state the efflux of ADP is higher than that of ATP, although the efflux rates may be equal, provided equal concentrations of free species. In the energized state the observed threefold preference of ATP to ADP would amount to a thirty-fold increase of specific efflux activity of ATP as compared to ADP when multiplied by the ratios of the free species $(ATP/ADP)_{free} \simeq 0.1$. Attempts to deplete the mitochondria from Mg^{++} by ionophore treatment in order to verify these assumptions resulted in inactivation of the phosphorylation and therefore lowering of ATP level.

The energy influence on the rate could be a modification of the affinities (K_m) for ADP and ATP such that the K_m for ATP is considerably decreased as that for ADP. This possibility must be considered in view of the rather low concentrations of free ADP and ATP in the mitochondria. The present data do not allow to decide whether energization influences

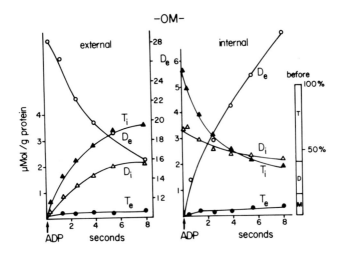

Figure 1. Differentiation between ADP and ATP in the efflux portion of the ADP,ATP exchange. Mitochondria in the controlled (A) and uncoupled (B) state.

Analysis of nucleotide movements inside and outside the mitochondria by rapid mixing, sampling, pressure filtration apparatus (Methods of Enzymology, in print). At time zero 50 μM ADP are added by the moving mixing chamber and a pair of samples of 350 μl withdrawn at the times indicated. This sample pair is spaced at 300 msec, first one through Millipore filter, second one unfiltered. The difference between both samples is used to calculate the intra- and extramitochondrial nucleotides.

The sampling vessels contain 0.5 ml 1 M $HClO_4$, which is neutralized immediately after sampling. The neutralized extracts are passed through small anion exchange columns to separate ATP, ADP and AMP, and the fractions were counted for ^3H- and ^{14}C-nucleotides. In a sample withdrawn before addition of ADP, the content of ATP,ADP and AMP in the mitochondria is determined by enzymatic assays. Rat liver mitochondria with internal nucleotides, prelabeled by ^{14}C, are incubated (30 mg protein per 15 ml) at $16°C$ with 4 mM ketoglutarate, 2 mM malate, 2 mM P_i for 1 min, 5 μg oligomycin/mg protein are added, after another 2 min the exchange is started by addition of ADP.

A.) Mitochondria in the energized (controlled) state.
B.) Mitochondria in the uncoupled state with additional 0.2 μM FCCP, added 1 min after oligomycin, before the start of the exchange.

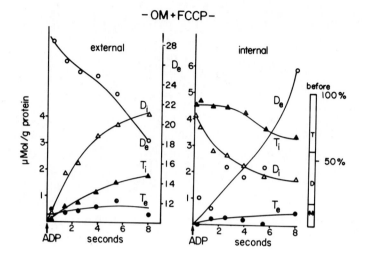

Fig. 1B

the K_m or the maximum translocation rate since changes of the free concentration of the ADP and ATP are difficult to accomplish. A mixed effect on K_m and translocation rate appears to be feasible (Klingenberg, 1976) similar to that found for the influx. With external ATP and ADP no single K_m could be defined, but a K_m for a high affinity and low affinity exchange reaction were found. Therefore, K_m for ATP appeared to be increased about three-fold by energization (Pfaff et al., 1969; Klingenberg, 1976).

Electrogenic versus Electrophoretic ATP, ADP Exchange

The extrusion of ATP against ADP uptake has been interpreted to be driven by the membrane potential in the energized state of the mitochondria in an electrophoretic manner. In the over-all process of the exchange reaction one net negative charge is extruded of the charge difference between ATP^{4-} and ADP^{3-}. The exchange can be visualized such that the carrier binding site exposes three positive charges for combination with ADP giving an electroneutral ADP carrier complex. Conversely ATP would give a complex with one negative charge in excess. There are several good reasons to assume that the binding center has a minimum of three positive charges.

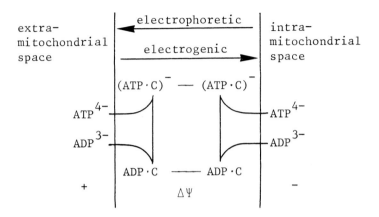

The electrical movement of ATP might cause a change of the membrane potential and possibly be made visible by corresponding membrane potential probes or even by the redox state of the respiratory carriers.

The membrane energization or membrane potential in mitochondria has been made visible by using fluorescent indicator 3,3'-dipropylthiocarbocyanine iodide (Laris, 1977) or absorbing dyes such as safranine (Akerman and Wikström, 1976). For the electrophoretic case, where mitochondria are energized to give a high endogenous ATP pool, the addition of ADP however did not cause the expected change in the fluorescence or absorbance signal. This was first shown by Laris (1977) and also observed in our hands. Only when mitochondria are pretreated with ATP, a corresponding change can be observed, although this is not very reproducible. The argument may be brought forward that the amount of ATP exchanged by ADP is too small to compete with the rapid regeneration of the membrane potential. A very small signal on the energy drain however might be visible in the respiratory chain. Thus a very small transient oxidation of NADH on addition of ADP has been reported by Out et al. (1977) reflecting the electrophoretic ADP-ATP exchange.

The electrical nature of the ADP,ATP exchange can be more clearly demonstrated by using the electrogenic reaction where ATP is taken up in exchange for ADP, moving a negative charge inside. Under anaerobic condition, thus from a low level the membrane potential can be transiently increased. It is of course important under these conditions to eliminate ATP driven H^+ transfer by ATPase action through the addition of oligomycin. In Fig. 2 safranine is used as an indicator of membrane energization. The effect can be fully suppressed in the presence of carboxyatractylate (not shown). The fast transient is in accordance with the fact that the ATP-ADP

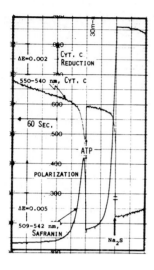

Figure 2. The polarization of mitochondrial membrane by electrogenic ATP-ADP exchange. Indicators for energization are safranine distribution, associated spectral shifts and cytochrome c oxidation by reversed electron transport. Conditions: Rat liver mitochondria (4 mg protein/ml) incubated at 20°C, pH 7.2, with 0.4 mM Na_2S, 20 μg/ml oligomycin, 5 μg/ml safranine, 600 μM ATP added. (Experiment by A.Bracht)

exchange is only of a short duration and limited by the endogenous ADP pool. The membrane potential generated is then caused by the transient charge conductance (e.g. by H^+, Ca^{++}, etc.).

In parallel recording the response of cytochrome c to the energization is correlated to that of the membrane probe. Cytochrome c is oxidized by the addition of ATP by reversed electron transfer utilizing the energization of the membrane. The oxidation of cytochrome c by ATP in the absence of oligomycin by reversal of ATPase and reversed electron transfer, as discovered by us first in 1961 (Klingenberg and Schollmeyer, 1961; Klingenberg, 1961), was shown to be a most sensitive indicator to small amounts of ATP. Furthermore, the response of the respiratory chain under these anaerobic conditions requires smaller phosphorylation potentials than that under aerobic conditions (ΔG = 12.8 kcal) versus ΔG = 14.7 kcal (Klingenberg, 1969). It is therefore understandable that a relatively small energization of the membrane even without interference of ATPase, by electrogenic ATP,ADP exchange, is sensed by cytochrome c.

The electrogenic ATP versus ADP exchange is further underlined by using ATP analogues which cannot be utilized by the ATPase but which have been shown to be also exchanged such as AMP-PNP and AMP-PCP (Klingenberg, 1976). The exchange rate with these compounds is considerably lower, about 0.2 and 0.1, of that of ATP respectively. These compounds permit to examine whether the fully dissociated triphosphonucleotide is responsible for the electrogenic effect since the pK of

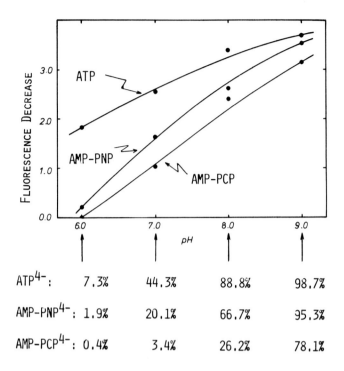

Figure 3. The electrogenic effect of adenosine triphospho-analogues in dependence on pH. Rat liver mitochondria (0.5 mg/ml) incubated at various pH with 10 μg oligomycin, 0.3 mM AMP-PCP, AMP-PNP and 0.6 mM ATP, 5 μg 3,3'-dipropylthiocarbocyanine iodide. Excitation at 622 nm, emission at 670 nm. Plot of the maximum emission intensity. (Experiment by A. Bracht)

the last group ($XH^{3-} \rightleftharpoons X^{4-} + H^+$) differs strongly from that of ATP. It therefore might be expected that the pH dependence of the electrogenic effect as followed by fluorescence decrease is different with these compounds from that of ATP. This in fact is shown in Fig. 3, where the maximum of the fluorescence response is plotted versus pH for the three compounds. At the same time the degree of dissociation of the four anionic forms is given. For example at pH 7, ATP^{4-} is present at 44% whereas $AMP-PNP^{4-}$ at 20% and AMP-PCP at 3%. Accordingly the electrogenic effects of the latter two are considerably smaller than that of ATP and disappear nearly completely at pH 6.

These data are best interpreted such that only X^{4-} participates in the exchange which results in the electrogenic effect whereas XH^{3-} probably is inactive. The results further substantiate that the electrogenic effect is not due to an eventual ATPase action which bypasses the oligomycin block.

CONCLUSIONS

The <u>electrogenic</u> effect of the ATP,ADP exchange can be seen as one of three ways of energizing the mitochondrial membrane, in addition to the energization by electron transport and by ATPase action. In the mitochondrial membrane there is no other electrogenic transport of comparable high activity. This third pathway of electrogenic effect may even have some physiological significance. It is postulated that in cells the existence of mitochondria requires energization (polarization) of the membrane, which depends on a continuous electrogenic source because of the final electrical resistance of any membrane. This question may be best illustrated with yeast mutants (see also Kovac, 1974). In ρ^- mutants and in promitochondria electron transport as a source of membrane polarization is missing and the membrane polarization might be sustained by ATP hydrolysis utilizing ATP from glycolysis. This requires the existence both of the ADP,ATP carrier and an intact ATPase. This process is probably operative also in promitochondria. In double mutants where both the respiratory chain and the ATPase are damaged, e.g. oligomycin insensitive, a polarization of the membrane is still possible by the electrogenic exchange of ATP against ADP as a last, although weak resource. ATP taken up is hydrolyzed by the uncoupled ATPase rendering ADP for the continuing exchange against external ATP. This, of course, is a wasteful way of utilizing cytosolic ATP and therefore growth of these mutants should be weak.

This reasoning shows that the mitochondria may dismiss of respiratory chain or ATPase but never can exist without the ATP,ADP carrier, which in these mutants came to the aid of energy supply necessary for mitochondrial survival.

In contrast, the <u>electrophoretic</u> action of ATP,ADP exchange has an obvious physiological role, which was first discussed by us in 1969 (Klingenberg et al., 1969b). The following points were made: (a) Increase of the phosphorylation potential of ATP in the cytosol versus in the mitochondria. (b) Consequent lowering of the P/O ratio in conforming with the needs of the eukaryotic cell of ATP maintaining a higher phosphorylation potential in the cytosol than in the mitochondria and therefore ATP is available in smaller amounts. Still another decade earlier, the difference of the phosphorylation potential in the cytosol to that of the intramitochondrial space was discussed by us (Bücher and Klingenberg, 1958). It was shown that the phosphorylation potential in the cytosol in mitochondria are inversely related to the redox potential of the protein nucleotide system in both compartments (Bücher and Klingenberg, 1958; Klingenberg, 1965). If in the cytosol the redox potential of the NAD system is low, phosphorylation potential of ATP,ADP is high, whereas in the mitochondria the coupling between the respiratory chain and the ATPase results in the opposite correlation. Therefore, the "symbiosis" of mitochondria with the cytosol requires the function of the electrophoretic ADP-ATP exchange. Also the electrophoretic function makes the ADP,ATP carrier again a *conditio sine qua non* of the mitochondria.

ACKNOWLEDGMENTS

The excellent collaboration of Maria Appel and Adelar Bracht in the experiments is gratefully acknowledged.

REFERENCES

Akerman, K., and Wikström, M. (1976). FEBS Lett. 68:91.
Bücher, Th., and Klingenberg, M. (1958). Angew. Chemie 70:552.
Heldt, H.W., Klingenberg, M., and Milovancev, M. (1972). Eur. J. Biochem. 30:434.
Heldt, H.W., and Pfaff, E. (1969). Eur. J. Biochem. 10:494.
Klingenberg, M. (1961). Biochem. Z. 335:263.
Klingenberg, M. (1965). In "Control of Energy Metabolism" (B. Chance et al., eds.), p. 149. Academic Press, New York
Klingenberg, M. (1969). In "The Energy Level and Metabolic Control in Mitochondria" (S. Papa et al., eds.) p. 185. Adriatica Editrice, Bari.
Klingenberg, M. (1972). In "Mitochondria: Biomembranes" p.147. Elsevier, Amsterdam.
Klingenberg, M. (1975). In "Energy Transformation in Biological Systems" p. 105. Associated Scientific Publ., Amsterdam.
Klingenberg, M. (1976). In "The Enzymes of Biological Membranes: Membrane Transport" Vol. 3 (A.N. Martonosi, ed.), p. 383. Plenum Publishing Corp., New York/London.
Klingenberg, M., and Pfaff, E. (1966). In "Structural and Functional Compartmentation in Mitochondria" (J. M. Tager et al., eds.), p. 180. Elsevier, Amsterdam.

Klingenberg, M., and Rottenberg, H. (1977). Eur. J. Biochem. 73:125.
Klingenberg, M., and Schollmeyer, P. (1961). Biochem. Z. 335:243.
Klingenberg, M., Heldt, H.W., and Pfaff, E. (1969a). In "The Energy Level and Metabolic Control in Mitochondria" (S. Papa et al., eds.), p. 237. Adriatica Editrice, Bari.
Klingenberg, M., Wulf, R., Heldt, H.W., and Pfaff, E. (1969b). In "Mitochondria: Structure and Function" (L. Ernster and Z. Drahota, eds.), p. 59, Academic Press, New York.
Kovac, L. (1974). Biochim. Biophys. Acta 346:101.
LaNoue, K., Mizani, S.M., and Klingenberg, M. (1978). J. Biol. Chem. 253:191.
Laris, P.C. (1977). Biochim. Biophys. Acta 459:110.
Out, T.A., Krab, K., Kemp, A., Jr., and Slater, E.C. (1977). Biochim. Biophys. Acta 459:612.
Pfaff, E., and Klingenberg, M. (1968). Eur. J. Biochem. 6:66.
Pfaff, E., Heldt, H.W., and Klingenberg, M. (1969). Eur. J. Biochem. 10:484.
Wulf, R., Kaltstein, A., and Klingenberg, M. (1978). Eur. J. Biochem. 82:585.

CATION FLUX ACROSS BIOMEMBRANES

PROTON AND POTASSIUM TRANSLOCATION BY THE PROTEOLIPID OF
THE YEAST MITOCHONDRIAL ATPase[1]

Richard S. Criddle
Richard Johnston

Department of Biochemistry and Biophyics
University of California at Davis
Davis, California U.S.A.

Lester Packer
Paul K. Shieh
Tetsuya Konishi[2]

Membrane Bioenergetics Group
University of California at Berkeley
Berkeley, California U.S.A.

INTRODUCTION

According to chemiosmotic theory, proton gradients established by the respiratory chain of the inner mitochondrial membrane are coupled to the production of ATP by discharge of the potential gradient through the membrane-bound ATPase enzyme complex (1). Recent studies have suggested a prominent role for a 7,500 dalton proteolipid subunit of the ATPase complex in linking proton translocation to synthesis ATP (2-4). This polypeptide appears to be present in all energy transducing ATPases studied (2-7) and has been identified as the binding site of the ATPase inhibitors oligomycin and dicyclohexyl carbodiimide, DCCD (8,9).

[1]This research was supported in part by the Department of Energy.
[2]Present address: Department of Radiochemistry and Biology Niigata College of Pharmacy, Niigata, Japan.

Previous studies from our laboratories (3,10) reported the use of reconstituted liposomes containing purified yeast mitochondrial proteolipid to examine proteolipidmediated proton translocation and collapse of electrical potentials. Activity of the proteolipid was demonstrated in liposomes in suspension and in association with lipid impregnated Millipore filters. This activity was inhibited by oligomycin and DCCD. Early studies of proton translocation using lipid vesicles required addition of valinomycin and K^+ for observation of electrogenic H^+/K^+ exchange. Curiously, some studies, using liposome associated with Millipore filter membranes bathed in K^+ containing solutions, indicated that valinomycin addition was not required. This discrepancy has now been resolved by showing that the differences in valinomycin requirements were due to the levels of proteolipid used in vesicle formation. When a high concentration of proteolipid was used, a new specificity developed that allowed translocation of K^+ or Rb^+ in exchange for H^+.

MATERIALS AND METHODS

Proteolipid was isolated from yeast mitochondria as described previously (3,11). basically by extraction with chloroform-methanol and precipitation with ether. This procedure, when repeated 4-5 times, yields material that shows only a single protein band when chromatographed on thin-layer silica plates. K^+-loaded proteolipid vesicles were prepared using purified soybean lecithin or mitochondrial phospholipid (3). Solutions of proteolipid (about 100 μg) and 20 mg of lipid were mixed and solvent removed with N_2. Then 3 ml of 20 mM Tris-Cl buffer, pH 7.5 containing 0.16 M sucrose, 0.2 M choline chloride and 0.6 M KCl were added. After allowing this mixture to stand for 2 hr under an N_2 atmosphere at room temperature to hydrate the lipid, the pH was adjusted to 7.5 and the mixture was sonicated in a Bronson sonifier for 2 min in a waterbath at 15° (power setting of 75 W output, 50% post mode). To prepare vesicles that were not K^+-loaded a similar procedure was employed except that the KCl was omitted from the solution. In some cases, $^{35}SO_4^=$ was included in the suspension buffer and ^{35}S-loaded liposomes were separated from excess $^{35}SO_4^=$ by gel filtration. This allowed simple assay of vesicle concentration in cation translocation studies. Following sonication, the vesicles suspension was ready for assay of the H^+ translocation activity.

RESULTS

Proteolipid-dependent Proton Translocation

The K^+-loaded vesicles were suspended in a 1 mM Tris-HCl buffer, pH 7.5, 8 mM KCl at a concentration of 0.33 mg lipid/ml and the pH was monitored (Fig. 1). After a stable base line has been achieved, usually within a minute, 2 μl of valinomycin (100 mg per ml in ethanol) was added to initiate the reaction. pH changes are created by the valinomycin-K^+ diffusion potential that drives counter movement of H^+. Note that carbonyl

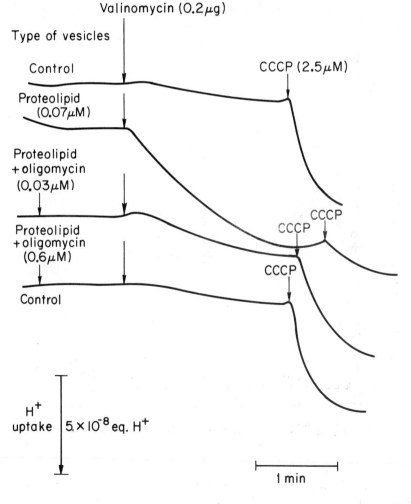

Fig. 1. Proton uptake by K^+-loaded proteolipid vesicles.

cyanide m-chlorophenylhydrazone (CCCP), a proton ionophore, maximizes the extent of the H^+ change. This also shows that proton translocation dependent upon the presence of proteolipid is inhibited by oligomycin.

Proteolipid-dependent K^+ Translocation

The dependence of proton translocation upon the ratio of proteolipid to lipid used in vesicle formation was studied by measuring the rate of efflux of H^+ from liposomes following addition of KCl into the external medium. Figure 2 shows that

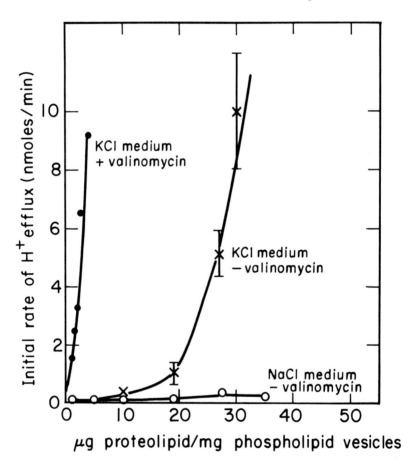

Fig. 2. Proton extrusion in proteolipid vesicles in the absence and presence of valinomycin. Just prior to the start of experiments KCl (or NaCl) was added to the suspending medium.

the initial rate of H^+ translocation in the presence of valinomycin increases, as proteolipid concentration is increased. At higher concentration, H^+ efflux no longer requires the presence of valinomycin though specificity for K^+ is retained. This was confirmed by studies of K^{42} influx into proteolipid containing vesicles (Fig. 3). The rate of K^{42} uptake was dependent upon the concentration ratio of proteolipid to lipid added prior to sonication to prepare vesicles. The amount of proteolipid was 40 μg/mg phospholipid. No valinomycin was present in this experiment. Figure 3 illustrates that the rate of K^{42} influx at high proteolipid concentration shows a saturation effect with increasing KCl concentration. Similar results were obtained in studies with vesicles bound to planar lipid membranes (not shown). K^+ transport from the outer compartment to the inner through the lipid membrane was measured with a K^+-sensitive electrode in the inner compartment. When the observed potentials were measured 10 min after KCl addition to the outer compartment, they were found also to be dependent on the concentration of proteolipid in the vesicles.

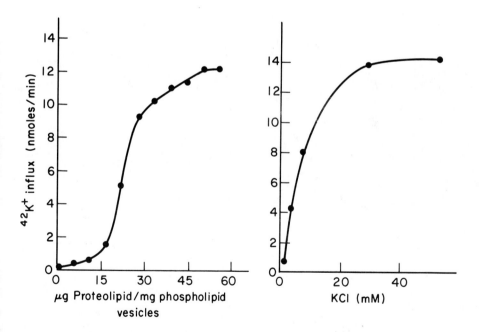

Fig. 3. Dependence of $^{42}K^+$ uptake into vesicles, on the concentration of proteolipid used to prepare vesicles, and on the external KCl concentration

Specificity of cation transport is illustrated in Figure 4. Rb^+, but not Na^+, Ca^+, Li^+ or Ca^{++} competed in ^{42}K uptake. Similarly, Rb^+ and K^+ can sustain photo-induced membrane potentials generated by bacteriorhodopsin across planar membranes containing proteolipid. The other cations tested had no effect. No anion specificity could be detected.

K^{42} uptake is inhibited by oligomycin (Figure 5). The levels of inhibitor required are identical with those for inhibition of H^+ translocation (3). The pH dependence of both K^+ translocation by proteolipid liposomes and conductance across planar membranes with attached proteolipid liposomes was also investigated. Figure 6 shows that electrical conductance across planar membranes decreases at pH values about 6.5, whereas K^+ translocation rates in vesicles increase over the pH range 3 to 6.5 and then they decrease. Both methods indicate the optimal pH for K^+ translocation is around neutrality.

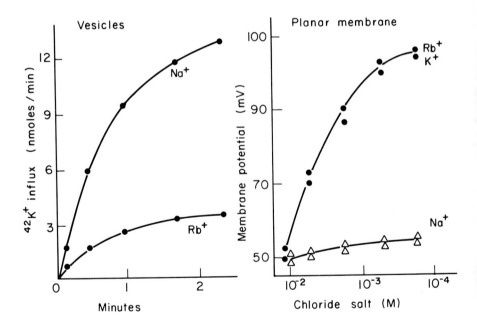

Fig. 4. Effect of monovalent cations on $^{42}K^+$ uptake into proteolipid vesicles and on the membrane potential changes in proteolipid vesicles attached to lipid impregnated planar membranes. Li^+, Cs^+ and Na^+ all gave essentially identical curves.

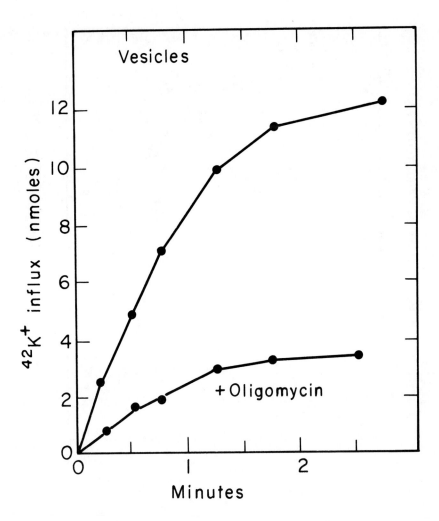

Fig. 5. Inhibition of $^{42}K^+$ uptake into proteolipid vesicles by oligomycin.

DISCUSSION

This study shows some unexpected properties of the oligomycin-sensitive proteolipid from the yeast mitochondrial ATPase reconstituted into liposomes. As the concentration of the proteolipid is progressively increased relative to that of the lipid prior to sonication and the preparation of lipid vesicles, a new ionic specificity for cation translocation arises: K^+ and Rb^+ as well as H^+ appear to be exchanged. To

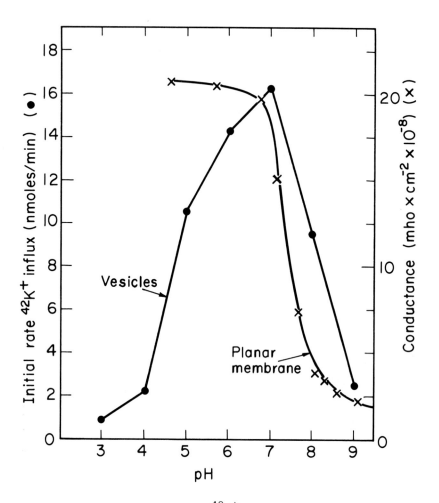

Fig. 6. pH dependence of $^{42}K^+$ uptake into proteolipid vesicles and conductance changes across planar membranes having proteolipid vesicles attached.

explain these results one can speculate that as proteolipid concentration is increased, more than one molecule is incorporated within a single vesicle. Assuming an average size of the vesicles near 300 Å, we calculate that oligomycin-sensitive translocation of K^+ and Rb^+ occurs in the concentration range where the number of proteolipid molecules averages 4-8 molecules per vesicle. Also it might not be coincidental that 4-6 moles of proteolipid are present per mole of the native ATPase complex (12) and that these appear to form a 45,000 dalton aggregate (11). Furthermore, other studies <u>in vitro</u> show that the levels of certain inhibitors blocking proton translocation by proteolipid in reconstituted systems and levels causing inhibition of ATPase activity is less than 1. Thus Azzi reports (personal communication) one mole of spin-labeled DCCD per 6 moles of proteolipid. These findings suggest an aggregated or oligomeric structure exists when liposomes with high proteolipid/phospholipid ratios are prepared and that cooperative interactions between the polypeptides may be important for determining the specificity of cation-translocation activity by proteolipid.

The significance of these new findings for the mechanism of oxidative phosphorylation is as yet unclear. Reconstitution studies may not have a direct correlation with in vivo function. Although the proteolipid exhibits specificity for K^+, Rb^+ and H^+ in our reconstitution assay, cation translocation in vivo might be exclusively for H^+. Nevertheless, in view of the present findings, it will be of interest to re-examine the possible importance of a role of K^+ in modulating the activity of the reversible mitochondrial ATPase with regard to regulation of ATP synthesis.

REFERENCES

1. Mitchell, P., FEBS Lett. 43;189 (1974).
2. Nelson, N., Eythan, E., Notsani, B., Sigrist, H., Sigrist-Nelson, K., and Gitler, C., Proc. Natl. Acad. Sci. USA 74:2375 (1977).
3. Criddle, R. S., Johnston, R. F., Packer, L., and Shieh, P., ibid., 74:4306 (1977).
4. Sone, N., Yoshida, M., Hirata, H., and Kagawa, Y., J. Biol. Chem. 250:7917 (1975).
5. Fillingame, R. H., ibid., 241L:6630 (1976).
6. Altendorf, K., FEBS Lett. 73:271 (1977).
7. Sebald, W., and Wachter, E. in: Energy Conservation in Biological Membranes (G. Schaffer and M. Klingenberg, eds.) Springer-Veerlag, Berlin (1978).
8. Enns, R. and Criddle, R. S., Arch. Biochem. Biophys. 182: 587 (1977).
9. Cattle, K. J., Lindop, C. R., Snight, I. G., Beechey, R. B.,, Biochem. J. 125:169 (1971).
10. Konishi, T., Packer, L., and Criddle, R. S., Methods in Enzymol. 55:415 (1979).
11. Sierra, M. F. and Tzagoloff, A., Proc. Natl. Acad. Sci. USA 70:3155 (1973).
12. Pedersen, P. L., Bioenergetics 6:243 (1975).

APPROACH TO THE MEMBRANE SECTOR OF THE CHLOROPLAST
COUPLING DEVICE[1]

Nathan Nelson[2]
Esther Eytan

Department of Biology
Technion-Israel Institute of Technology
Haifa, Israel

I. INTRODUCTION

The coupling device of energy transduction is a reversible proton translocating ATPase. It contains an ATPase section that can readily be solubilized and a membrane section that functions in energy transduction. The key role of the membrane sector is to conduct protons across the membrane under specific conditions that will bring about the phosphorylation of ADP to ATP on the level of the ATPase enzyme (1). It was suggested that the protons are conducted across the membrane via a specific channel composed of a proteolipid molecule (2). A purified proteolipid in a state allowing reconstitution of a native proton channel was widely sought for in various organelles that are active in phosphorylation. Recently it was reported on the isolation of functional proteolipid from three different systems (3-5).

The function of energy transduction across the membrane seems to be more complex than the mere proton conductance. Therefore, we deliberate some studies on the properties of the membrane sector (CF_O) of the chloroplast coupling device.

---

[1]This research was supported by a grant from the United States-Israel Binational Science Foundation (BSF), Jerusalem, Israel.
[2]Present address: Biocenter, Department of Biochemistry, Klingelbergstrasse 70, CH-4056 Basel, Switzerland.

II. METHODS

Lettuce or pea chloroplasts were prepared as previously described (6) with the addition of 0.15 M NaCl wash to eliminate the presence of ribulose diphosphate carboxylase (7). Chlorophyll (8), protein (9), photophosphorylation (10), and proton uptake (11) were assayed by published procedures.

Sodium dodecyl sulfate (SDS) gel electrophoresis was performed as described by Weber and Osborn (12). Chloroplasts and CF_1 depleted chloroplasts, containing about 0.5 mg chlorophyll per ml, were incubated for about 2 hours at room temperature in the presence of 2% SDS, 2% mercaptoethanol and about 10% sucrose. Incubation at elevated temperatures brings about polymerization of chloroplast polypeptides (7). Samples of 50 µl were applied on the gels. After electrophoresis for 5 hours at 7 mA per tube, the gels were fixed, stained, destained and scanned as previously described (13).

III. RESULTS

Treatment of chloroplasts with dilute solution of EDTA results in a partially CF_1 depleted preparation. This preparation is widely used for reconstitution studies with purified CF_1. The main drawback of EDTA particles is that considerable amounts of CF_1 remain on the membranes after the treatment. Table I shows that upon addition of purified CF_1 the degree of reconstitution of photophosphorylation in EDTA particles is dependent on the amount of CF_1 that was retained on the membranes after the EDTA treatment. On the other hand, treatment of chloroplasts with 2 M NaBr yielded a preparation which was essentially free of CF_1 with intact electron transport properties but without reconstitution activity (6). An aggregation phenomenon, which prevents the CF_1 from reaching its binding sites upon the membranes, has been proposed to explain the failure to obtain reconstitution with particles that have been effectively depleted of CF_1 (14).

Chloroplast preparation which is essentially free of CF_1 and exhibiting a high reconstitution degree can be obtained as the following: A solution of 5 M NaBr is added to chloroplast suspension in a medium containing 0.4 M sucrose, 10 mM NaCl, 5 mM dithiothreitol (DTT) and 10 mM Tricine (pH 8), to give a final concentration of 2 M NaBr and a chlorophyll concentration of 1 mg per ml. After incubation at 0°C for 30 min equal volume of water is added and the suspension is centrifuged at 20,000 g for 10 min. The pellet is thoroughly homogenized by

TABLE I. The dependence of the reconstitution activity on the amount of CF_1 that was retained on the EDTA particles

Concentration of EDTA during the treatment	% CF_1 retained on the membranes	Photophosphorylation µmoles ATP/mg Chl/h	
		$- CF_1$	$+ 50$ µg CF_1
2 mM EDTA particles	46	2	17
3 mM EDTA particles	77	8	130

glass-Teflon homogenizer at 10 times the original volume of 10 mM Tricine (pH 8). The suspension is centrifuged at 20,000 g for 10 min and the resulting pellet is homogenized in a medium containing 0.4 M sucrose, 10 mM NaCl, 5 mM DTT and 10 mM Tricine (pH 8) at chlorophyll concentration of about 2 mg/ml. The inclusion of DTT during the NaBr treatment and in the suspension medium and the thorough homogenization in 10 mM Tricine were found to be necessary to get reconstitutable particles.

The depleted particles are essentially free of CF_1 as judged by the lack of ATPase activity. Figure 1 shows SDS gels of chloroplast membranes and NaBr treated chloroplast. In the control chloroplasts the α, β and γ subunits of CF_1 are prominent while in the treated chloroplasts they cannot be detected. Since residual 10% of subunits α and β will not escape detection, it seems as if the particles are over 90% depleted of CF_1 and if there is a residual enzyme it is not active.

The depleted particles possess a negligible photophosphorylation activity. Upon addition of purified CF_1, in the

TABLE II. Reconstitution of photophosphorylation and proton uptake in NaBr treated chloroplasts

	Photophosphorylation µmoles ATP/mg Chl/h	Proton uptake n equ H^+/mg Chl	
		-DCCD	+DCCD
NaBr particles	2.5	-102	181
NaBr particles + 170 µg CF_1	235	107	232

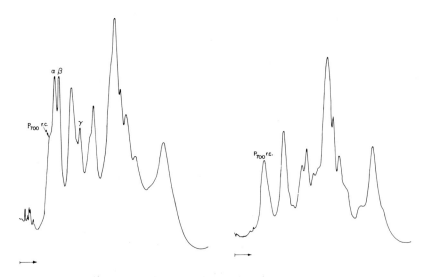

Figure 1. The SDS gel pattern of control chloroplasts (left) and NaBr-treated chloroplasts (right).

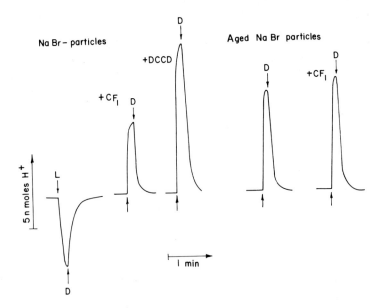

Figure 2. Proton uptake activities of NaBr-treated chloroplasts and similar particles that were incubated for 7 days at 4°C.

TABLE III. Effect of aging on proton uptake and reconstitution of photophosphorylation in NaBr treated chloroplasts

	Photophosphorylation µmoles ATP/mg Chl/h	Proton uptake n equ H^+/mg Chl
NaBr particles	0.5	-110
" + CF_1	235	107
NaBr particles after 5 days incubation at 4°	0.8	37
" + CF_1	198	202
NaBr particles after 8 days incubation at 4°	0.7	206
" + CF_1	95	239

presence of Mg^{++}, fairly high phosphorylation activities can be obtained (Table II).

Figure 2 shows the light-induced pH changes in NaBr treated particles and in particles which were reconstituted with CF_1. In the resolved particles light-induced proton release is observed. The properties of this effect were described previously (6). Upon addition of CF_1 uncoupler-sensitive light-induced proton uptake is observed. This represents about 60% restoration of the proton uptake activity in control chloroplasts. When the NaBr-particles were stored in the presence of 1% bovine serum albumin for 5 days at 4°C, a spontaneous restoration of the uncoupler sensitive proton uptake activity took place. Table III shows that the resolved particles are surprisingly stable. Even though their proton uptake properties were markedly modified, they still maintain remarkable reconstitution activity.

IV. DISCUSSION

Chloroplast preparation totally lacking CF_1 and exhibiting a high reconstitution degree was badly needed for the study of phosphorylation with chemically modified CF_1 and for better understanding of the membrane sector of the coupling device. To the best of our knowledge this work provides, for the first time, particles which are suitable for these purposes. The remarkable stability of the resolved particles should be valuable for future studies.

The beneficial effect of thorough homogenization stresses the assumption that aggregation of highly depleted chloroplast particles, is one of the main obstacles in successful reconstitution of these particles (14). The presence of DTT during the NaBr treatment is not required for the preservation of intact electron transport properties (6). The fact that DTT is obligatory for the preparation of reconstitutable particles, might suggest that either, a free SH group in the membrane sector is necessary for the binding of CF_1 or that the membrane sector is involved directly in the process of photophosphorylation.

The effect of aging of the NaBr-treated chloroplast on their proton uptake activity has several interesting features. The phenomenon that proton uptake is restored upon prolonged incubation could be explained by inactivation of the chloroplast proteolipid which is responsible for the high proton conductance across the chloroplast membrane (15). This cannot be the case because upon addition of CF_1 photophosphorylation activity of the particles is restored. Second explanation might be that the CF_1 molecule sits within a cleft in the chloroplast membrane. Upon removal of the CF_1 and during prolonged storage the cleft is covered by lipids and the proton conductivity across the membrane is markedly decreased. This situation resembles the ATPase depleted chromatophores in which the proton uptake activity is barely affected (16).

REFERENCES

1. Mitchell, P., Biol. Rev. 41:445 (1966).
2. Racker, E., "A New Look at Mechanisms in Bioenergetics". Academic Press, New York, 1976
3. Okamoto, H., Sone, N., Hirata, H., Yoshida, M., and Kagawa, Y., J. Biol. Chem. 252:6125 (1977).
4. Nelson, N., Eytan, E., Notsani, B., Sigrist, H., Sigrist-Nelson, K., and Gitler, C., Proc. Natl. Acad. Sci. USA 74:2375 (1977).
5. Criddle, R. S., Packer, L., and Shien, P., ibid. 74: 4306 (1977).
6. Kamienietzky, A., and Nelson, N., Plant Physiol. 55:282 (1975).
7. Nelson, N., and Notsani, B., in "Bioenergetics of Membranes" (L. Packer et al., eds.), p. 233. Elsevier, Amsterdam, 1977
8. Arnon, D. I., Plant Physiol. 24:1 (1949).

9. Lowry, O. H., Rosebrough, N. J., Farr, A. L., and Randall, R. J., J. Biol. Chem. 193:265(1951).
10. Nelson, N., and Broza, R., Eur. J. Biochem. 69:203 (1976).
11. Neumann, J., and Jagendorf, A. T., Arch. Biochem. Biophys. 107:109 (1964).
12. Weber, K., and Osborn, M., J. Biol. Chem. 244:4406 (1969).
13. Nelson, N., Deters, D. W., Nelson, H., and Racker, E., J. Biol. Chem. 248:2049 (1973).
14. Nelson, N., Biochim. Biophys. Acta 456:314 (1976).
15. Nelson, N., Eytan, E., and Julian, C., in "Proceedings of the Fourth International Congress on Photosynthesis" (D. Hall et al., eds.), p. 559. Printed in Great Britain, 1977
16. Melandri, B. A., Baccarini-Melandri, A., San Pietro, A., and Gest, H., Proc. Natl. Acad. Sci. USA 67:477 (1970).

PROTON TRANSLOCATION BY BACTERIORHODOPSIN[1]

Lester Packer
Tetsuya Konishi[2]
Peter Scherrer
Rolf J. Mehlhorn
Alexandre T. Quintanilha
Paul K. Shieh
Irmelin Probst
Chanoch Carmeli

Membrane Bioenergetics Group
University of California
Berkeley, California

Janos K. Lanyi

NASA Ames Research Center
Moffett Field, California

INTRODUCTION

The Nobel Prize in Chemistry was awarded this year to Dr. Peter Mitchell for having developed the chemiosmotic theory, a conceptual framework for relating to the development of proton gradients and electrical potentials across membranes, expressed as the protonmotive force (PMF) as PMF = $\Delta\Psi$ + 60mVΔ pH, to other energy-linked cellular processes (1). An ideal experimental system for studying the relationship of ion

[1] Supported by the United States Department of Energy and a University of California - NASA Interchange No. NCR2-ORO50-607.
[2] Present address: Department of Radiochemistry and Biology Niigata College of Pharmacy, Niigata, Japan.

gradients to each other is found in the halobacteria which live in concentrated salt solution and hence divert much of their energy to salt pumping. These bacteria derive their energy either from oxidation of substances scavenged from their environment or from sunlight (Figure 1). Bacteriorhodopsin in the purple membranes of Halobacteria halobium (1) evolved as a light energy converter (2), establishing a gradient of protons across this membrane as the primary energy harvesting event. In our laboratory we have been interested in the mechanism of light energy conversion by bacteriorhodopsin. Specifically we want to find out how protons move in time and space across the membranes spanned by this protein. Unlike photosynthetic membranes, which contain complex structures of oxidation and reduction catalysts, the halobacteria achieve conversion of light into electrical and osmotic energy

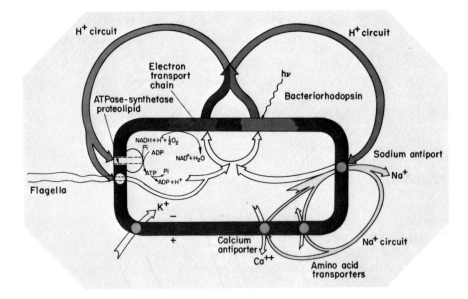

Fig. 1. Overall scheme of energy conversion by Halobacterium halobium. Direct conversion of sunlight into electrical current (proton movement) occurs within the "purple patches", crystalline arrays of the protein bacteriorhodopsin. Proton gradients are collapsed by being coupled to salt pumping and chemical energy synthesis in the form of adenosine triphosphate (ATP) in the cell envelope membrane.

by means of a photocycle operated by a single, relatively simple protein, bacteriorhodopsin.

To comprehend the light energy conversion system of bacteriorhodopsin the amino acid arrangement of the protein and how this structure is coupled to photoexcitation events and proton conduction must be understood. The resolution of this problem is feasible because of several advantages unique to this protein: there is only a single chromophore, retinal, in this protein; the protein can be isolated in large quantities and in a pure form with ease because it is the only protein present in the purple "patches", i.e, in large crystalline-like arrays of 75% bacteriorhodopsin and 25% lipid, and the lipid in the patches is tightly associated with the protein so that it is not susceptible to the rapid oxidative degradation that occurs upon fragmentation of photosynthetic membranes of higher plants (3).

The bacteriorhodopsin molecules occur in clusters of trimeric units. Because of its purity and crystalline-like arrangement the purple membrane can be more readily studied by physical methods than complex membranes containing more randomly dispersed components.

The small molecular size (about 26,000 Daltons) of bacteriorhodopsin has made it possible for a large team of Moscow scientists to determine its amino acid sequence to the last twenty amino acids (4). Thus molecular modeling of the mechanism of action of the protein has now become feasible.

There are compelling advantages for using chemical modifications to study bacteriorhodopsin. The protein is stable to drastic alterations in the pH and ionic strength of its environment and has proven to retain activity after extensive reaction with several chemical reagents. The reactive amino acids cysteine and histidine are not found in bacteriorhodopsin so cross reaction of many reagents with different amino acids in this protein is not a serious problem. An additional advantage of bacteriorhodopsin for chemical modification studies is that the retinal chromophore can be replaced in the protein after selective removal, thus allowing the binding site environment to be probed.

Bacteriorhodopsin is probably the simplest and most stable naturally occurring light energy converter presently known. One aspect of the program in the Membrane Bioenergetics Group for study of the proton conductance of this protein has involved chemical modification studies of specific amino acid residues to obtain molecular information about the groups essential for proton conduction and light absorption. These studies have led us to advance a tentative hypothesis (described herein) for the mechanism of proton conduction which can be tested and perfected (5).

To elucidate the mechanism of light driven proton movements which forms an electrochemical gradient it is important to know which amino acids in bacteriorhodopsin are involved in the proton movement. Our working hypothesis is that light induced conformational changes and/or isomerization of the retinal chromophore causes movement of the proton attached to the Schiff base nitrogen of the chromophore and that this is the primary step in the proton pump. This proton movement very likely occurs over a small distance relative to the transmembrane dimension over which protons must finally traverse, estimated to be overall about 45 Å based on the dimensions of the protein (8). There is considerable interest in how the protein is organized to effect the remaining charge separation. One goal of our current research is to elucidate the amino acid arrangement along this proton "channel".

Chemical Modification of Aromatic Amino Acid Residues in Bacteriorhodopsin

Last year we used the tryptophan reactive reagent, N-bromosuccinamide to demonstrate that even though only one or two tryptophan residues were located in close proximity to the chromophore, all four of the tryptophans in the molecule appear to be involved in photocycling activity since progressive loss of all tryptophan absorbance by N-bromosuccinamide coincided with progressive loss of activity (9). During the past year we have shown that at least one of the eleven tyrosines in the molecule was directly involved in the light dependent proton movement across the protein (5). Iodination was used to alter tyrosine residues which were accessible to this reagent. In analogy with the downward pK_a shift of iodinated tyrosine molecules in water, the kinetics of photocycling of iodinated bacteriorhodopsin were observed to undergo a downward shift in their pH dependence relative to the control protein. Of particular interest is the observation that the decay of the 412 nm intermediate became significantly prolonged after iodination (Fig. 2). Since it has been established previously that the 412 nm decay is due to reprotonation of the Schiff base linkage of the chromophore, we have inferred that tyrosine is a source of the protons which are donated to the chromophore. These studies, as well as others (10), indicate that aromatic hydrophobic amino acids play an important role in maintaining a proper environment in the vicinity of the chromophore and elsewhere in the molecule to promote proton conductance.

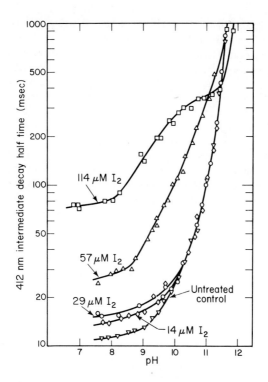

Fig. 2. Effect of iodination on the pH dependence of the decay of the 412 nm photocycle transient of bacteriorhodopsin, as measured by the flash photolysis technique.

Studies of the Proton Channel within Bacteriorhodopsin

We have assumed that the chromophore retinal fulfills the function of pumping protons in a light driven reaction and that the protons subsequently are conducted along a proton "channel" to reach the bulk aqueous phases. Another approach towards understanding the nature of this proton "channel" is to remove the chromophore and study the residual protein for proton conductivity. The chromophore can be removed by a bleaching process whereby the retinal is displaced from the protein by illuminating bacteriorhodopsin in the presence of a large excess of hydroxylamine (11). This process is reversible, i.e. the light driven proton pump can be recovered upon addition of retinal to the bleached protein. Using this procedure bleached bacteriorhodopsin was prepared and incorporated into lipid vesicles which were loaded with potassium and then suspended in sodium ion medium (12). Normally only a negligible proton leakage will occur in these vesicles since no permeability mechanism is available for movement of the ions. However, upon adding the potassium specific ionophore valinomycin to the vesicles, a diffusion potential was established and H^+ was taken up in electroneutral exchange of H^+ for K^+ ions. The addition of retinal to the bleached bacteriorhosopsin led to a substantial decrease in this proton movement. Thus the experiment suggests qualitatively that protons may be able to move through the bleached bacteriorhodopsin molecule in the dark. To obtain a measure of the capability of bacteriorhodopsin to promote proton fluxes, lipid vesicles containing equivalent amounts but native bacteriorhodopsin were illuminated in the presence of valinomycin. The light-induced and the passive proton fluxes were comparable to one another.

Photoelectrical Studies

It has been shown that a photovoltaic cell can be constructed from bacteriorhodopsin incorporated into Millipore filter membranes (6,7). The filter membrane serves as a support for biological lipids which comprise the permeability barrier of the membrane. Bacteriorhodopsin was incorporated into the filter membranes by a multi-step absorption and fusion process. The current-voltage characteristics of the photocell were studied at high light intensity. A maximum open circuit voltage of 300 millivolts and a maximum short circuit current of 0.9 milliamp were obtained. Relative to the incident light intensity the power output of the cell was determined to be 0.07%. Under continuous high intensity illumination the activity of the cell declined, and no photovoltage could be observed after ninety minutes. One goal of future research is to improve the stability and efficiency of the photocell.

Conductance Studies

Planar membranes are a useful tool for discriminating between channel conductance mechanisms. Channels which are open intermittently can lead to discrete conductance jumps across lipid bilayer membranes provided that the opening and closing process is long enough to be observed. Using planar membranes painted across a small Teflon orifice we have attempted to obtain discrete conductance jumps when bleached bacteriorhodopsin was incorporated into the membranes. We incorporated bleached bacteriorhodopsin into such lipid bilayer by means of the calcium absorption and fusion technique which was developed earlier for the Millipore filter system (6). Discrete conductances are indeed observed. Alterations in the concentrations of sodium, potassium or calcium did not significantly alter the conductance magnitude.

The principal conclusion to be drawn from these experiments is that bleached bacteriorhodopsin is not a continuously open channel for protons or other ions when an electrical potential is applied across the membrane bearing the protein. Rather, at infrequent intervals an occasional bacteriorhodopsin molecule seems to open up to protons. We have attempted to define further the mechanism whereby this channel opening occurs, by varying the composition of the planar membranes and the aqueous phase so as to induce changes in the conductance jump frequency. Thus far the parameters we have found which affect the jump frequency are the concentration of bleached bacteriorhodopsin and the temperature. Another finding of interest was that such current fluctuations in the bleached molecule were only exhibited by ions having a similar hydrated radius as K^+ and Rb^+, suggesting an effective "pore" radius of about 3-4 Å.

Computer Modeling

Knowledge of most of the amino acid sequence of bacteriorhodopsin from the work in Moscow (4) has made it feasible to attempt to construct a molecular model of the protein. The electron density map with seven Angstrom resolution suggests that the molecule consists of a nearly cylindrical array of seven helical coils with their axes oriented almost perpendicular to the plane of the membrane. Studies with digestive enzymes have shown that the amino terminus of the sequence faces the extracellular space while the carboxyl terminus lies within the cytoplasm. These factors imply that the protein contains six nonhelical regions where the seven helices are connected at the membrane interfacial region.

The overall symmetry of the protein suggests that the helical regions are of nearly equal lengths. To identify the amino acids directly involved in the bend portions several criteria should be satisfied: polar or charged groups will be stabilized at the polar aqueous interface, and certain amino acids, particularly proline residues interrupt alpha-helical structure. In addition it is known that bacteriorhodopsin is an intrinsic membrane protein so the majority of residues in the helical regions should be hydrophobic to promote thermodynamically favorable interaction with the hydrocarbon chains of lipids. Because the amino acid sequence contains eleven proline residues it appears that there must be nonhelical regions of the protein embedded within the protein as well as within the interfacial region. Moreover the charged amino acids are distributed in a manner that is inconsistent with their exclusive location at the polar interface of the membrane. This suggests that some ionic bridges are located within the interior of the protein and that ionic bridges may be responsible for the protein interactions which stabilize the crystalline array.

Using the above criteria a preliminary model of bacteriorhodopsin has been constructed (Fig. 3). A space filling model constructed from plastic CPK atoms was used to estimate

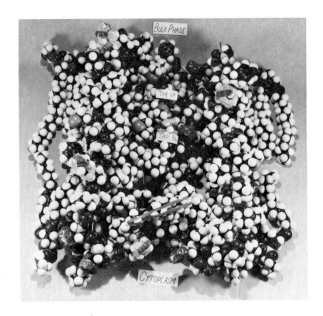

Fig. 3. Photograph of the three dimensional space-filling model of bacteriorhodopsin assembled from the amino acid sequence of Ovchinnikov et al. (4). The model shows the distribution of aromatic tryptophan and tyrosine residues in the vicinity of the retinylidene lysine chromophore (marked with two light parallel strips of tape). The view is of the interior of the bacteriorhodopsin molecule and shows four of the seven helices.

bond angles within the peptide backbone. These angles were
fed into a computer which has been programmed for molecular
model building at the NASA Ames Research Center (13). At present
only the graphic capabilities of the Ames program are
being utilized to examine some of the probable conformations
of the proteins. In the future we intend to carry out energy
calculations for the more probable conformations so as to eliminate
some of them from consideration. A particular goal of
these studies is to understand protein conformational changes
that can result from the possible photo-isomerizations and
rotations of the chromophore. A further goal is to identify
which amino acids are directly involved in protein conduction.

Photodesalination Studies with Cell Envelopes of Halobacterium Halobium

Ion movements across cell envelope vesicles of H. halobium
are being assayed with a newly developed spin probe method.
The method uses spin labeled amines (see Fig. 4) or carboxylic
acids to monitor transmembrane pH gradients (14). The basic
assumption is that pH gradients are achieved rapidly due to
the limited buffering capacity within the vesicles. Subsequent
proton movement across the membrane in response to this
gradient must be balanced by the movement of other ions (see
Fig. 1). The secondary ion movements result from an exchange
mechanism whereby a cation moves in one direction while a proton
moves in the opposite direction. A Na^+ gradient is thus
coupled to the development of the pH gradient which can be
observed with the spin probes. As shown in Fig. 4, a rapid
initial H^+ extrusion is followed by a reversal of the proton
movement when sodium ions are present within the cell
envelopes. The interpretation of this experiment is that
there is a proton-sodium exchange mechanism in the membranes
which allows protons to move into the vesicles and sodium
ions move out and that this mechanism is only triggered after
a threshold value of the initial proton gradient has been
achieved (15,16). As depicted in Fig. 1, the extrusion of
sodium ions is compensated for by potassium ion uptake (and
chloride ion efflux, depending on the concentration of potassium
ions added).

Tentative Hypothesis for Mechanism of Proton Conduction

A scheme showing a proposed mechanism of proton conductions
by bacteriorhodopsin is shown in Fig. 5. Although this
model is almost certainly wrong in its details, it provides
a hypothesis which can be tested and modified. The model

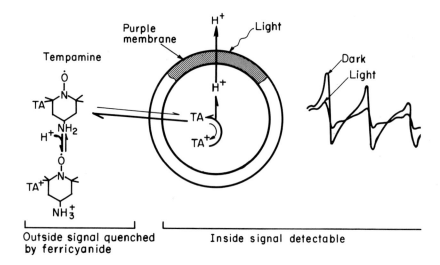

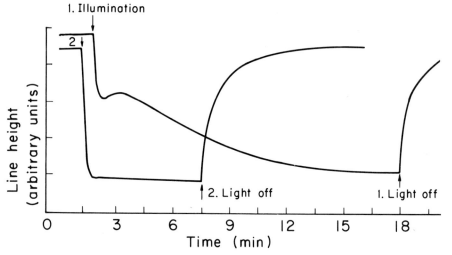

Fig. 4. Light-induced proton release from H. halobium cell envelopes loaded with sodium ions, assayed with the spin labeled amine, tempamine. The probe distributes across the membrane in direct proportion to the proton concentrations. Quenching of the signal outside the vesicles with ferricyanide allows the intravesicle signal to be observed. In the first trace of the line height a delay in proton extrusion is observed during illumination as sodium ion movement allows protons to reenter the vesicles. On subsequent illumination only rapid efflux of protons is observed since no sodium is left within the vesicles. Conditions: 2.7 mg Protein/ml vesicles, 3M KCl + 1.0 M NaCl + 1 mM Tempamine + 100 mM Potassium ferricyanide, pH ∼ 6.5.

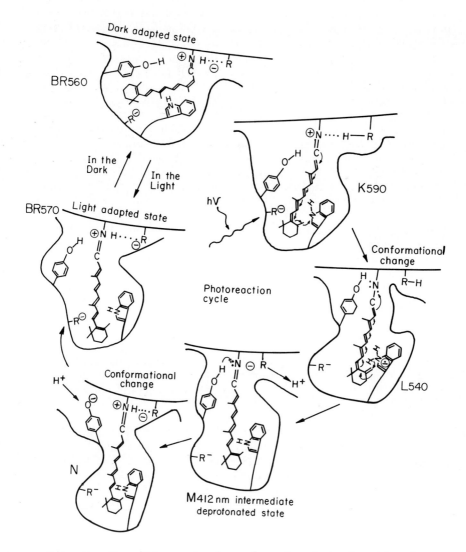

Fig. 5. Possible mechanism of proton translocation by bacteriorhodopsin.

shown is consistent with the chemical modification studies of
lysine (17), tryptophan (9), and tyrosine (5). Crosslinking
of amino groups of lysl residues in dark adapted and light
adapted samples (18), and hydrogen exchange studies (19) have
demonstrated different conformations of the dark and light
adapted forms and the essentiality of conformational changes
to the process of proton conduction. In constructing this
scheme a tentative assignment to the major transients of the
photocycle has been made. Several assumption have been made:
that charges in the vicinity of the retinal chromophore are
involved in the spectral and chemical changes, that dark
adapted bacteriorhodopsin exists in a more open configuration,
and that the chromophore is a 50-50 mixture of cis and trans
forms in the dark. Upon illumination all-trans retinal forms,
and the configuration of the light adapted molecule is commensurate with more rapid proton translocation (12). Perhaps
a more compact hydrophobic pocket with aromatic amino acid
residues like tryptophans is formed under these conditions,
four of which appear to be required but two residues are more
closely associated with light driven proton translocation.
Photon absorption by the light adapted molecule results in an
electron delocalization at the Schiff base nitrogen, i.e.,
so-called bathochromic shift). This promotes a charge transfer complex to arise with a nearby negatively charged amino
acid (perhaps aspartate 38). Then conformational change
results in vectorial movement of the proton away from the
retinal. The electron then returns to the Schiff base nitrogen. A nearby negative charge aids in the transfer of a proton from a nearby tyrosine, resulting in reprotonation of the
Schiff base. Formation of the phenolate ion results in a pK
and a conformation change leading to realignment and subsequent reprotonation of the tyrosine, leading to formation of
the original 570 nm chromophore. During the occurrence of
the photocycle a transmembrane conduction of protons occurs
resulting in the establishment of a gradient. The origin of
the released protons is still in dispute. Some of the protons
appear directly linked to the photocycle, but others may
originate from the protein itself and from cooperative interactions (20) between the adjacent molecules in the trimeric
cluster in the plane of the purple membrane.

REFERENCES

1. Stoeckenius, W., in "Light Transducing Membranes (D. W. Deamer, ed.), p. 127. Academic Press, New York, 1977.
2. Oesterhelt, O., and Stoeckenius, W., Nature New Biol. 233:149 (1971).

3. Constantopoulos, G., and Kengon, C. N., Plant Physiol. 43:531 (1978).
4. Ovchinnikov et al., The Russion Journal of Bio-organic Chemistry 4:979 (1978).
5. Konishi, T., and Packer, L., FEBS Lett 92:1 (1978).
6. Packer, L., Konishi, T., and Shieh, P. K., in "Living Systems as Energy Converters" (R. Buvet and M. J. Massue, eds.), p. 119. North Holland Publishing, Amsterdam, 1977.
7. Shieh, P. K., and Packer, L., Biophys. J. 17:2, 257a (1977).
8. Henderson, R., and Unwin, P.N.T., Nature 257:28 (1975).
9. Konishi, T., and Packer, L., FEBS Lett. 79:369 (1977).
10. Bogomolni, R. A., Stubbs, L., and Lanyi, J. K., Biochem. 17:1037 (1978).
11. Becker, B., Cassim, J. Y., Biophys. J. 19:285 (1977).
12. Konishi, T., and Packer, L., FEBS Lett. 89:333 (1978).
13. Coeckelenbergh, Y., Hart, J., Rein, R., and MacElroy, R.D., Computer Graphics 3:9 (1978).
14. Quintanilha, A. T., and Mehlhorn, R. J., FEBS Lett. 91:104 (1978).
15. Lanyi, J. K., "Gating Effects in <u>Halobacterium halobium</u> Membrane Transport" (manuscript submitted).
16. Probst, I., Mehlhorn, R., Quintanilha, A. T., Lanyi, J., and Packer, L., in Abstract submitted, Biophysical Society 23rd Annual Meeting, Atlanta, Georgia, February 25-28, 1979.
17. Konishi, T., Tristram, S., and Packer, L., Photochem. Photobiol. (in press).
18. Packer, L., Konishi, T., and Shieh, P., Fed. Proc. 36: 1819 (1977).
19. Konishi, T., and Packer, L., FEBS Lett. 80:455 (1977).
20. Hess, B., Korenstein, R., and Kuschmitz, D., in "Energetics and Structure of Halophilic Microorganisms" (S. R. Capland and M. Ginzburg, eds.), p. 89. Elsevier-North Holland, Amsterdam, 1978.

AUTHOR INDEX

Numbers in italics refer to pages where references appear.

A

Abe, H., 180, 181, 182, 187, 188, 189, *190*
Ahmed, K., 29, *39*
Akerman, K., 392, *396*
Albers, R. W., 3, *18*, 29, *38*
Alexandre, A., 322, *328*, *329*, 334, *342*, 344, 345, 348, 351, 352, 353, *353*, *354*
Allen, J. C., 3, *19*
Alonzo, G. L., 99, *100*
Altendorf, K. H., 316, *319*, 399, *407*
Anderson, J. P., 78, *87*
Andreo, C. S., 264, *272*
Anraku, Y., 294, 296, *298*, 366, *370*
Antunes-Madeira, M. C., 105, *116*
Anzai, K., 105, 106, *116*
Arai, K., 120, *124*
Arakaki, N., 355, *356*
Arnon, D. L., 262, *272*, 410, *414*
Arrio, B., 175, *177*
Artzatbanov, V. Yu., 326, 327, *329*
Asada, K., 263, *272*
Asimakis, G. K., 376, *376*
Atkinson, M. R., 45, *47*
Avron, M., 243, *248*
Azzi, A., 348, *354*
Azzone, G. F., 322, *328*, 331, 332, 333, 336, *342*, 344, 348, 351, 352, *354*

B

Baccarini-Melandri, A., 157, *160*, 209, 210, 212, *218*, 220, 221, 222, 226, 227, *228*, 243, 246, *248*, 414, *415*
Backman, B., 291, *298*
Baltscheffsky, H., 210, 212, *218*, 243, *248*
Baltscheffsky, M., 210, 212, *218*, 220, 222, 225, 226, *228*, 243, 246, *248*
Banerjee, S. P., 29, *39*
Bangham, A. D., 167, *177*
Barber, J., 230, *242*
Barnes, E. M., 301, *302*
Barter, M. J., 378, *385*
Bartlett, G. R., 76, *76*
Baskin, R., 194, *204*
Bastide, F., 71, 72, *76*, 78, *87*, 119, 122, 124, *124*
Bäuerlein, E., 357, 358, *363*
Becker, B., 421, *429*
Beechey, R. B., 365, *370*, 399, *407*
Belyakova, T. N., 310, *319*
Benz, R., 176, *177*
Berger, H., 76, *76*
Berriman, J. A., 226, *228*
Berzborn, R. J., 261, *272*
Bhalla, R. C., 189, *190*
Bhattacharyya, P., 301, *302*
Birdsall, N., 105, 106, 108, 111, *116*, *117*, 197, *204*
Blanck, J., 187, *190*
Blondin, G. A., 366, *370*
Blostein, R., 70, *76*
Bogomolni, R. A., 420, *429*
Boland, R., 111, *117*
Bond, G. H., 68, *76*
Bonting, S. L., 4, *18*
Borle, A. B., 371, *376*
Bornet, E. P., 187, 189, *190*
Boyer, P. D., 41, *47*, 78, *86*, 102, *102*, 127, *128*, 250, 258, *258*, 279, *290*
Bragadin, M., 336, *342*, 351, 352, *354*

Brand, M. D., 321, 322, 324, 327, *328,* 334, *342,* 343, 344, 345, 346, 349, 350, 351, 352, *353*
Brasie, K., 203, *204*
Bray, G. A., 263, *272,* 378, *385*
Brierley, G. P., 348, *354*
Briggs, R. N., 78, 84, *86*
Brinley, F. J., Jr., 164, *164*
Broun, I. I., 309, 311, 313, 314, 315, *319*
Brown, M., 203, *205*
Broza, R., 410, *415*
Bücher, Th., 396, *396*
Buschlen-Boucly, S., 78, *87,* 119, 122, *124*
Butlin, J. D., 294, *298*
Bygrave, F. L., 371, *376*

C

Calvayrac, R., 175, *177*
Capaldi, R., 201, *204,* 285, *290*
Capano, M., 375, *376*
Carafoli, E., 322, 328, *329,* 375, *376*
Carbalho, A. P., 105, 113, *116*
Carlson, R. D., 162, *162*
Carmeli, C., 209, 216, 217, *218,* 250, 251, 256, *259,* 338, *340*
Carston, A., 250, *258*
Casadio, R., 157, *160,* 220, 221, 222, 226, 227, *228*
Case, G. D., 239, *242*
Cassim, J. Y., 421, *429*
Cattle, K. J., 399, *407*
Cavieres, J. D., 3, 17, 18, *19*
Champeil, P., 78, *87,* 119, 122, *124*
Chance, B., 250, *259,* 279, *290,* 331, 333, *342,* 377, 378, 379, 380, 382, 384, *385*
Chang, H. H., 55, 56, 57, 58, *65, 66*
Chao-Huei, W., 141, 142, *159*
Chen, A., 41, *47*
Chen, C. H., 334, *342,* 344, 351, *353*
Chevallier, J., 175, *177*
Chipperfield, A. R., 3, *19*
Clough, D. L., 68, *76*
Coan, C. R., 78, *86,* 119, *124*
Cockrell, R. S., 344, *354*
Coeckelenbergh, Y., 425, *429*
Coffey, P., 111, *117*
Colombo, G., 262, 269, *272*
Conner, G., 193, *204*
Constantopoulos, G., 419, *429*
Copenhaver, J. H., Jr., 340, *342*
Cornish-Bowden, A., 43, *47,* 183, *190*
Cox, G. B., 291, 294, 296, *298*
Crago, S., 53, *65*
Cranston, A., 209, *218*
Criddle, R. S., 285, *290,* 400, *407,* 409, *414*
Crofts, A. R., 157, *160,* 237, 238, *242,* 331, *342*
Crompton, E., 375, *376*
Cunarro, J., 366, *370*

D

Dahms, A. S., 41, *47*
Dailey, D., 53, 55, *65*
Dalton, L. A., 111, *117*
Dalton, L. R., 107, 111, 112, *116, 117*
Danchin, A., 250, *258*
Datta, D. B., 263, *272*
Davidson, M. M. L., 189, *190*
Davies, T., 189, *190*
Davis, D. G., 105, *116*
Davoust, J., 107, *116*
Deamer, D. W., 157, *160,* 194, *204*
Dean, W. L., 121, *124*
Degani, C., 41, *47*
De Gier, J., 167, *177*
de Kruijff, B., 167, *177*
Delbrück, M., 113, *117*
Dell'Antone, P., 336, *342*
del Valle Tescon, S., 220, 227, *228*
deMeis, L., 77, 78, *86*
Demel, R. A., 167, *177*
Deters, D. W., 261, 269, *272,* 410, *415*
Devaux, P. P., 107, *116*
De Weer, P., 3, *19,* 42, *47*
De Witt, W., 304, *318*
Dilley, R. A., 220, 221, *228*
DiPolo, R., 164, *165*
Dittrich, F., 37, *39*
Di Virgilio, F., 322, *328,* 351, 352, *354*
Dockter, M. E., 293, *298*
Dorwart, W., 45, *47*
Dose, K., 248, *248*
Dougherty, J. P., 78, *86,* 103, 104, *104,* 129, *137*
Douzou, P., 377, *385*
Downie, J. A., 296, *298*
Downs, J., 78, *87*
Drachev, L. A., 305, *319*

Author Index

Duggan, P. F., 167, *177,* 194, *205*
Dunn, S. D., 292, 293, 296, *298*
Dupont, Y., 78, *87,* 119, *124,* 139, *140,* 143, *160*
Duysens, L. M. N., 220, 225, *228*

E

Ebashi, S., 89, *100,* 139, *140,* 163, *164*
Edsall, J. T., 266, *273*
Edwards, P. A., 210, 217, *218,* 270, *273*
Eilam, Y., 36, *39*
Eisenthal, R., 43, *47*
Eletr, S., 78, *86,* 105, *116*
Ellman, G. L., 80, *87*
Endo, M., 139, *140,* 161, 162, *162,* 177, *177*
Enns, R., 399, *407*
Entman, M. L., 187, 189, *190*
Epstein, W., 312, *319*
Erecińska, M., 377, *385*
Ernster, L., 279, *290*
Eytan, E., 285, *290,* 399, *407,* 409, 414, *414, 415*

F

Fabbri, E., 209, *218*
Fabiato, A., 177, *177*
Fagan, J., 250, *258*
Fahn, S., 29, *39*
Fairfield, J., 366, *370*
Farr, A. L., 22, *28,* 262, *272,* 410, *415*
Farrance, M. L., 68, *76*
Farron, F., 209, *218,* 250, *258,* 262, 263, 264, *272*
Fayle, D. R. H., 296, *298*
Ferguson, S. J., 321, *329,* 344, *354*
Fillingame, R. H., 285, *290,* 399, *407*
Finegan, J. M., 183, *190*
Fischer, E. H., 262, 263, 269, *272*
Fishkes, H., 299, *300,* 301, *302,* 317, *319*
Fiske, C. H., 81, *87*
Fleischer, S., 71, 72, *76,* 193, 197, 199, 201, 202, 203, *204, 205*
Foder, B., 68, 70, *76*
Forry, A. W., 263, *272*
Forte, J. G., 53, 54, *65*
Fraser, A., 162, *162*

Froehlich, J. P., 101, 102, *102,* 129, *137,* 184, *190*
Fuhrmann, G. F., 167, 175, 176, *177*
Fujihara, Y., 22, *28*
Fujita, M., 22, *28*
Fukushima, Y., 12, 17, *19*
Futai, M., 292, 293, 294, 295, 296, 297, *298*

G

Gache, C., 29, *38*
Galmiche, J. M., 250, *259*
Ganser, A. L., 53, 54, *65*
Garay, R. P., 3, 7, *19*
Garrahan, P. J., 3, 7, *19,* 38, *39,* 68, 69, 70, 72, 73, 74, 75, *76,* 99, *100*
Gary-Bobo, C., 78, *87,* 119, 122, 124, *124,* 168, *177*
George, J. N., 189, *190*
George, P., 45, *47*
Gepshtein, A., 251, *259*
Gerger, G., 76, *76*
Gergely, J., 105, 111, *116*
Gerretsen, W. J., 167, *177*
Gest, H., 414, *415*
Gibson, F., 291, 294, 296, *298*
Ginzburg, B. Z., 303, *318*
Ginzburg, M., 303, *318*
Girault, G., 250, *259*
Gitler, C., 285, *290,* 399, *407,* 409, *414*
Glagolev, A. N., 309, 310, 311, 313, 314, 315, *319*
Glasser, P., 366, *370*
Glynn, I. M., 3, 12, *19,* 29, 38, *39*
Good, N. E., 220, 221, *228*
Gottschlich, R., 317, *319*
Gouglas, M. G., 293, *298*
Govindjee, 270, *273*
Graham, N., 377, *385*
Green, D. E., 366, *370*
Green, N. M., 78, *87*
Greengard, P., 180, *190*
Greville, G. D., 167, *177*
Griffith, O., 201, *204*
Grinius, L. L., 309, 311, 313, 314, 315, *319*
Grinvald, A., 141, 142, *159*
Gromet-Elhanan, Z., 248, *248*
Guarraia, L. J., 276, *277*
Guillory, R. J., 248, *248*
Gulik-Krzywcki, T., 105, *116*

Gutman, M., 250, 251, *259,* 353, *354*
Gutnick, D. L., 296, *298*

H

Hagihara, M., 371, *376*
Halliwell, H. F., 45, *47*
Hanstein, W. G., 366, *370*
Hara, Y., 22, *28*
Hariyaga, S., 180, *190*
Harold, F. M., 291, *298,* 301, *302,* 303, 312, 316, *318, 319*
Harris, D. A., 250, *258*
Harris, E. J., 344, *354*
Hart, J., 425, *429*
Hartmann, R., 317, *319*
Hartree, E. F., 251, *259*
Haslam, R. J., 189, *190*
Hasselbach, W., 77, 78, *86,* 89, *100,* 139, *140,* 151, *160*
Hastings, D. F., 22, *28,* 29, *39*
Hatefi, Y., 366, *370*
Hattori, A., 355, *356*
Hauska, G. A., 261, *272*
Hayashi, Y., 21, 28, *28,* 30, *39,* 49, *50*
Hegyvary, C., 28, *28,* 29, *38,* 42, *47*
Heldt, H. W., 387, 388, 391, 395, *396, 397*
Henderson, R., 420, *429*
Henke, W., 37, *39*
Heppel, L. A., 295, 296, 297, *298*
Herbette, L., 203, *204*
Herz, R., 151, *160,* 168, *177*
Hesketh, T. R., 111, *117*
Hess, B., 250, *259,* 428, *429*
Heytler, P. G., 365, *370*
Hidalgo, C., 105, 111, *116, 117*
Higashida, M., 262, 269, 270, 271, *272, 273*
Higuti, T., 355, *356*
Hille, B., 14, *19*
Hinkle, P. C., 241, *242,* 322, 325, 326, *328, 329*
Hirata, H., 219, 226, *228,* 279, 280, 281, 282, 283, 284, 288, *290,* 292, *298,* 399, *407,* 409, *414*
Hobbs, A. N., 42, *47*
Hoch, G., 209, *218*
Hochman, Y., 250, 256, *258, 259*
Hodge, A. J., 234, *242*
Hoffman, J. F., 70, *76,* 141, 142, *159*
Hoggle, J. H., 29, *39*
Hokin, L. E., 29, *39,* 41, *47*

Holland, P. C., 193, *204*
Hollunger, G., 380, *385*
Homareda, H., 21, *28,* 30, *39,* 49, *50*
Hope, A. B., 234, *242*
Hori, H., 78, *86*
Horio, T., 210, *218,* 243, *248*
Horiuti, T., 210, *218,* 243, *248*
Hosoi, K., 243, *248*
Houslay, M. D., 108, 111, *116, 117*
Hui, Bon Hoa, G., 377, *384*
Hunter, M. J., 176, *177*
Hvidt, A., 110, *117*
Hyde, J. S., 107, 111, 112, *116, 117*

I

Ierokomas, A., 78, *87,* 121, *124*
Iida, S., 46, *47*
Ikegami, A., 115, *117*
Ikemoto, N., 77, 79, 81, 82, 85, *86, 87,* 105, 111, 115, *116, 117,* 119, *124,* 194, *204*
Iles, G. H., 176, *177*
Inagaki, C., 37, *39*
Inesi, G., 77, 78, 84, *86, 87,* 105, *116,* 119, *124*
Iorio, J. M., 180, *190*
Ishimoto, M., 275, *277*
Itask, F., 377, 378, *385*
Ito, I., 250, *259*
Itoh, S., 230, 233, *242*
Izumi, F., 42, *47*

J

Jackson, J. B., 210, 217, *218,* 226, 227, *228,* 237, 238, *242,* 270, *273*
Jacobs, E. E., 345, *354*
Jagendorf, A. T., 209, *218,* 219, *228,* 250, *258,* 263, 271, *272, 273,* 410, *415*
Jakabova, M., 189, *190*
James, T. L., 29, *39*
Janig, G. R., 76, *76*
Jensen, J., 28, *28,* 29, *39*
Jilka, R. L., 111, *117,* 167, *177*
Johansson, B. C., 210, 212, *218,* 243, 246, *248*
John, P., 322, *329*
Johnston, R. F., 400, *407*
Jorgensen, P. L., 3, 12, *19*
Jost, P., 201, *204*

Judah, J. D., 29, *39*
Julian, C., 414, *415*
Jullien, M., 175, *177*

K

Kagawa, Y., 219, 226, *228,* 279, 280, 281, 282, 283, 284, 288, *290,* 292, *298,* 399, *407,* 409, *414*
Kajdos, I., 61, *66*
Kakuno, T., 243, 244, *248*
Kalbitzer, H. R., 139, *140*
Kaltstein, A., 344, *354,* 387, *397*
Kamen, M. D., 243, *248*
Kametani, F., 366, *370*
Kamienietzky, A., 261, *272,* 410, 413, 414, *414*
Kanazawa, T., 46, *47,* 77, 78, *86,* 89, 96, *100, 102,* 114, *117,* 120, *124,* 127, *128,* 129, 136, *137,* 184, 187, *190,* 294, 295, 296, 297, *298*
Kanemasa, Y., 366, *370*
Kaniike, K., 21, *28,* 29, *39,* 49, *50,* 187, 189, *190*
Kanner, B. I., 296, *298*
Karlish, S. J. D., 3, 12, *19,* 29, 36, *39*
Kasai, M., 142, 143, *159,* 167, 168, 176, *177*
Kaser-Glanzmann, R., 189, *190*
Katsumata, K., 243, *248,* 371, *376*
Katsura, E., 275, *277*
Katz, S., 70, *76,* 104, *104,* 179, 180, 183, 187, 188, 189, *190*
Kaulen, A. D., 305, *319*
Kawai, K., 22, *28*
Kawakita, M., 120, *124*
Kawato, S., 115, *117*
Kaziro, Y., 120, *124*
Keister, D. L., 225, *228*
Kell, D. B., 322, *329*
Kemp, A., Jr., 392, *397*
Kengon, C. N., 419, *429*
Kent, A. B., 262, 269, *272*
Kessler, R. J., 366, *370*
Kiehl, R., 357, *363*
Kimura, I., 189, *190*
Kimura, M., 21, *28,* 30, *39,* 189, *190*
Kinoshita, N., 180, 182, 187, 188, 189, *190*
Kirchberger, M. A., 179, 180, 187, 188, 189, *190*
Kirino, Y., 105, 106, 107, *116*
Klingenberg, M., 344, 354, 387, 388, 391, 393, 395, *396, 397*

Knauf, P. A., 70, *76,* 167, 175, 176, *177*
Kobayashi, H., 301, *302*
Kobayashi, K., 191, *192,* 275, *277*
Kobayashi, S., 189, *190*
Koh, Y., 293, *298*
Kok, B., 230, *242*
Kometani, T., 142, 143, *159,* 168, *177*
Konings, A. W. T., 248, *248*
Konishi, T., 400, *407,* 419, 420, 428, *429*
Konstantinov, A. A., 326, 327, *329*
Korenstein, R., 428, *429*
Koshland, D. E., Jr., 271, *273*
Kovac, L., 395, *397*
Koval, G. J., 29, *39*
Krab, K., 322, 324, 325, 327, *328, 329,* 347, 351, *354,* 392, *397*
Krebs, E., 262, 269, *272*
Kubota, S., 280, *290,* 366, *370*
Kume, S., 16, *19,* 28, *28,* 29, *38,* 42, *47*
Kuo, J. F., 180, *190*
Kuroda, Y., 191, *192*
Kurzmack, M., 84, *87*
Kuschmitz, D., 428, *429*

L

Lacaz-Vieira, F., 51, *52*
Landgraf, W. C., 78, *86*
Lane, L. K., 29, *39,* 49, *50,* 187, 189, *190*
Lange's Handbook of Chemistry (1973), 233, *242*
Lanir, A., 250, 256, 258, *259*
LaNoue, K., 387, *397,* 344, *354*
Lanyi, J. K., 303, *304,* 317, *318,* 420, 425, *429*
LaRaia, P. J., 187, *190*
Lardy, H. A., 330, *342,* 375, *376*
Laris, P. C., 392, *397*
Larson, R. J., 292, *298*
Lazdunski, M., 29, *39,* 271, *273*
Lea, C. H., 106, *117*
Leavis, P. C., 79, *87*
Lee, A. G., 105, 106, 111, *116, 117,* 197, *204,* 331, 332, 337, *342, 343*
Lee, J., 53, *65*
Legallias, V., 377, *385*
Lehninger, A. L., 321, 322, 324, 327, *328, 329,* 332, 333, 334, *342,* 343, 344, 345, 348, 349, 350, 351, 352, *353, 354,* 371, *376*
Lehrer, S., 79, *87*
LeMaire, M., 193, *204*
Lentz, B., 45, *47*

Lerbs, V., 227, *228*
Levitzki, A., 271, *273*
Levy, H. R., 262, 269, *272*
Lewin, M., 54, 58, *65, 66*
Liberman, E. A., 366, *370*
Lieb, W. R., 36, *39*
Lien, S., 250, *259*, 262, 270, *272*
Lifshitz, Y., 209, 217, *218*, 250, 251, *259*
Lindenmayer, G. E., 3, *19*, 21, *28*, 29, 37, *39*, 49, *50*
Lindrop, C. R., 399, *407*
Lipmann, F., 89, *100*, 209, *218*
Love, J., 230, *242*
Low, K. B., 291, *298*
Lowry, O. H., 22, *28*, 262, *272*, 410, *415*
Lundin, A., 210, *218*, 220, 222, 225, 226, *228*
Luscher, E. F., 189, *190*
Lustorff, J., 349, *354*
Lynham, J. A., 189, *190*

M

McCarty, R. E., 250, *258*, 271, *273*
McClean, J. D., 234, *242*
McClenaghan, M. D., 189, *190*
MacDonald, R. E., 301, *302*, 303, 317, *318*
MacElroy, R. D., 425, *429*
McFarland, B. H., 77, *86*
McGill, K. A., 111, *117*
McIlwain, H., 191, *192*
McIntyre, J., 199, *204*
McKinley, D., 142, 143, *159*, 167, *177*
McLaughlin, S., 230, 231, *242*
MacLennan, D. H., 106, *116*, 139, *140*, 176, *177*, 183, *190*, 193, *204*
Madeira, V. M. C., 105, 113, *116*, 157, *160*
Maeda, M., 294, 296, *298*
Makinose, M., 77, *86*, 89, 93, *100*, 151, *160*, 163, *164*
Malstrom, K., 375, *376*
Manuck, B. A., 113, *117*
Marling, E., 77, *86*
Martin, I., 209, *218*
Martonosi, A., 77, 78, *86, 87*, 105, 111, *116, 117*, 121, *124*, 167, 193, 194, *204*
Masoro, E. J., 78, *87*
Massari, S., 222, *228*, 331, 332, *342*, 344, *354*
Matsui, H., 21, *28*, 29, 30, *39*, 49, *50*
Matsuno, A., 262, 269, 270, *272, 273*
Matsuura, K., 237, 238, 241, *242*
Maurel, P., 377, *385*

Mehard, C., 199, *204*
Mehlhorn, R., 425, *429*
Meissner, G., 71, 72, *76*, 78, *86*, 96, *100*, 142, 143, *159*, 167, *177*, 193, *194*, 197, 199, *204*
Melandri, B. A., 151, 157, *160*, 209, 210, 212, *218*, 220, 221, 222, 226, 227, *228*, 243, 246, 248, 414, *415*
Mercer, F. V., 234, *242*
Metcalfe, J. C., 105, 106, 108, 111, *116, 117*, 197, 201, *204*
Michel, H., 317, *319*
Michelangeli, F., 62, *66*
Miconi, 322, *328*
Mihas, A. A., 53, *65*
Mikulecky, D. C., 167, *177*
Milhausen, M., 262, 268, 269, *272*
Millman, M., 105, *116*
Mills, J. D., 230, *242*
Milovancev, M., 387, *396*
Minton, N. J., 227, *228*
Mitchell, P., 217, *218*, 219, *228*, 229, 241, *242*, 279, *290*, 299, 300, 301, *302*, 303, 304, 317, *318, 319*, 321, 322, 324, 325, 326, 327, *328*, 329, 331, 333, 334, 335, *342*, 344, 348, 349, 350, 351, *353, 354*, 397, 399, 409, *414*
Mitsui, T., 78, *86*, 105, *116*
Miyamoto, E., 167, 168, 176, 177, 180, *190*
Mizani, S. M., 344, *354*, 387, *397*
Moller, J. V., 78, *87*
Morales, M. F., 78, *86*
Morgan, J. F., 79, 81, 82, *87*
Morkin, E., 187, *190*
Morton, R. K., 45, *47*
Moyle, J., 217, *218*, 321, 322, 324, 325, 326, 327, *328, 329*, 331, 334, 335, *342*, 344, 349, 350, 351, *353, 354*
Muhle, H., 230, *242*
Mukohata, Y., 262, 264, 265, 266, 267, 269, 270, 271, *272, 273*
Muller, H. W., 248, *248*
Mullins, L. J., 3, *19*, 164, *164*
Muraoka, S., 365, 366, *370*
Murcus, F., 262, 269, *272*
Murphy, A. J., 77, 78, *86, 87*, 111, *117*, 119, 120, 122, *124*

N

Nagai, K., 42, *47*
Nakajima, Y., 177, *177*
Nakamura, A., 194, *204*
Nakamura, H., 78, *86, 87*, 111, *117*, 121, *124*

Author Index

Nakamura, Y., 137, *137*
Nakanishi, M., 105, 106, 110, *116*
Nakao, M., 22, *28*
Nakao, T., 22, *28*
Nakase, Y., 377, *385*
Nakasima, S., 355, *356*
Navon, G., 139, *140*
Nelson, N., 251, *259*, 261, 269, *272, 273,*
  290, 296, *298,* 399, *407,* 409, 410, 413, *414,*
  415
Neumann, J., 410, *415*
Nielsen, A. O., 110, *116*
Nishi, N., 243, 244, *248*
Nishigaki, I., 45, *47*
Nishikawa, K., 210, *218,* 243, *248*
Nishimura, M., 237, 238, 241, *242,* 331, *342*
Noack, E., 84, *87*
Nolan, C., 263, *272*
Noltman, E. A., 270, *273*
Norby, J. G., 28, *28,* 29, *39*
Notsani, B., 285, *290,* 399, *407,* 409, 410, *414*
Nyburg, S. C., 45, *47*

O

Oerlemans, A., 167, *177*
Oesterhelt, D., 316, 317, *319*
Ogawa, Y., 139, *140,* 161, *162*
Ohkuma, T., 105, 107, *116*
Ohmori, F., 180, 181, 182, 187, 188, 189, *190*
Ohnishi, T., 163, *164,* 355, *356*
Ohno, K., 219, *228,* 279, *290*
Ohta, S., 105, 106, *116*
Okamoto, H., 279, 280, 282, 283, 284, *290,*
  409, *414*
Okuoka, Y., 275, *277*
Olsgaard, R., 263, *272*
Oren, R., 248, *248*
Orii, Y., 115, *117*
Ort, D. R., 220, 221, *228*
Ortanderl, F., 262, *272*
Osborn, M. J., 410, *415*
Ostroy, F., 29, *39*
Out, T. A., 392, *397*
Ovchinnikov, Yu. A., 419, *429*
Overbeak, J. T. G., 230, 231, *242*

P

Packer, L., 199, *204,* 285, *290,* 400, *407,* 409,
  414, 419, 420, 422, 425, 428, *429*

Packham, N. K., 226, *228*
Padan, E., 220, *228*
Pang, D., 78, 84, *86*
Papa, S., 322, *328,* 344, *354*
Papinean, D., 316, *319*
Parson, W. W., 239, *242*
Pearl, L., 356, *356*
Peck, H. D., Jr., 276, *277*
Pedersen, P. L., 406, *407*
Penefsky, H. S., 120, *124*
Penttila, T., 328, *329*
Perutz, M. F., 16, *19*
Peterson, E. A., 262, *272*
Petrack, B., 209, *218,* 250, *258*
Pfaff, E., 387, 388, 391, 395, *396, 397*
Pfleiderer, C., 262, *272*
Pittman, P. R., 271, *273*
Pontremoli, S., 122, *124*
Post, R. L., 12, *13,* 16, *19,* 28, *28,* 29, 37, 38,
  39, 41, 42, 45, 46, *47,* 71, 72, *76*
Postma, P. W., 291, *298*
Pouchan, M. I., 68, *76*
Pozzan, T., 322, *328,* 336, *342,* 351, 352, *354*
Pressman, B., 344, *354*
Price, B., 331, *342*
Prichard, W. W., 365, *370*
Prince, R. C., 157, *160*
Probst, I., 425, *429*
Proverbio, F., 62, *66,* 70, *76*

Q

Quintanilha, A. T., 425, *429*
Quist, E. E., 74, *76*

R

Rabon, E., 53, 55, 56, 57, 58, 61, *65, 66*
Racker, E., 61, *66,* 142, 158, *159,* 197, *204,*
  219, *228,* 243, *248,* 250, *258,* 259, 261, 262,
  263, 264, 269, 270, *272,* 279, *290,* 291, *298,*
  409, 410, *414, 415*
Radik, J., 296, *298*
Randall, R. J., 22, *28,* 262, *272,* 410, *415*
Rapoport, S. M., 76, *76*
Rasmussen, H., 371, *376*
Recktenwald, D., 250, *259*
Reed, P. W., 375, *376*
Rega, A. F., 68, 69, 70, 72, 73, 74, 75, *76,* 99,
  100
Reich, R., 230, 237, *242*

Rein, R., 425, *429*
Reisse, I., 151, *160,* 168, *177*
Renthal, R., 303, 317, *318*
Repke, K. R. H., 37, *39*
Repke, D. I., 179, 180, 187, *190*
Requena, J., 164, *164*
Reynafarje, B., 321, 322, 324, 327, *328, 329,* 334, *342,* 343, 344, 345, 348, 351, 352, *353, 354,* 371, *376,* 388, 395, *397*
Reynolds, J., 193, *204*
Rhoads, D. B., 312, *319*
Rhodes, D. N., 106, *116*
Rich, G. T., 167, 168, *177*
Richards, D. E., 69, 70, 74, *76*
Richardson, A., 210, *218*
Richman, L., 45, *47*
Rippa, M., 122, *124*
Rizzolo, L., 193, *204*
Robinson, B. H., 111, *117*
Robinson, J. D., 29, 37, *39*
Robles, E. C., 106, *117*
Roelofsen, B., 67, *76*
Rogers, F. N., 41, 45, *47*
Rogowski, R. S., 78, 84, *86*
Romualdez, A., 167, 168, *177*
Rosebrough, N. J., 22, *28,* 262, *272,* 410, *415*
Rosen, B. P., 294, *298,* 299, *300,* 301, *302*
Rosenthal, A. S., 29, *39*
Rossi, B., 29, *39*
Rossi, C. S., 333, *342*
Rossi, E., 333, *342*
Rossi, J. P. F. C., 74, 75, *76*
Rothstein, S. S., 167, 175, 176, *177*
Rothstein, A., 167, 175, 176, *177*
Rottenberg, H., 220, *228,* 353, *354,* 387, *397*
Roufogalis, B. D., 74, *76*
Rugolo, M., 375, *376*
Rumberg, B., 230, *242*
Ryrie, I. J., 209, *218,* 250, *258,* 263, 271, 272, 273

S

Saari, H. T., 322, 324, 325, 327, 328, *328, 329,* 344, 347, 348, 351, 352, *354*
Saccomani, G. J., 53, 54, 55, 56, 59, 60, *65, 66*
Sachs, G., 53, 54, 55, 56, 57, 58, 59, 60 61, *65, 66*
Sachs, L., 303, *318*
Saffman, P. G., 113, *117*

Saito, A., 197, 201, *204*
Saito, M., 191, *192*
Saito, S., 294, 295, 296, 297, *298*
San Pietro, A., 414, *415*
Sanadi, D. R., 345, *354*
Saphon, S., 227, *228*
Saraste, M., 328, *329*
Sarau, H. M., 53, *65*
Sarrif, S., 29, *39*
Scarpa, A., 164, *164,* 203, *204*
Scatchard, G., 36, *39*
Schackmann, R., 55, *55,* 56, 58, 59, 60, *66*
Scharff, O., 68, 70, *76*
Schatz, G., 293, *298*
Schatzmann, H. J., 67, *76*
Schenackerz, K. D., 270, *273*
Schenk, F. J., 102, *102*
Schlimme, E., 349, *354*
Schmidt, S., 230, 237, *242*
Schneider, E., 248, *248*
Scholes, P., 53, *65,* 217, *218*
Schollmeyer, P., 393, *397*
Schon, R., 37, *39*
Schonfeld, W., 37, *39*
Schuldiner, S., 299, *300,* 301, *302,* 317, *319*
Schwartz, A., 3, *19,* 29, 30, 37, *39,* 49, *50,* 59, *66,* 187, 189, *190*
Schwartz, W., 14, *19*
Schwulera, U., 248, *248*
Sebald, W., 285, *290,* 399, *407*
Seelig, J., 78, *86,* 203, *204*
Seeman, P., 176, *177*
Seki, Y., 275, *277*
Sen, A. K., 29, 38, *39*
Senior, A. E., 291, *298*
Sha'afi, R. I., 167, 168, *177*
Shah, G., 54, *65*
Shallenberger, M. K., 294, *298*
Sharaf, A. A., 366, *370*
Shaw, D., 54, *65*
Sheppy, F., 209, *218,* 250, *258*
Shieh, P., 285, *290,* 409, *414,* 422, 428, *429*
Shigekawa, M., 78, *86,* 103, 104, *104,* 129, *137,* 183, *190*
Shimizu, H., 105, 106, 116
Shorr, E. J., 263, *272*
Shurayh, F., 45, *47*
Sierra, M. F., 285, *290,* 400, *407*
Sigel, E., 322, 328, *329,* 375, *376*
Sigrist, H., 285, *290,* 399, *407,* 409, *414*
Sigrist-Nelson, K., 285, *290,* 399, *407,* 409, *414*

Author Index

Siliprandi, D., 375, *376*
Siliprandi, N., 375, *376*
Simoni, R. D., 291, 294, *298*
Simpson, G., 53, *65*
Sims, P. J., 141, 142, *159*
Singh, D., 189, *190*
Sireix, R., 377, *385*
Skou, J. C., 3, *17*, 29, *39*
Skulachev, V. P., 304, 305, 306, 309, 310, 311, 313, 314, 315, 317, *318*, 319, 322, 326, 327, *328, 329*, 366, *370*
Slater, E. C., 250, 258, *258*, 279, *290*, 293, *298*, 392, *397*
Smettan, G., 187, *190*
Smith, G. A., 111, *117*
Smith, J. B., 292, 293, *298*, 399, *407*
Snyder, E. R., 262, 269, *272*
Sober, H. A., 262, *272*
Soe, G., 243, 244, *248*
Solomon, A. K., 167, 168, *177*
Sone, N., 219, 226, *228*, 279, 280, 281, 282, 283, 284, 289, *290*, 292, *298*, 399, *407*, 409, *414*
Sordahl, S. A., 376, *376*
Sorgato, M. C., 322, *329*, 344, *354*
Soumarmon, A., 61, *66*
Spanio, L., 122, *124*
Spenney, J. G., 54, *65*
Spisni, A., 55, *65*
Spitzer, H. L., 55, *65*
Sreter, F., 194, *204*
Stark, G., 176, *177*
Stehlik, D., 139, *140*
Stein, W. D., 36, *39*
Steinhart, R., 292, *298*
Sternweis, P. C., 292, 293, 295, 296, 297, *298*
Stewart, H. B., 54, *65*
Stock, A., 262, *272*
Stoeckenius, W., 219, *228*
Stubbs, L., 420, *429*
SubbaRow, Y., 81, *87*
Sugiyama, Y., 262, 264, 265, 266, 267, 269, 270, *272, 273*
Sumida, M., 101, *102*, 115, *117*
Sykes, B. D., 113, *117*

T

Tada, M., 119, *124*, 179, 180, 181, 182, 183, 184, 187, 188, 189, *190*, 193, *204*
Takahashi, M., 263, *272*

Takeda, H., 377, *385*
Takeuchi, Y., 279, *290*
Takisawa, H., 129, 130, 133, 134, *137*
Tanford, C., 119, *124*, 193, *204*
Tani, I., 355, *356*
Taniguchi, K., 16, *19*, 37, *39*, 41, 42, 46, *47*
Taupin C., 119, 124, *124*
Taussky, H., 263, *272*
Taylor, A. L., 291, *298*
Taylor, E. W., 101, *102*, 129, *137*, 184, *190*
Terada, H., 365, 366, *370*
Tew, W. P., 371, *376*
Thomas, D. D., 107, 111, 112, *116, 117*
Thore, A., 210, *218*, 222, 226, *228*
Thorley-Lawson, D. A., 78, *87*
Tiffert, T., 164, *164*
Tillack, T. W., 167, *177*
Toda, G., 16, *19*, 37, *39*, 41, 45, *47*
Toniello, A., 375, *376*
Tonomura, Y., 16, 17, *19*, 46, *47*, 77, 78, *86*, 89, 96, *100*, 101, *102*, 108, 114, 115, *116, 117*, 119, 120, *124*, 125, *126*, 129, 130, 133, 134, 136, *137*, 157, *160*, 179, 183, 184, 187, *190*, 193, *204*
Toon, P. A., 106, 111, *116, 117*, 197, *204*
Trachtman, M., 45, *47*
Trasch, H., 357, 358, *363*
Travers, F., 377, *385*
Tristram, S., 428, *429*
Tsopanakis, A. D., 378, *385*
Tsuboi, M., 105, 106, 110, *116*
Tsuchiya, Y., 271, *273*, 299, *300*, 301, *302*
Tume, R. K., 78, *86*
Tyson, C. A., 366, *370*
Tzagoloff, A., 285, *290*, 400, *407*

U

Uda, M., 366, *370*
Unwin, P. N. T., 420, *429*
Urano, M., 263, *272*
Uribe, E., 219, *228*

V

Vallejos, R. H., 264, *272*
Vambutas, V. K., 243, *248*, 250, *258*, 261, *272*
Van Brunt, J., 301, *302*

Van Deenen, L., 167, *177*
van den Berg, D., 106, *116*
Vanderkooi, G., 78, *87,* 121, *124,* 282, *290*
Vandermeullen, D. L., 270, *273*
van der Sluis, P. R., 293, *298*
van Grondelle, R., 220, 225, *228*
van Winkle, W., 187, 189, *190*
Varanda, W., 51, *52*
Vercesi, A., 344, 351, *353,* 371, *376*
Verjovski-Almeida, S., 84, *87*
Verschoor, G. J., 293, *298*
Viale, A., 264, *272*
Vincent, S. P., 378, *385*
Vincenzi, F. F., 68, *76*
Vogel, G., 292, *298*
vonStedingk, L.-V., 210, *218*

W

Wachter, E., 399, *407*
Waggoner, A. S., 141, 142, *159*
Wagner, G., 316, 317, *319*
Wahlquist, H., 107, *116*
Wakabayashi, T., 280, *290*
Wallick, E. T., 29, *39,* 49, *50*
Wang, C.-T., 141, 142, *159,* 197, 199, 200, 201, 203, *204*
Warren, G. B., 106, 108, 111, *116,* 197, 201, *204*
Webb, R. C., 189, *190*
Weber, A., 77, *86,* 151, *160,* 168, *177,* 410, *415*
Webster, G. D., 227, *228*
Weiner, M. W., 366, *370*
Werber, M. M., 250, *258*
West, I. C., 299, *300,* 301, *302,* 317, *319*
Wharton, D., 78, *87*
Whittam, R., 3, *19*
Wikstrom, M. K. F., 322, 324, 325, 327, *328,* 329, 331, 334, 336, 337, *342,* 344, 347, 348, 351, 352, *354,* 392, *396*
Will, H., 187, *190*

Williams, G. R., 382, *385*
Williams-Smith, D. L., 378, *385*
Williamson, R. L., 365, *370*
Winget, G. D., 261, 262, *272*
Witonsky, R. J., 45, *47*
Witt, H. T., 227, *228,* 230, 237, *242*
Wollenberger, A., 187, *190*
Wong, S. M. E., 29, *39*
Wu, C., 45, *47*
Wulf, R., 344, *354,* 387, 395, *397*
Wyman, J., 266, *273*

Y

Yagi, T., 262, 270, *272,* 273
Yamada, S., 77, 79, 81, 82, 85, 86, *87,* 96, *100,* 114, *117,* 120, *124,* 129, 130, 136, *137,* 157, *160,* 180, 184, 187, 189, *190*
Yamaguchi, A., 366, *370*
Yamamoto, T., 77, *86,* 89, 96, *100,* 107, 114, *116, 117,* 119, 120, *124,* 129, 130, 134, *137,* 179, 183, 184, 187, *190,* 193, *204*
Yamashita, J., 243, *248*
Yates, D. W., 10, 12, *19,* 29, *39*
Yearwood-Drayton, V., 303, 317, *318*
Yip, C. C., 176, *177*
Yokota, M., 355, *356*
Yon, J., 175, *177*
Yoshida, M., 219, 226, *228,* 279, 280–284, 288, *290,* 292, *298,* 399, *407,* 409, *414*
Yoshida, S., 42, *47,* 78, *86,* 115, *117*
Younis, H. M., 261, 262, *272*
Yu, B. P., 78, *87*

Z

Zahler, W., 199, *204*
Zande, H. V., 366, *370*
Zimniak, P., 142, 158, *159*
Zoccarato, F., 375, *376*

SUBJECT INDEX

A

Acridine orange, 58
Adenine nucleotide transport, 387–397
Adenylylimidodiphosphate (AMPPNP), 139, 293
ADP carrier, 391
Alkylamines, 357
  as uncouplers in mitochondria, 357–363
9-aminoacridine, 299
Amphipathic anion, 355
Amphipathic cation, 355
Anilinonaphthalene sulfonate, 289, 311
Antimycin A, 222, 337
Antiporters
  in bacterial systems, 303–319
  in *E. coli*, 299–300
  in *H. halobium*, 417–429
  in *S. faecalis*, 301–302
ATP–ADP exchange activity, 10, 11, 98
ATP analogs, 119, 393
ATP synthesis
  in chloroplasts, 261–273
  in *R. capsulata*, 219–228
ATPase, 29
  activity, 157
  calcium
    in erythrocytes, 67
    in sarcoplasmic reticulum, 77, 89–117, 119–137, 139–140
  in chloroplasts, 249–259, 261–273
  in *D. vulgaris*, 275–279
  in *E. coli*, 77–87
  proton-potassium, 55
  in *R. rubrum* chromatophores, 209–218, 243–248
  sodium-potassium
    overview, 3
    p-nitrophenylphosphate hydrolysis, 41
    sodium binding, 49
    sodium-sensitive electrode, 21
    in thermophilic bacteria, 179–290
    in yeast mitochondria, 399–407
Atractyloside, 352

B

β-anilinonaphthalene sulfonate, 289
β-hydroxybutyric dehydrogenase, 199–205
b-c_1 complex, 338
Bacteriorhodopsin, 304, 417–429

C

Ca^{2+}-ATPase, 249–259, 268, 275, 276
Caffeine, 139, 163
Calcium ion, 67–87, 89–117, 119–126, 129–137, 141–165, 193–205, 243–248, 275–276, 377–385
Calcium pump protein, 193
Carboxyatractylate, 392
Carotenoid, 226, 238
CF_0, 409
Chemiosmotic hypothesis, 229, 321
Chloroplasts, 249–259, 261–273, 409–415
Chromatium, 239
Chromatium vinosum, 249
Chromatophores, 209–228
Chymotrypsin, 61
Conformers, 15
Correlation time, 113
Coupling factor 1 (CF_1), 249–259, 261–273
Creatine phosphotransferase, 308
Cu^{2+}, 275
Cu^{++} phenanthroline, 55
Cyanine dyes, 142, 365–370
Cyclic AMP, 179–192
Cyclic AMP-dependent protein kinase, 179

Cyclic electron flow, 243
Cytochrome b, 384
Cytochrome c, 322, 384
Cytochrome oxidase, 115, 321–329, 349

D

D. gigas, 276
DCCD-binding protein, 285
Depolarization-induced calcium release, 163
Desuphovibrio vulgaris, 275–290
Diaminodurene, 325
Dicyclohexylcarbodiimide, 13, 210, 276, 279, 399–407
Diethyloxodicarbocyanine, 57
Dimethyl sulfoxide, 377
3,3′-dipropylthiocarbocyanine iodide, 392
Diuresis, 5
Dodecyl octaoxyethyleneglycol monoether, 130

E

EDTA, 2, 49, 68, 255
EDTA particles, 411
Electrogenic, 142
  calcium transport, 141
  transport, 395
  uniport, 53
Electroneutral
  exchanger, 53
  Na^+/H^+ antiport, 318
Electrophoretic
  ADP–ATP exchange, 392
  K^+ transport, 318
Emulgen 810, 275
Erythrocytes, 6, 67–76
Escherichia coli, 291–298, 299, 300, 303–319
ESR, 112, 256
Ethylene glycol, 380
N-ethylmaleimide, 85, 106, 324, 346

F

Flagellar motor, 308
Formycin nucleotides, 12
Freeze-fracture electron microscopy, 194

G

Gastric vesicles, 53–66
Gel–liquid crystalline phase transition, 105, 108
Glycerophospholipids, 57
Glycoprotein, 54, 58
Gouy–Chapman theory, 230

H

^{3}H-ouabain, 30–31
H^+/e^- ratio, 321, 331
H^+/site ratio, 344
Halobacterium halobium, 303–319, 417–429
Hemoglobin, 16
Hydrogen–deuterium exchange, 105
d-β-hydroxybutyrate apodehydrogenase, 200

I

Immunodiffusion, 296
Ionophores, 212, 223

L

Li^+, 8, 49
Light scattering, 167
Lineweaver–Burk plot, 90, 95
Luciferase, 210

M

Magnesium ion, 3–19, 41–47, 49–50, 89–100, 105–117, 129–137, 243–248, 270, 275
Manganese ions, 139–140, 243–249, 270, 275
Membrane potential, 157, 370, 387
Mercuric N-dansyl cysteine, 79
Mitochondria, 321–329, 343–363, 365–385, 387–397, 399–407
Motility, 311
Mutants, 291

Subject Index

N

Negative staining, 197
Nigericin, 57, 215, 221
p-Nitrophenyl phosphatase, 13
p-nitrophenylphosphate, 41

O

Oligomycin, 356, 366, 390, 399–407
Ouabain, 5, 29, 52, 53
Ouabain-sensitive K^+ binding, 31
Ouabain-sensitive Na^+ binding, 34
Oxalate, 151
Oxygen pulse technique, 331

P

Phosphoenzyme, 10, 15, 31, 37, 44, 45, 54, 70, 72, 77, 96, 101, 103
    formation, 45, 93, 95, 162
    intermediate, 103, 179
Phospholamban, 180
Phospholipase A_2, 57
Phospholipids, 46, 55
Photophosphorylation, 249, 263
Photosynthetic units, 227
Picric acid, 359
Potassium ion, 3–19, 21–28, 29–39, 41–47, 49–50, 53–66, 399–407
Presynaptic terminals, 191
Promitochondria, 395
Propranolol, 84, 85
Prostaglandin, 165
Proteolipid, 399–407
Protonmotive force, 221, 299, 302
Protons
    by bacteriorhodopsin, 417–429
    by cytochrome c oxidase, 321–329
    by gastric vesicles, 53–66
    by mitochondria, 331–356, 399–407
Pyridoxal phosphate, 119, 262

R

Reconstitutuion
    ATPase of *E. coli*, 291–298

ATPase of thermophilic bacteria, 279–290
    chloroplast vesicles, 409–415
    sarcoplasmic vesicles, 193–205
Rheogenic Na pump, 51
Rhodopseudomonas capsulata, 219–228
Rhodopseudomonas sphaeroides, 229–242
Rhodospirillum rubrum, 209–218, 243–248
Ruthenium red, 373

S

Saccharomyces cereviseae, mitochondria, 399–407
Safranine, 392
Sarcoplasmic reticulum, 141–165, 167–177, 179–190, 193–205, *see also* ATPase, calcium, in sarcoplasmic reticulum
Saturation transfer electron spin resonance, 105, 111
SITS, 176
Sodium ion, 3–19, 21–28, 29–39, 41–47, 49–52
Sodium pump, 51–52
Sodium sensitive electrode, 21
Spin labeling, 111
Streptococcus faecalis, 301–302
Surface potentials, 229–242
Symport, 344
Symporter, 321
Synaptic transmission, 191–192

T

Tannic acid, 198
Tetrachlorsalicylanilide, 57
Tetradecylamine, 359
Tetramethylmurexide, 161
Tetramethyl-p-phenylendiamine, 337
TF_1, 279–290
TF_0-vesicles, 284
Thallous, 8
Thermophilic bacteria, 279–290
Thiols, 81
Thyroxine, 375
Toad skin, 51–52
Tolbutamide, 371
$TPMP^+$, 62

Triiodothyronine, 375
Trypsin, 11, 55, 58, 61, 296

U

Ubiquinones, 372
Uncoupler, 212, 366
Uncoupling, 359–363, 365–370

V

Valinomycin, 57, 59, 171, 215, 220, 324, 333, 345, 355, 387
Volume changes
  in gastric vesicles, 371–376
  in mitochondria, 53–66
  in sarcoplasmic reticulum, 167–177